# From Nuclear Transmutation to Nuclear Fission, 1932–1939

## Also by Per F Dahl

Ludvig Colding and The Conservation of Energy Principle:
Experimental and Philosophical Contributions (1972)

Superconductivity: its Historical Roots and Development from Mercury
to the Ceramic Oxides (1992)
American Institute of Physics

Flash of the Cathode Rays: A History of J J Thomson's Electron (1997)
Institute of Physics Publishing

Heavy Water and the Wartime Race for Nuclear Energy (1999)
Institute of Physics Publishing

## Related Titles published by
## Institute of Physics Publishing

Cockcroft and the Atom
G Hartcup and T E Allibone

The Origin of the Concept of Nuclear Forces
L M Brown and H Rechenberg

The Defining Years in Nuclear Physics, 1932–1960s
M Mladjenović

Operation Epsilon: The Farm Hall Transcripts
Edited by Sir Charles Frank

Radar Days
E G Bowen

Echoes of War: The Story of $H_2S$ Radar
Sir Bernard Lovell

Boffin: A Personal Story of the Early Days of Radar
and Radio Astronomy and Quantum Optics
R Hanbury Brown

Technical and Military Imperatives:
A Radar History of World War II
L Brown

# From Nuclear Transmutation to Nuclear Fission, 1932–1939

Per F Dahl

I*o*P

Institute of Physics Publishing
Bristol and Philadelphia

*British Library Cataloguing-in-Publication Data*
A catalogue record of this book is available from the British Library.

ISBN 0 7503 0865 6

*Library of Congress Cataloging-in-Publication Data are available*

Commissioning Editor: James Revill
Production Editor: Simon Laurenson
Production Control: Sarah Plenty
Cover Design: Frédérique Swist
Marketing: Nicola Newey and Verity Cooke

Published by Institute of Physics Publishing, wholly owned by The Institute of Physics, London

Institute of Physics Publishing, Dirac House, Temple Back, Bristol BS1 6BE, UK

US Office: Institute of Physics Publishing, Suite 1035, The Public Ledger Building, 150 South Independence Mall West, Philadelphia, PA 19106, USA

Typeset by Academic + Technical, Bristol
Printed in the UK by MPG Books Ltd, Bodmin, Cornwall

# Contents

# Preface

The present volume deals with a particular phase of the early history of experimental nuclear physics: what in effect became a race, *circa* 1930, between four laboratory teams to be the first to achieve the transmutation of atomic nuclei with artificially accelerated nuclear projectiles (protons) in high-voltage discharge tubes or vacuum chambers. (Experiments 15 years earlier under Ernest Rutherford had relied on alpha-particles from radium sources in the disintegration of nitrogen nuclei.) The laboratories and their team leaders were John D Cockcroft at the Cavendish Laboratory in Cambridge, England; Merle A Tuve at the Department of Terrestrial Magnetism (DTM) of the Carnegie Institution of Washington; Ernest O Lawrence at the Radiation Laboratory and Department of Physics of the University of California in Berkeley; and Charles C Lauritsen at the Kellogg Radiation Laboratory, California Institute of Technology in Pasadena, CA.

As is generally known, the 'race' was won by the English team in 1932 — the 'annus mirabilis' of nuclear physics with a number of important breakthroughs in nuclear and particle physics that year. Less well known are the details of the competing accelerators, the personalities of the team members, their variegated experiments, and certain external factors that played a role in the British victory.

I start, after some preliminaries, with the early days of the Cavendish under Rutherford and James Chadwick, with certain provocations from the Radium Institute in Vienna, and the arrival of John Cockcroft and his partner Ernest Walton in Cambridge. Next, we meet Ernest Lawrence and Merle Tuve, boyhood friends in South Dakota. We also meet George Gamow and Gregory Breit, theorists who would profoundly affect the upcoming competition. Early geophysics and high-speed electron experiments at the DTM led, oddly enough, to a successful proton accelerator in Tuve's laboratory, taking advantage of Robert Van de Graaff and his clever generator. As a young professor at Berkeley, meanwhile, Lawrence came across an accelerator scheme by the young Norwegian physicist Rolf Wideröe, and magically transformed it into his cyclotron. At Caltech, Charlie Lauritsen had his own ideas for the design of a high-voltage installation for various purposes. With prototype accelerators running after a fashion in Cambridge, in Washington, and in Berkeley and in Pasadena, the race was on in the midst of poor economic times. We learn of the success at Cambridge in more ways than one, and how the competition met the challenge.

By way of rounding out the volume, we follow subsequent developments in nuclear science, culminating in the discovery of fission and its aftermath, and learn how the accelerator laboratories responded to the discovery, each in their own way. While focusing on a particular theme in early nuclear physics, the book thus also provides an overview of the history of modern physics, from Rutherford's experiment at the end of World War I to prospects for nuclear energy on the eve of World War II.

A 'sub-plot' running through much of the narrative, and not so well known, is connected with the divergent ambitions of the two friendly rivals, Merle Tuve and Ernest Lawrence. Once ensconced in Berkeley, Lawrence emphasized ever-more-powerful accelerators, at the expense of the underlying physics made possible with the new atomic artillery. Tuve, on the other hand, had his eye on the basic physics from the start. At the same time he found himself in the awkward position of chasing down a serious blunder in physics interpretation by his friend in Berkeley, while honing his own high-precision accelerator technology in Washington to the point of allowing him to perform the world's first fundamental experiments bearing directly on the internal constitution of the atomic nucleus. In a sense Tuve, less well known than Cockcroft or Lawrence to most, is the unsung hero of our story.

My primary sources include the wealth of archival material in Cambridge, England, in Berkeley, and in Washington, DC. In Cambridge, I had access to the papers and notebooks of Cockcroft (and Ernest Walton) at Churchill College, as well as the Rutherford archives at Cambridge University Library. Not to be overlooked is the museum collection of original apparatus, as well as the photographic archives, at the Cavendish Laboratory nowadays located in the rolling fields of west Cambridge. At Berkeley I availed myself of the E O Lawrence papers, in the Bancroft Library, University of California, including the Berkeley notebooks of M Stanley Livingston. In Washington, DC, I had access to the Merle Tuve papers, all held in the Madison Building, Library of Congress, as well as archives in the DTM Library of the Carnegie Institution of Washington. As always important were the extensive holdings of the Niels Bohr Library, American Institute of Physics, aided by their excellent *Guide to the Archival Collections*. The Niels Bohr Library holdings include a copy of the Archives for the History of Quantum Physics, which I also accessed at the Office for History of Science and Technology of the University of California in Berkeley.

Among the secondary sources pertinent to the subject at hand, I have profited particularly from Guy Hartcup and T E Allibone's *Cockcroft and the Atom,* Herbert Child's *An American Genius: The Life and Times of Ernest Orlando Lawrence,* and John L Heilbron and Robert W Seidel's *Lawrence and His Laboratory: A History of the Lawrence Berkeley Laboratory.* As for Tuve's program, I had the benefit of three unpublished manuscripts. The first, *'Respectfully Submitted, Merle A. Tuve,' Reports from a Golden Age of Physics,* edited, with notes and comments by Louis Brown, contains Merle Tuve's monthly reports to John A Fleming, DTM director, from June 1928 to September 1940. The second manuscript, written by Louis Brown, is a draft for DTM's Centennial History, prepared by Brown as part of a larger history of the Carnegie Institution, which will celebrate its centennial in 2002. The third manuscript was the PhD dissertation by Thomas D Cornell, entitled *Merle A Tuve and His Program of Nuclear Studies at the Department of Terrestrial Magnetism: The Early Career of a Modern American Physicist.*

I am indebted to the following individuals and institutions for archival assistance: Louise King (Churchill Archives Centre, Churchill College, Cambridge), Godfrey Waller (Cambridge University Library), Keith Papworth (Cavendish Laboratory, Cambridge), Fred Bauman (Manuscript Reading Room, Library of Congress), Joseph Andersen and his staff (Niels Bohr Library, American Institute of Physics, College Park, MD), Shaun Hardy (DTM Library, Carnegie Institution of Washington), Rita Labrie (Research Library, Lawrence Berkeley National Laboratory), and the staff of the Bancroft Library (University of California, Berkeley).

Along the way, I have been helped directly and indirectly by many individuals, including Laurie Brown, Louis Brown, Cathy Carson, John Heilbron, Roger Stuewer, Diana Wear, and Spencer Weart.

I thank my publisher, Jim Revill, and his editorial staff at Institute of Physics Publishing, including Simon Laurenson, my editor, for their splendid support and cooperation in all stages of manuscript processing and production of the volume.

Finally, the book depended, as always, on Eleanor, my personal editor, companion, and wife. Much of the work on the name index was done by her; she has a keen sense in copy-editing, and her word processing expertise was essential. All in all, she is truly a partner in my literary and historical endeavors.

**Per F Dahl**                                                                 **January 2002**

# Acknowledgments

The author gratefully acknowledges permission to quote excerpts from the following works.

Peierls R 1985 *Bird of Passage: Reflections of a Physicist* (Princeton: Princeton University Press). Copyright 1985 Princeton University Press. Reprinted with permission from Princeton University Press.

Frisch O R 1979 *What Little I Can Remember* (Cambridge: Cambridge University Press). Copyright 1979 Cambridge University Press. Reprinted with permission from Cambridge University Press.

Alvarez L 1987 *Adventures of a Physicist* (New York: Basic Books). Copyright 1987 Basic Books. Reprinted with permission from Perseus Books Group.

Stuewer R H (editor) 1979 *Nuclear Physics in Retrospect* (Minneapolis: University of Minnesota Press). Copyright 1979 Regents of the University of Minnesota. Reprinted with permission from the University of Minnesota Press.

Heilbron J L and Seidel R W 1989 *Lawrence and his Laboratory: A History of the Lawrence Berkeley Laboratory* (Berkeley: University of California Press). Copyright 1989 The Regents of the University of California. Reprinted with permission from the University of California Press.

Roberts R B 1979 *Autobiography.* Unpublished manuscript, DTM Archives, Carnegie Institution of Washington. Reprinted with permission from Carnegie Institution of Washington, Dept. of Terrestrial Magnetism.

Brown L, "*Respectfully Submitted, Merle A Tuve*," *Reports from a Golden Age of Physics.* Draft manuscript, DTM Archives, Carnegie Institution of Washington. Reprinted with permission of Louis Brown.

Gamow G 1970 *My World Line: An Informal Autobiography* (New York: Viking Press). Copyright 1970 the Estate of George Gamow. Reprinted with permission from Elfriede Gamow.

Letters from L Meitner to O Hahn in Meitner–Hahn correspondence, MPG-Archiv, Archiv zur Geschichte der Max-Planck-Gesellschaft, Berlin-Dahlem. Reprinted with permission from Archiv zur Geschichte der Max-Planck-Gesellschaft.

Letters from N Bohr to colleagues, Bohr Scientific Correspondence, Niels Bohr Archive, Copenhagen.

E O Lawrence Papers, The Bancroft Library, University of California, Berkeley. Portions reprinted with permission from The Bancroft Library.

R T Birge Papers, The Bancroft Library, University of California, Berkeley. Portion reprinted with permission from The Bancroft Library.

G Lewis Papers, College of Chemistry and The Bancroft Library. University of California, Berkeley. Portions reprinted with permission from The Bancroft Library.

P Harteck Papers, Folsom Library, Rensselaer Polytechnic Institute, Troy, NY. Portion reprinted with permission of Folsom Library.

The author gratefully acknowledges permission to use the following illustrations:

Figure 2.1 by courtesy of Cambridge University Press.

Figure 4.1 by courtesy of AIP, Emilio Segrè Visual Archives and the University of Cambridge, Cavendish Laboratory, Madingley Road, Cambridge, UK.

Figures 6.1, 12.1, and 13.1 by courtesy of the Department of Terrestrial Magnetism, Carnegie Institution of Washington.

Figure 10.7 by courtesy of the Department of Terrestrial Magnetism, Carnegie Institution of Washington and the American Physical Society.

Figures 4.2, 5.2, 7.1, 7.2, 10.5, and 10.6 by courtesy of the American Physical Society.

Figures 9.1, 9.3, 9.4, and 11.1 by courtesy of the Royal Society.

Figure 9.5 by courtesy of the University of Cambridge, Cavendish Laboratory, Madingley Road, Cambridge, UK.

Figures 10.1 and 10.2 by courtesy of Lawrence Berkeley National Laboratory and AIP, Emilio Segrè Visual Archives.

# List of Illustrations

# Chapter 1

# PROLOGUE

## 1.1.  1932: Wings over Europe, and other upheavals

In early March of 1932, a play opened at the Globe Theatre in London's West End. *Wings over Europe,* by the poet Robert Nicholes and the actor-producer Maurice Browne, was remarkably prescient with respect to certain scientific developments under way just then, only an hour by train north of London [1-1].

When the curtain rises, a young physicist of twenty-five, Francis Light-foot, has discovered how to release the energy locked in the atom. By happenstance, Lightfoot is also a nephew of the Prime Minister, who has summoned a Committee of the Cabinet to consider how to deal with this ominous discovery of potentially devastating threat to humanity and the earth itself. The hours tick by toward a noontime doomsday, with Lightfoot, arrogant and confused, resolved to face annihilation along with the despairing Committee members, by activating his material disintegration process. Alas, Lightfoot is shot dead in the nick of time by the Secretary of War. Too late, nevertheless! The P.M. is informed that control of the atom has also been discovered in Yugoslavia. As the Cabinets of all the nations are summoned by the Guild of United Scientists of the World, aeroplanes capable of dropping atomic bombs are heard droning overhead.

1932 was indeed a turbulent year in many areas of world affairs—both in the East and West. Central Europe was still reeling from the Great Depression, precipitated by the failure of the Austrian credit market the year before; by the beginning of 1932, the number of unemployed was already more than six million [1-2]. In the United States, the depression following the Stock Market crash of 1929 was just as evident, with wide-spread unemployment, bank failures, and business calamities. In the Far East, matters took a different, albeit equally disastrous turn. January of 1932 saw Sino-Japanese hostilities in full swing. Large-scale Japanese forces attacked Woosung Forts below Shanghai, and warships of the

1

Western powers gathered in Shanghai waters to protect citizens from the onslaught. In Russia, a severe famine swept the land, in the wake of governmental agrarian policies run amok.

On the purely political front, much was happening. In Bombay, the year 1932 began with Mohandas K Gandhi's arrest on the eve of his new civil disobedience campaign, and the Congress was promptly outlawed by the British India government. March 1932 saw presidential elections in Germany, with Paul von Hindenburg falling short by a narrow margin of the required majority over Adolf Hitler and Ernst Thälmann (the latter representing the Communists). In a run-off election the next month, Hindenburg secured a plurality of 6,000,000 out of a total of 36,000,000 votes. Late in the year, with the Reichstag already dissolved on a Communist motion of 'no confidence' in the Ministry of Franz von Papen, Hitler rejected a conditionally proffered Chancellorship. In the United States, the same month of November, Franklin Delano Roosevelt, Democrat, was elected president over Herbert Hoover, Republican, by 472 to 59 electoral votes.

Among human events of 1932, two stand out, concerning as they did an aviator and aviatrix of kindred fame. The baby of Charles A Lindbergh was kidnapped in March; two months later Amelia Earhart became the first woman to fly solo across the Atlantic, from Newfoundland to Londonderry, Ireland, in 13.5 hours. August saw another aerial record, when Auguste Piccard ascended 17.5 miles above Zürich in a stratospheric balloon. On terra firma, records were broken one after another, following the opening of the Third Olympic Winter Games at Lake Placid, NY, that February.

1932 was a banner year for physical science as well. It began in January with the announcement by Harold Urey at Columbia of his discovery, in late 1931, of a heavy isotope of hydrogen, which he named 'deuterium' [1-3]. Just about the same time came word from Frédéric and Irène Joliot-Curie in Paris of their detection of high-speed protons from the collision of energetic $\gamma$-rays from beryllium with hydrogen nuclei in paraffin. Following up on the latter, startling communication at Cambridge, England, James Chadwick soon had an alternative explanation for the process at hand: the strange effect was actually due to protons being knocked out of paraffin by a neutral particle, of mass about equal to the proton, emanating from the beryllium source. Chadwick's discovery of the neutron was announced in mid-February. Six months later, an American physicist, Carl D Anderson, again made news, announcing his discovery among tracks of cosmic-ray particles of the 'positron,' so named to his displeasure by the editors of *Science News Letters*. An unstable, short-lived particle identical to the electron, except for carrying a positive charge, the 'antiparticle' to the electron is never found in ordinary matter. It had been predicted three years earlier by

Paul A M Dirac as one of two solutions to the equation forming the basis of his own theory of the electron. (The theoretical *pièce de résistance*, Enrico Fermi's theory of $\beta$-decay, came two years later, in January 1934. It heralded the last of the four forces of nature — gravity, electromagnetism, the 'strong' nuclear force, and now Fermi's 'weak' force — the latter responsible for radioactive decay.)

However, arguably the most important development in physics in 1932 was one anticipated by only a few days, in a manner of speaking, by the opening of the play in London. A few days after the inspiring yet somber affair staged in the West End, two physicists at Cambridge, John Douglas Cockcroft and Ernest T S Walton, the first ten years older and the second only four years senior to the pertinacious young Lightfoot, submitted a letter to *Nature* [1-4]. It announced the first nuclear disintegration of an atom by artificially accelerated particles — the splitting of the lithium nucleus, under proton bombardment, into two $\alpha$-particles (helium nuclei). More startling, wrote Waldemar Kaempffert in the *New York Times*, was the energy release of 16,000,000 volts for a maximum energy expenditure of 400,000 volts in firing ten million proton shots for every hit scored [1-5]. On 21 April Ernest Rutherford, the Cavendish Professor at Cambridge, in writing to Niels Bohr about the recent excitement over the neutron, mused that

> ...it never rains but it pours, and I have another interesting development to tell you about of which a short account should appear in *Nature* next week. [1-6]

Describing the work of Cockcroft and Walton, under way since 1928, he cautiously added that their results 'may open up a wide line of research in transmutation generally.' Kaempffert was more emphatic as well as pessimistic in contemplating the portents of the Cambridge experiment. Echoing, perhaps unbeknownst, the situation aired on the stage of the Globe Theatre, he paraphrased Sir Oliver Lodge to the effect that

> ...man is not yet spiritually ripe for the possession of the secret of atomic energy.... Technically we are demi-gods, ethically still such barbarians that we would probably use the energy of the atom much as we used the less terrible forces that almost destroyed civilization during the last war. [1-7]

Thirteen years and one month earlier, Rutherford himself had smashed the atom, using the reverse reaction. He had fired $\alpha$-particles from a radioactive source at nitrogen atoms, and driven out protons — particles vaguely foretold in still earlier experiments at Cambridge. Now, his young assistants had fired protons, accelerated in a high-tension apparatus, into lithium and obtained $\alpha$-particles. Whereas Rutherford, in 1919, had provided the first example of artificially induced transmutation of a chemical element (as distinct from the naturally occurring, spontaneous transmutations of radioactivity), Cockcroft and Walton had achieved a more far-reaching goal: the first nuclear transmutations with projectiles artificially accelerated to high energies in laboratory apparatus. They thereby set the stage, if we go along with the pun, for manipulating the forces of nature under precisely controlled conditions.

The Cambridge physicists were, however, not the only ones to have devised an electrical method of hurling particles at atomic nuclei. Physicists in Berkeley, in Pasadena, and in Washington, DC, had also been hard at work on the same problem since 1928 — the year, incidentally, when *Wings over Europe* made its debut in New York. How four far-flung teams happened to focus on a central problem in atomic research at the very same time, will be explored in what follows.

# Chapter 2

---

# THE ENGLISH STAGE IS SET

## 2.1. Manchester, 1919

The guns of World War I fell silent in November of 1918, on the eleventh hour of the eleventh day of the eleventh month of the year. The Treaty of Versailles was signed seven months later, in June 1919. That June, marking the formal end of the Great War, also marked the completion of a series of experiments in Manchester, England, by Ernest Rutherford. The publication of his epoch-making results that very month, also signaled the last piece of scientific work that Rutherford undertook at Manchester University; later in the year he succeeded Joseph John (J J) Thomson as Cavendish Professor at Cambridge. From then on, administrative duties and other affairs would claim much of his time.

Preoccupied as he was with wartime matters, Rutherford had nevertheless found time for the work that led to his crowning experimental discovery, when he could scrape up 'an odd half-day' for it. That work had its genesis in early 1915, as an extension of an experiment carried out by one of his students, Ernest Marsden. Marsden's experiment, in turn, was an outgrowth of the classic investigation by Hans Geiger and himself in 1909 under Rutherford's guidance at Manchester: the scattering of $\alpha$-particles off thin metallic foils. Their results revealed unexpectedly large $\alpha$-deflections, quite at variance with the small-angle, multiple scattering expected on the basis of J J Thomson's model of the atom making the rounds just then. Rutherford's pondering over these results during the winter of 1910–11 had culminated in his inspired nuclear atom of 1911 [2-1]. Featuring a central, positively charged nucleus with discrete electrons forming the periphery of an almost empty atom, Rutherford's atom contrasted sharply with Thomson's 'plum pudding' model of distributed positive electricity (the pudding) embedded with negative corpuscles (the plums).

The experiment of Geiger and Marsden involved collisions between alphas from a radium source and atomic nuclei massive enough for recoils of the target nuclei, typically gold, to be ignored. But what might happen in a close encounter between an $\alpha$-particle and a much lighter nucleus, say one of hydrogen with one-quarter the mass of the incident $\alpha$-particle? Charles Galton Darwin, grandson of the great Charles Darwin and a young mathematical physicist on Rutherford's Manchester team, had given the matter some thought. He concluded that head-on collisions in hydrogen should give rise to 'H' particles recoiling with nearly twice the velocity and four times the range of the $\alpha$-projectiles [2-2].

Marsden had attacked the problem experimentally on the eve of the Great War. Preliminary results were about as predicted for collisions in hydrogen, showing recoiling H-particles with a range of nearly 100 cm. However, under certain conditions, among them collisions in air, something was evidently amiss: now and then H-particles with a range much too great to be accounted for as recoils showed up. To Marsden's chagrin, the outbreak of World War I prevented him from pursuing the anomaly; the experiments were hurriedly written up for publication as he dashed off to war. The paper ended on a tantalizing note, expressing 'a strong suspicion that the H-particles are emitted from the radioactive atoms themselves' [2-3].

Rutherford was intrigued by the possibility of some new particle (besides $\alpha$-, $\beta$-, and $\gamma$-rays [2-4]) emitted from the radioactive atoms of the source. Making sure he was not stepping on Marsden's turf [2-5], he set about following up on the experiment, though getting started was sooner said than done, under the prevailing conditions. Only in 1917 was he able to devote much time to the problem. It is remarkable that he could tackle it at all, not to speak of bringing the investigation to a brilliant conclusion, with the war raging on the Western Front, and the struggle claiming nearly everybody, in or out of uniform. About the only one left to assist him at Manchester was William Kay, his trusted laboratory steward. Moreover, Rutherford, too, was soon heavily involved in the war effort, diverting most of his remaining scientific resources to anti-submarine warfare — submarine detection, in particular. He also became an active member of the short-lived, disastrous Board of Inventions and Research. Important despite its shortcomings for setting a precedent in harnessing science for wartime needs, the BIR was spearheaded by Lord Fisher, the eccentric Admiral, to render expert technical assistance to the Admiralty by screening inventions and sponsoring research. Rutherford's little-known legacy in the latter connection, his secret reports advising on patterns for underwater warfare, remains as applicable today as when he wrote them in 1915 [2-6].

Rutherford has also, happily, left an extensive, if disorganized, record of his investigation of the 'H-particle problem,' in his laboratory notebooks for this period — records still extant, among his archives at Cambridge University Library. The first entries for the problem at hand

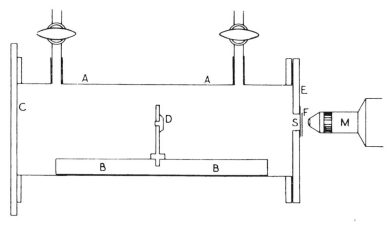

**Figure 2.1.** *Apparatus used by Rutherford in observing the first nuclear disintegrations. Rutherford, Chadwick, and Ellis, 1931,* Radiations from Radioactive Substances *(Cambridge: Cambridge University Press).*

are for March of 1917, and the notebooks culminate with records of a final experiment exactly two years later [2-7]. The experimental arrangement, similar to Marsden's, is shown in figure 2.1. D is a RaC (polonium 83) source of $\alpha$-particles near the center of the brass vessel AA that could be evacuated or filled with the gas to be studied. The opening in the brass plate E was covered with absorbing metal foils of various thicknesses, presenting different stopping powers for alphas when the vessel was under vacuum. A zinc sulphide screen F just outside the vessel was viewed through the microscope M. (Much of the counting of scintillation flashes was done by Kay alone, since Rutherford had limited patience for the drudgery of getting dark-adapted for over a half-hour and then counting for all of a few minutes, due to the strain.) The screen was located far enough from the source that scintillations observed could not be due to the alphas emitted by the source, since they were absorbed by the gas between the source and the screen. The apparatus was placed in a strong magnetic field to keep lighter $\beta$-particles (swift electrons) from striking the screen. (Some of the atoms transformed by $\alpha$-emission, the rest by $\beta$-decay.)

With dry oxygen or carbon dioxide admitted, scintillations from what appeared to be hydrogen-like particles identical in mass and range to Marsden's H-particles decreased with increasing gas concentration. This was as expected from the stopping power of the column of gas between source and screen. But wait! Admitting dry air brought a 'surprising effect': the number of scintillations increased. Rutherford's reaction is an understatement, to say the least. Having shown that oxygen alone decreased the count, the increase could only be due to the nitrogen in the dry air. The $\alpha$-particles had collided with nitrogen atoms and knocked hydrogen ions out of their nuclei. A new phenomenon, glimpsed but not grasped by Marsden, was at hand: the first artificially induced disintegration and transmutation of a chemical atom. The experiment established something else, long suspected, as well: that hydrogen ions or *protons* (as Rutherford was soon to call them [2-8]) are among the constituents of atomic nuclei generally, not only of hydrogen.

The first explicit reference to 'disintegration of nitrogen' only appears on page 39 of the 40-page notebook covering the final round of these experiments, recorded in mid-March of 1919 [2-9]. Marsden, too, participated in this mopping-up work, having returned to Manchester from overseas for a brief stint as Major E Marsden of the New Zealand Division Signals Company; he and Kay alternated as observers, and he assisted in changing absorption screens, filling the gas-chamber, and performing control experiments. By 18 March, the date for the last sparse entry for Notebook 25, Rutherford was no doubt busy writing his formal report on the full set of experiments, arranged in the form of four consecutive manuscripts under the common heading 'Collision of $\alpha$ particles with light atoms.' They were completed in April, and published as back-to-back papers in the *Philosophical Magazine* for June of 1919. The 'Discussion of results' in Part IV, the centerpiece of the quartet with the sub-title 'An anomalous effect in nitrogen,' leads off with the following observation:

> From the results so far obtained it is difficult to avoid the conclusion that the long-range atoms arising from collision of $\alpha$ particles with nitrogen are not nitrogen atoms but probably atoms of hydrogen, or atoms of mass 2. If this be the case, we must conclude that the nitrogen atom is disintegrated under the intense forces developed in a close collision with a swift $\alpha$ particle, and that the hydrogen atom which is liberated formed a constituent part of the nitrogen nucleus. [2-10]

In the largely typewritten draft manuscript of Part IV, headed curiously, by 'An anaomalous [sic] effect in nitrogen,' what is surely no more than a second typo in the last sentence of the above citation has been corrected in Rutherford's hand as follows:

nitrogen

'...formed a constituent part of the ~~hydrogen~~ nucleus' [2-11].

Rutherford's picture of the nuclear process involved, we might add, viewed the 14 particles inside the nitrogen nucleus as arranged in three $\alpha$-particles, each of mass 4, plus two other particles, one of which is a proton [2-12]. These two particles orbit in the outer regions of the nucleus, and are, consequently, easily dislodged if struck by an $\alpha$-particle. Since the neutron was not yet known in 1919 (but would be vaguely anticipated by Rutherford scarcely a year later, as we shall see), Rutherford considered these two outer particles as 'either two hydrogen nuclei or one of mass 2.' The actual equation for the reaction,

$$_2\mathrm{He}^4 + {_7}\mathrm{N}^{14} \longrightarrow {_8}\mathrm{O}^{17} + {_1}\mathrm{H}^1$$

was only confirmed by P M S Blackett, a junior colleague of Rutherford, in 1925. An expert on the cloud chamber (invented years before at the Cavendish by C T R Wilson), Blackett's study of 400,000 $\alpha$-tracks in photographic emulsions revealed eight cases in which two tracks emerged from the point of impact between an $\alpha$-particle and a nitrogen nucleus: a light, lengthy track of the proton, and a short, heavy track left by the oxygen-17 nucleus [2-13].

Part IV ends on the following upbeat note (inserted, curiously, as a handwritten addition to the original manuscript):

> The results as a whole suggest that, if $\alpha$ particles — or similar projectiles — of still greater energy were available for experiment, we might expect to break down the nucleus structure of many of the lighter atoms. [2-14]

Rutherford himself harbored no doubt as to the importance of the experiments just completed. (This, oddly enough, was in contrast to his earlier discovery of the nuclear model of the atom, which he failed initially to appreciate [2-15]). His famous apology, for having arrived late one day for an international antisubmarine warfare meeting, also underscores his disdainful attitude towards the war: 'I have been engaged in experiments which suggest that the atom can be artificially disintegrated. If it is true, it is of far greater importance than a war!' [2-16]

## 2.2. From Manchester to Cambridge: off to a poor start

Ernest Marsden was not the only former student of Rutherford to return, if only for a few weeks, to Manchester following the Armistice in late 1918. So did James Chadwick, albeit with an altogether different wartime experience behind him. Raised in Bollington near Macclesfield, the son of John Joseph Chadwick who ran a laundry business in Manchester, Chadwick had obtained his undergraduate degree in physics under Rutherford in 1911, and his Master's degree in 1913. Rutherford then recommended him for an Exhibition of 1851 Senior Research Studentship. A condition for the scholarship being that its recipient carry out research in a laboratory removed from where he had earned it, Chadwick cast about for a suitable place to carry on his work in radioactivity. What better place than the Physikalisch-Technische Reichsanstalt in Berlin/ Charlottenburg? Among other things, Hans Geiger had recently, following six years in England, first under Arthur Schuster and then Rutherford at Manchester, taken up the newly appointed position as director of the laboratory for radium research at the PTR, and there Chadwick went.

In the event, the outbreak of war put an untimely end to what had begun as a promising stint at the Reichsanstalt, studying $\beta$-radiation under Geiger's 'genial tuition' [2-17], while exposed to a roster of scientific luminaries at the PTR generally, and at Heinrich Rubens's weekly Colloquium in particular. Chadwick was interned for the duration of the war in a camp for civilian prisoners at Ruhleben, near the Berlin suburb of Spandau. Through his own resilience and the kindness and generosity of colleagues at the Reichsanstalt as well as camp authorities, plus some bribery, he even managed some token research in molecular physics, with improvised apparatus in a makeshift laboratory. In this he was joined by Charles Drummond Ellis, a few years younger and a cadet at the Royal Military Academy at Woolwich. Ellis had been caught summering in Berlin when the war began.

As for Geiger, he wound up serving as artillery officer on the Western Front, across the lines from his former colleague Marsden, who served with a Sound-Ranging Section of the Signal Corps in France (as did Darwin), steadily advancing in grade and eventually wounded in action. While at the front, Marsden received a letter from Geiger, forwarded through Niels Bohr (back in Copenhagen after spending 1914–15 at Manchester), congratulating him on his Wellington appointment, put off by the war.

Following the Armistice, Chadwick made his way home to Manchester, and set about picking up the pieces of his interrupted career. Suffering from malnutrition, he promptly penned a handwritten report on his ragtag experiments at Ruhleben to the Commissioners of the 1851 Exhibition. He also paid a call on Rutherford, who was just then putting the finishing touches on the disintegration experiments. Pleased to have him back, Rutherford took him on as a teaching assistant, and brought him along when he departed for Cambridge in mid-1919 as the new Cavendish Professor.

The first task of Rutherford after settling in at Cambridge was to reorganize the Cavendish Laboratory, renowned though it may have been as one of the world's leading centers in experimental physics research [2-18]. The Cavendish was 42 years old when Rutherford began his appointment. The Laboratory, and a Professorship in Experimental Physics, were both named after the family of the seventh Duke of Devonshire, who founded them both in the 1870s. Lord Kelvin (Sir William Thomson at the time) was approached to occupy the first Cavendish Professorship, but declined, not wishing to relinquish his private researches and professorship at Glasgow; so did his German counterpart, Hermann Helmholtz. Instead the position went to James Clerk Maxwell, another Scot, who supervised the construction of the original Cavendish building on Free School Lane in central Cambridge; the building formally opened in 1874. After Maxwell's untimely death from cancer in 1879, the Professorship was again offered to Kelvin, who declined once more. Instead, the post was accepted by John William Strutt, the third Lord Rayleigh, on his proviso that his appointment be limited to five years. Rayleigh was succeeded by J J Thomson in 1884, who held the post for all of 35 years, until he stepped down in favor of Rutherford.

Despite its fame the Cavendish was perennially a poor institution, and 1919 was no exception. More laboratory space and money were desperately needed, both for teaching and research. With regard to space, the venerable Laboratory had been updated twice, in 1896 and again during 1906–08; now Rutherford demanded three new, well-equipped laboratory additions as well, earmarked for applied physics, optics, and for what Rutherford called 'General Properties of Matter.' Money, for the new building addition, for equipment and for salaries for new teaching staff, proved to be a daunting proposition, however. Though Rutherford requested £200,000, the majority for equipment and salaries, the ponderous university machinery for allocating funds moved so slowly that significant money for scientific research only became available in the second half of the 1920s [2-19]. Rutherford himself didn't help matters with his ingrained reluctance of following through on money initially demanded — or refusing to act with the 'importunate mendacity' as Sir Joseph Larmor, the Lucasian Professor of Mathematics at Cambridge, assured Rutherford was the strong point among the higher-ups in Vice-Chancellor Sir Arthur Shipley's administration [2-20]. Believing in simplicity, Rutherford hated *spending* money as well as raising it.

The lack of funds did have one saving grace, in forcing the Cavendish to concentrate on nuclear physics — albeit with a paucity of equipment and lack of singleness of purpose among the laboratory staff. The one who felt the frugality and lack of dedication most strongly was Chadwick, in whom Rutherford had entrusted day-to-day running of the laboratory; Chadwick eventually assumed the post of Assistant Director of the Cavendish. While thus preoccupied, Chadwick also found time to pursue a PhD degree, a degree only formally established at Cambridge in 1920, long after the Doctorate of Philosophy had become the standard second degree at Oxford and elsewhere in the UK.

Chadwick attacked his thesis problem, on the size and shape of the $\alpha$-particle, by the scattering of $\alpha$-particles in hydrogen sheets of paraffin wax. As such, it was an auxiliary investigation in his renewed researches under Rutherford's direction, addressing two general subjects: the scattering of $\alpha$-particles in hydrogen, and the disintegration of light elements under $\alpha$-bombardment. The disintegration experiments, carried out during the first half of the 1920s, demonstrated the transmutation in a range of elements besides nitrogen — namely nuclei of all elements up to potassium, except lithium, carbon, and oxygen. The experimental arrangement was much like that for the nitrogen disintegration setup of 1917–19 (figure 2.1), featuring a movable source of $\alpha$-particles in the middle of a cylindrical brass vessel. The target elements to be exposed to $\alpha$-rays were deposited as powders on a gold foil placed immediately in front of the source. One end of the vessel was pierced with a small hole and covered with a silver foil. A zinc-sulphide screen was fixed next to the foil, leaving a narrow slot between them in which absorbing sheets could be inserted. They were thin sheets of mica equivalent in their stopping power of 5, 10, 20 or 50 cm of air. The scintillations were observed through a microscope with a right-angle bend in it to shield the observer from direct $\gamma$-rays from the target. To reduce the illumination of the screen by $\beta$-rays, the whole apparatus was placed between the poles of a large electromagnet, which bent the rays away before they struck the screen.

Rutherford and Chadwick found that recoil-hydrogen ions produced by collisions with $\alpha$-particles had a range of about 30 cm in air. By considering only H ions with a range greater than 32 cm, they could be sure that the ions came from the disintegration of the target element, and not from hydrogen impurities in it. All six elements that could be disintegrated had an atomic mass which could be expressed as $4n + 2$ or $4n + 3$, lending credence to Rutherford's pet scheme of the nucleus consisting of $n$ helium nuclei tightly bound, plus 2 or 3 hydrogen nuclei ('H satellites'), less tightly bound, orbiting the central core [2-21].

The radioactive sources, as well as the thin absorbing strips of mica, were prepared by Rutherford's personal laboratory steward, George Crowe, who succeeded the equally admirable William Kay who had preferred to remain in Manchester. Crowe has left a vivid description of these tedious experiments, and the Professor's impatient quest for new phenomena.

> He was shooting protons out of light atoms by means of $\alpha$-parti-cles; carbon would not play. The next day was the turn of alumi-num, and on some theoretical grounds Rutherford thought that the protons would have a high velocity and a long range. 'Now, Crowe, have some mica absorbers ready tomorrow with stopping power equivalent to 50 cm of air.' 'Yessir.'
>
> On the next day: 'Now, Crowe, put in a 50 cm screen.' 'Yessir.' 'Why don't you do what I tell you—put in a 50 cm screen.' 'I have, sir.' 'Put in 20 more.' 'Yessir.' 'Why the devil don't you put in what I tell you, I said 20 more.' 'I did, sir.' 'There's some damned contamination. Put in *two* 50s.' 'Yessir.' 'Ah, it's all right, that's stopped 'em! Crowe, my boy, you're always wrong until I've proved you right! Now we'll find their exact range!' [2-22]

The $\alpha$-scattering experiments, which addressed the fundamental question of the structure of nuclei in greater detail, were resumed during the second half of the decade — alas, a period of declining productivity in Rutherford's own scientific and personal life. The Professor had reached middle age, with little further progress to show in elucidating the fundamental problems at hand. His family life was troubled as well. His only daughter, Eileen, though married to Ralph Fowler, the mathematical physicist, led an unconventional life style contrary to her father's way of thinking. In December of 1930 she gave birth to a fourth child, and all was going well when she died of a blood clot. In addition, his wife Mary never adjusted to the transition from provincial New Zealand society to the mannered circles of Cambridge Dons and academicians. Fortunately, his grandchildren remained his one great delight to the end.

Another elusive research objective was the neutron, a hypothetical neutral particle postulated by Rutherford in 1920. He had proposed it on the basis of the two fundamental building blocks of matter known since his alchemical experiment in Manchester the year before, electrons and protons. In his Bakerian Lecture to the Royal Society on 3 June 1920, Rutherford had observed that, within experimental limits, the masses of all isotopes [2-23] hitherto studied were represented by whole numbers, when tabulated on a scale in which oxygen is taken as 16. The only exception appeared to be hydrogen, with a mass of 1.008. This anomaly suggested to Rutherford that 'it may be possible for an electron to combine much more closely with the H nucleus [than in the ordinary hydrogen atom], forming a kind of neutral doublet' [2-24]; that is, an 'atom' of mass unity and zero nuclear charge.

> Such an atom would have very novel properties. Its external field would be practically zero, except very close to the nucleus, and in consequence it should be able to move freely through matter. Its presence would probably be difficult to detect by the spectroscope, and it may be impossible to contain it in a sealed vessel. On the other hand, it should enter readily the structure of atoms, and may either unite with the nucleus or be disintegrated by its intense field, resulting possibly in the escape of a charged H atom or an electron or both. [2-25]

Note that the word *proton* had not yet been coined by Rutherford to denote the positively charged 'H-particles' discovered in 1919; he proposed the new term in a talk before the British Association at Cardiff later in the year. At any rate, soon after his Bakerian Lecture, Rutherford had asked J L Glasson, a research student, to search for evidence for his 'neutral doublets,' or what he still casually called 'atoms,' with a beam of positive hydrogen ions striking a lead target, but without success. A year later, in 1922, J K Roberts, another Cavendish researcher, looked for the missing energy due to the production of neutrons in an electrical discharge through hydrogen gas, and in so doing 'wasted a lot of time' on 'a crazy idea of the Prof. or Chadwick' [2-26]. Undeterred, the detection of *existing* neutrons, and *creating* them by firing protons into atoms, remained a vexing preoccupation of Rutherford and Chadwick. In 1923 Chadwick looked in vain for $\gamma$-radiation which could be produced in the process of combination of a proton and an electron, using an ioniza- tion chamber and a point counter as detectors. He also attempted to unite protons, accelerated to 200,000 volts with a Tesla coil, with the tightly bound inner electrons of the heavy elements, but the equipment and techniques then at the Cavendish were inadequate for the task [2-27].

Chadwick would indeed discover the neutron by a different pro- cedure, in the disintegration of atoms under $\alpha$-particle bombardment, a full decade later. For our purpose, however, the thing to note is that the need for higher voltages across gaseous discharge tubes at Cambridge was driven, not so much by the need for higher voltages in disintegration experiments, as by the somewhat misguided effort of producing Ruther- ford and Chadwick's pet construct, the neutron, by electrical means.

The hunt for the neutron languished as the prey stubbornly refused to reveal itself, and Rutherford, increasingly under pressure from Chadwick to spend more on equipment to get on with the search, grew more and more morose over the lack of progress in all of the various ongoing investigations — not merely searching for the neutron. In his Presidential Address to the British Association in 1923, he characterized the difference between the positive and negative electricity simply as 'an enigma,' and complained that 'in the present state of our knowledge it does not seem possible to push the enquiry further' [2-28]. The next year, speaking before the Franklin Institute, he was equally pessimistic, viewing the nucleus as 'a world of its own which is little if at all influenced by the phy- sical and chemical forces at our command' [2-29]; perhaps having to put up with stiff competition, out of the blue in Vienna, fueled his pessimism. Only in 1927, after five years of wasted effort, did he reveal his old self in proposing a fresh line of attack, in his Royal Society Anniversary Address on 30 November of that year. The background for his sudden optimism was a meeting with a young man named Allibone in March of 1926. First, however, we must deal with the Radium Institute in Vienna, and an unfortunate affair precipitated there by two young researchers.

### 2.3.  Provocations in Vienna

Just as Rutherford and Chadwick were getting started on the disintegra- tion of light elements by $\alpha$-bombardment, so were two young physicists in Stefan Meyer's Institut für Radiumforschung in Vienna. Rutherford had experienced kindly relations with his old friend Meyer and his Radium Institute in the war-ravaged former capital of the splintered

Austro-Hungarian Empire at the end of World War I. He dispatched reprints and copies of *Nature* so that Meyer's demoralized staff could keep up with their colleagues abroad, and even contrived among British officialdom to pay for the radium which the Institute had loaned the Cavendish before the war. Strictly speaking the radium was now 'enemy property.' However, Rutherford's modest grant of £500 was enough to ensure the survival of Meyer's Institute.

In the spring of 1922 a Swedish oceanographer named Hans Pettersson had arrived at the Radium Institute, intending to measure the radioactivity of deep-sea sediments [2-30]. Before long, however, he began a collaboration with Gerhard Kirsch, one of Meyer's assistants, on the artificial disintegration of elements in a prolonged investigation that would generate an international controversy and strain the patience of Rutherford and Chadwick. In September 1923 Pettersson and Kirsch reported their first results, that $\alpha$-particles from RaC' (an isotope of polonium) produced disintegrations in lithium, beryllium, magnesium, and silicon, yielding protons with maximum ranges between 10 and 18 cm in air. They concluded that 'the hydrogen nucleus is a more common constituent of the light atoms than one had hitherto been inclined to believe' [2-31]. In a private letter to Meyer, dated 24 November 1923, Rutherford was moved to inquire whether Pettersson and Kirsch were working under his direction; describing their experiments 'as a valiant piece of work,' he thought, nevertheless, on the basis of work carried out by two research students at the Cavendish, that 'one or two of the conclusions are doubtful' [2-32]. The two Cavendish students were L F Bates and J S Rogers, who argued that Pettersson and Kirsch were not observing disintegration protons at all, but ordinary $\alpha$-particles emitted directly from their RaC' source.

The difference in opinion came to a head as a result of a paper by Pettersson, read at a meeting of the Physical Society of London in 1924; in it, he disputed Rutherford's satellite model of the nucleus. Instead, he proposed 'an alternative hypothesis which assumes that the $\alpha$-particle communicates its energy to the nucleus as a whole, precipitating an explosion.' Not only that; artificial disintegration was not confined only to certain elements, as Rutherford would have it. Rather, the Viennese experiments indicated that it was 'a general property, common to the nuclei of *all* atoms' [2-33]. Rutherford and Chadwick struck back in March 1924 [2-34]. In their previous work, they had concentrated on detecting protons with ranges greater than 30 cm in air. Now they reported on observations at an angle of 90° with respect to the direction of the incident alphas. Assuming 'that the particles of disintegration are emitted in all directions,' this allowed them to detect short-range disintegration protons as well. By this arrangement, they were able to add seven more light elements to their previous six, including magnesium and silicon. Pettersson and Kirsch responded on a smug note, this time bolstering their claim by reporting 'that carbon, examined as paraffin, as very pure graphite, and finally as diamond powder,' yielded protons of about 6-cm range [2-35].

Rutherford, in turn, disputed the latest Viennese claim as 'in complete disagreement' with the Cambridge experiments; he and Chadwick had searched for disintegration protons of a range down to 2.6 cm in air and found none whatsoever. Rutherford also vented his opinion of Pettersson and his work in a letter to Bohr.

> He seems a clever and ingenious fellow, but with a terrible capacity for getting hold of the wrong end of the stick. From our experiments, Chadwick and I are convinced that nearly all his work published hitherto is either demonstrably wrong or wrongly interpreted. [2-36]

The controversy dragged on. With Rutherford out of touch for much of 1925, traveling abroad and with other matters claiming his attention, the affair was largely left for Chadwick to handle. Pettersson actually spent a few days in the lion's den in Cambridge, discussing their work, but neither side was prepared to yield on their respective positions. Chadwick, too, visited Vienna in December of 1927, and in so doing settled the affair, in a manner of speaking. Learning little from Pettersson, he focused on the scintillation counting, which was done by three young women. They were, wrote Chadwick to Rutherford,

> ...of what Pettersson called Slavic descent because he believed (I'm only repeating what he said to me) that ... Slavs had better eyes, secondly, and mainly, that women would be more reliable than men as counters of scintillations because they wouldn't be thinking while they were observing them. [2-37]

The problem was, the young women were seeing what they were expected to see. Adds Otto Frisch, who began his career at the Radium Institute under Karl Przibram, and left Vienna the same year, 1927:

> On the face of it, [using women students to do the counting] appears to be a very objective method because the student would have no bias; yet the students quickly developed a bias towards high numbers because they felt that they would be given approval if they found lots of particles. Quite likely this situation caused the wrong results along with a generally uncritical attitude and considerable enthusiasm over beating the English at their own game. [2-38]

Said Przibram to Frisch as he left for Berlin, with sadness in his voice, 'You will tell the people in Berlin, won't you, that we are not quite as bad as they think?' [2-39].

The matter was thrashed out in a tense meeting between Chadwick, Pettersson, and Stefan Meyer. Director Meyer accepted Chadwick's view that subjective bias was at work, and offered to make a public retraction to set the record straight. Without consulting Rutherford, and out of respect for Meyer, Chadwick declined his offer, recommending instead that they drop the matter then and there. And so ended a controversy in modern physics, with Rutherford and Chadwick remaining on good terms with both Meyer and Pettersson in the years to come.

In their tome, *Radiations from Radioactive Substances*, all Chadwick had to say about the matter was that 'it seems difficult to put forward any satisfactory explanation of these and other divergences between two series of investigations, and this is not the place to attempt it' [2-40]. And if not satisfactorily resolved, the protracted affair was of indirect benefit to the Cavendish, in drawing attention to the realization that scintillation counting could no longer be relied upon for acquiring charged-particle data. Fortunately, new counting techniques were in an early stage of development just then, among them electrical methods of detection, under development principally in Germany. Above all, there was the cloud chamber—resurrected in about 1925 by Patrick Blackett at the Cavendish, as noted earlier, and of which we shall hear more before long.

### 2.4. A million volts or more?

*Wings over Europe*, as noted earlier, was first performed by the New York Theater Guild in December of 1928. It seems unlikely that its two playwrights were aware that young physicists in America and in England were just then setting out to accomplish what Lightfoot failed to do. In fact, the physicists themselves scarcely knew of their forthcoming preoccupation with atom smashing when the year began, except possibly for one or two in the UK.

Ernest T S Walton, who happened to be 25, like Lightfoot, was the son of a Methodist Minister, born in County Waterford, Ireland. In 1915 he entered Methodist College in Belfast, where he excelled both in mathematics and science, and in 1922 he went to Trinity College, Dublin. Upon receiving his MSc degree in 1927, he also obtained an 1851 Exhibition Overseas Research Scholarship [2-41], for pursuing graduate studies abroad. Despite his mathematical aptitude, however, young Walton found himself equally at home behind a laboratory workbench, being gifted in crafting things with his hands. He arrived at the Cavendish Laboratory in October 1927, where he spent most of the first term gaining experience in glass blowing, operating high-vacuum apparatus, and performing radioactive measurements in the 'Nursery' (the attic) of the Cavendish. Before long, Rutherford sent for him, in order to discuss a research problem for him to work on [2-42].

17

The Old Man asked Walton if he had any suggestions to make, and Walton replied that 'he would like to try producing fast particles' [2-43]. Disintegration experiments were then in full swing at the Cavendish, though the potentials required to accelerate positive ions fast enough to smash atoms asunder were generally considered unattainable with apparatus at hand, except from radioactive sources. (Alpha particles emitted from a radium atom traveled with a velocity that could only be equaled if they were to be accelerated through an electric potential of roughly eight million volts.) Walton wondered, too, why other projectiles, besides $\alpha$-particles, were not also tried. However, the only beam of *protons* was to be found in the mass spectrograph, which featured an extremely low beam-current. Instead, Rutherford considered *electrons* more promising projectiles, being more likely to enter an atomic nucleus and be attracted to the positive nuclear core, than positive ions of modest energy. Besides, he was a firm believer in 'trying anything once, and see.'

Walton had given the matter some thought, and suggested a scheme for accelerating electrons in a circular electric field. Rutherford pointed out that the time required was too long and stray fields would have time to cause particles to fly off the orbit. Instead, he countered with a more practical method of producing the circular field — essentially the method developed by J J Thomson for his 'electrodeless discharge' in gases [2-44]. However, not getting anywhere with the induction method, Walton subsequently suggested to Rutherford an alternative scheme for accelerating heavy ions by what would be known as a linear accelerator. Such an accelerator, in the form of a linear array of cylindrical electrodes, had been first proposed by Gustaf Ising in Stockholm in 1924, and resurrected in a more practical form by the Norwegian accelerator physicist Rolf Wideröe in 1927 [2-45]. Wideröe's published paper of 1928 would prove of inestimable value for Ernest O Lawrence when he came across it in 1929, as we shall see. As for Walton, however, he seems not to have profited greatly by Wideröe's paper when a copy was circulated in Cambridge; at least he got bogged down in messy details having to do with the quenched-spark generator source and the focusing of the cesium ion beam.

> I am not an easily excited person and I have no recollection of feeling very annoyed [on seeing the Wideröe paper]. I was only mildly disappointed. No injustice had been done to me and so I had no cause to nurse any grievance or ill feeling. As far as I can recall I merely decided to try to be quicker off the mark next time and any slight feelings of disappointment vanished very quickly. [2-46]

Walton did not have to wait long for yet another opportunity. Unknown to him, about a week or so prior to his initial talk with the Prof., Rutherford had expanded on the importance of new methods for obtaining high voltages and high magnetic fields. The inspiration for Rutherford's newfound optimism seems to have been the technical progress of another research student, Thomas E Allibone.

Allibone's background was quite different from Walton's. He began his career as an electrical engineer with the Metropolitan-Vickers Company in Manchester. He learned of the transmutation experiments of Rutherford and Chadwick from a lecture in 1925 by Chadwick's old partner C D Ellis. Ellis spoke at Sheffield University, where Allibone was serving a Metro-Vick scholarship at the time [2-47]. After the lecture, he asked Ellis about chances for getting a post-graduate scholarship to Cambridge, and on his recommendation ended up on a Wollaston research scholarship at Caius College, supplemented by financial support from Sir Arthur P M Fleming, Research Director at Metro-Vick. More so than Walton's, his background prepared him well for a research project at Cambridge. Having worked in the High Tension Laboratory at Metro-Vick, Allibone was familiar with the latest techniques for generating high voltages and applying them to X-ray tubes. Thus, at the General Electric Company of America, William D Coolidge had reached 350 kV across a tube in 1926 by shielding the glass walls of the X-ray tube with copper. In so doing, he prevented the buildup of charge on the walls of the tube from backstreaming electrons driven into them. The tube was fitted with a thin metallic window through which fast electrons could stream out into the air, as Philipp Lenard had demonstrated in 1894 at a much lower voltage. Why not put similar apparatus to work in atomic transmutations?

The close association of the Metropolitan-Vickers Company with the Cavendish Laboratory had its origins in World War I [2-48]. At the beginning of the war, Mr. Fleming was the Superintendent of the Transformer Department of the Company, then called the British Westinghouse Company. In that capacity, he was responsible for testing not only the finished transformers, but their subcomponents as well, including magnetic sheet steel, and solid and liquid insulation. The Transformer Test Department in Trafford Park eventually mushroomed into one of the largest research laboratories in the UK. At any rate, when an anti-submarine committee was formed early in the war, Fleming was invited to join Ernest Rutherford, who had flung himself into the work on submarine detection with nearly superhuman energy, and share in the experimental work on the Firth of Forth and Lake Windermere. Before the end of the Great War, the Sheffield armaments firm of Vickers was considering how its huge steel-making capacity could be profitably used in peacetime, and decided to buy out the Westinghouse concerns in Trafford Park, some of which constituted the Metropolitan Carriage, Waggon and Finance Corporation. Thus the Metropolitan-Vickers Company was formed. The Company's renewed interaction with Cambridge came about chiefly through Allibone and John Cockcroft, as we shall see.

Director Fleming reacted positively to Allibone's suggestion of assembling a high-voltage installation at Cambridge and attempt to produce disintegrations with particles accelerated to 300 kV or more in a vacuum-discharge tube. Fleming not only sent him off with pocket money, but with the promise of Metro-Vick apparatus thrown in at moderate cost. Rutherford was equally receptive when Allibone met with him at the Cavendish on 29 March 1926. Rutherford took him down to the 'biggest room we have,' where he was introduced to John Cockcroft, another former Metro-Vick engineer, working in one corner of the room. 'Is this room high enough for you?' barked the Professor. Allibone could only assure him that he would try to generate the highest voltage possible in that height, about 500,000 volts.

Allibone formally entered the Cavendish in October 1926. As he did, he also consulted Brian L Goodlet, head of the High Voltage Laboratory at Metro-Vick. (Goodlet, who later became Chief Engineer at the British Atomic Energy Research Establishment at Harwell, was born in Russia and had shot his way down Nevsky Prospect in St Petersburg during the Russian Revolution. He ended up in Sheffield, where he had been taught mathematics by Allibone's father and engineering by the man who eventually became Allibone's father-in-law [2-49].) Goodlet advised Allibone to build a Tesla transformer coil for 500–600 kV — a device far more compact than a ponderous power frequency transformer of the same nominal voltage rating. By the fall of 1927 Allibone had his Tesla coil, and was experimenting with a discharge tube of the Coolidge-type capable of withstanding several hundred kilovolts. The high potential was applied across a pair of hollow electrodes in a long evacuated glass tube with an elliptical bulb blown at the center. Beams of electrons accelerated in the discharge tube were electrostatically focused through a slit orifice in the tube and were then deflected in a magnetic field to obtain a roughly monochromatic beam for scattering experiments. It seems that Rutherford was more than satisfied with Allibone's progress, judging by the tone of his Presidential Address at the anniversary meeting of the Royal Society on 30 November — even if he did not single out Allibone by name.

> It has long been my ambition to have available for study a copious supply of atoms and electrons which have an individual energy far transcending that of the $\alpha$- and $\beta$-particles from radioactive bodies. I am hopeful that I may yet have my wish fulfilled, but it is obvious that many experimental difficulties will have to be surmounted before this can be realized, even on a laboratory scale. [2-50]

Appropriately, W D Coolidge, prominently referred to in the talk, was in the audience that day, present to receive the Hughes Medal of the Royal Society for his 'invention and production of a new type of X-ray tube, called by his name, of great flexibility and power, which has proved of great service not only in Medical Radiology but in numerous scientific researches.' In a lecture to undergraduates in 1928, Rutherford was even more enthusiastic, looking forward to such high voltages that 'the machine would have to be housed in a cathedral.' In fact, there was a need, not only for accelerating voltages comparable to nuclear binding energies, or several million electron volts, but also for sources of higher currents. In the ongoing experiments of Rutherford and Chadwick, only one in a million $\alpha$-bullets hit their mark, precluding quantitative measurements.

A problem with the Tesla coil, in which a resonant primary coil of a few turns induces high alternating voltages in a multiturn secondary coil, was the extremely loud and unpleasant noise of the rotary sparkgap in the primary circuit, as well as the intense radio-interference which spoiled Ernest Walton's experiments in the nearby radio laboratory. More serious was the oscillatory character of the output potential, lasting only a small fraction of a second per second of operation, and generating all velocities of electrons up to some maximum determined by the peak voltage on the tube. What was really needed for energizing the discharge tube was a steady d.c. voltage, such as might be obtainable from a transformer with auxiliary equipment to rectify the voltage and deliver d.c. to the discharge tube. The man who would look into this approach was John Cockcroft, assisted by the hapless Ernest Walton. Nevertheless, Allibone deserves credit for having been the first at the Cavendish to try seriously to create disintegrations with accelerated particles [2-51], albeit with little success. About all he disintegrated was a rat, in an experiment designed to determine whether the Tesla frequency of 100,000 cycles per second was as lethal as the 50 cycles of the mains. Allibone passed a spark through a rat to earth, boring a hole half an inch in diameter [2-52].

John Douglas Cockcroft was the oldest of five sons of a miller in the Todmorden region of the North Country. He matriculated at the top of his class in the Todmorden Grammar School, then won easily a scholarship to Manchester University in 1914. Though he entered for the BSc course in mathematics, his real interest was in atomic physics, after poring over a popular book by James A Crowther on the pioneering work of J J Thomson and Ernest Rutherford. Rutherford was then head of the Physics Department at Manchester, and at the height of his powers, surrounded by a galaxy of gifted scientists, some of whom we have met, including Geiger, Marsden, and Darwin. Unfortunately, the outbreak of war on the Continent scattered them all to the four winds, Cockcroft among them. After only a year at Manchester, Cockcroft, too, enlisted and endured the most bitter fighting on the Western Front, first as a non-commissioned signaller and then as a gazetted second lieutenant in the Royal Field Artillery. On one occasion he was the sole survivor at a forward post. Released from the Army in early 1919, he returned to Manchester to complete a BSc degree in electrical engineering under Miles Walker at the Manchester College of Technology.

With his degree in hand, Cockcroft joined Metropolitan-Vickers in 1920 as a College Apprentice working towards a MSc thesis. As with Allibone, Mr. Fleming took a close interest in his career, and saw to it that Cockcroft went to Cambridge 'to improve his mathematics'—that is, sit for the Mathematical Tripos—for future progress with Metro-Vick. Instead, Cockcroft wound up at St John's College as a Proper Sizer, starting in 1922, and remained in academia for the better part of his career. In his second year at Cambridge, he encountered Rutherford in one of the laboratories,

> ...sitting, as he so often did, on a stool. He received me very kindly and gave me authority to devote such time as I could spare from mathematics to work in the advanced practical class, then, looking at me with those penetrating eyes, he promised to take me into the Cavendish *if* I got a first. [2-53]

In addition to his income from St John's Sizarship and Exhibition, Metro-Vick, in their enlightened manner, continued to provide partial support, on the understanding that he would continue to carry out some research work for the company. Cockcroft also made himself useful at the Cavendish as a 'spare-time, honorary electrical engineer.'

Cockcroft selected for his PhD program studies of the surface properties of vacuum-deposited metallic films. Chadwick, recognizing his engineering flair, asked him to develop a method for counting large numbers of $\alpha$-particles from a strong radioactive source, for which he came up with an elegant solution involving a motor-driven wheel with a hole in it. By that time he had completed his apprenticeship under Chadwick in the 'Nursery,' and he caught the eye of Peter Kapitza, the Russian electrical engineer in charge of the Magnetic Laboratory. Formerly a lecturer in electrical engineering in the Petrograd Polytechnic Institute, Kapitza was sent to England in 1921 to purchase scientific equipment to refurbish devastated Russian laboratories after the Revolution. In England he took the opportunity of visiting Rutherford at the Cavendish, and was so impressed with what he saw and heard that he stayed on as research worker. Rutherford, for his part, came to consider Kapitza 'a man of marked skill, exceptional ability and great originality of mind, who has very unusual qualifications for pioneer work of this character.' It has been claimed that, for the next fifteen years after Kapitza's arrival at the Cavendish, Kapitza remained the most important single figure in Rutherford's life [2-54].

Kapitza's initial preoccupation at the Cavendish was the generation of intense magnetic fields for the study of the electronic lattice in metallic crystals. As such, the lasting value of the project lay in the technical approach, not in the scientific results *per se*. Discharging a large battery through a copper coil proving insufficient, he conceived of the idea of short-circuiting an AC generator through the coil for a single half-cycle (one hundredth of a second) — thus gaining energy from the loss of momentum of the heavy rotor during the short. Cockcroft pointed out to Kapitza that the Metro-Vick Company made just such equipment. He also helped him draw up specifications for the apparatus, as well as calculating the enormous forces in the coil when subjected to the very high transient current. In due course the machinery, tons of switchgear and a very large generator with specially reinforced end-windings to withstand the great forces generated during the short circuit, was purchased with funds from the Department of Scientific and Industrial Research (DSIR), which administered a fund to promote industrial research. The equipment was assembled in a shed in the Cavendish courtyard formerly occupied by chemistry students. To everybody's satisfaction, the installation worked just fine, producing a field of 320,000 gauss over a volume of about 3 cc without bursting the coil, as Rutherford proudly noted in his Royal Society address of 1927 [2-55]. While Cockcroft had been officially nominated as one of Kapitza's assistants, Kapitza, in turn, sharpened Cockcroft's intellectual powers as well.

> [Kapitza] was a man who had wide interests, was extremely sociable and enjoyed discussion. While he had great respect for the Cambridge tradition, he found the undergraduates, on the whole, tended to be treated too much like schoolboys, and when abroad he had observed that young English research workers, who were usually more talented than their continental contemporaries, were afraid to express opinions for the fear that they would appear ignorant or stupid. Kapitza therefore founded what became known as the 'Kapitza Club' which met every week during term in one of the colleges. On the first occasion Kapitza broke down the silence by deliberately making howlers so obvious that even the shyest young research worker put up his hand to correct them. In this way Kapitza raised the 'standard of general discussion of physics and engendered a freer exercise of the scientific intellect.' [2-56]

While Cockcroft collaborated with Kapitza in the Magnetic Laboratory, and completed his PhD work, the mainstream of the Cavendish Laboratory's orientation continued to center on nuclear scattering and transmutation studies. Despite his repeated statements to the contrary, in his later years, Rutherford was far from keen on building machinery for accelerating particles. He was intolerant of the delay it invariably caused from obtaining results, and felt that such apparatus was the task of his successor as Cavendish Professor. Nevertheless, he recognized the need to 'have a supply of electrons and atoms of matter in general, of which the individual energy of motion is greater even than that of the α-particle.' By 'atoms' he undoubtedly meant α-particles or protons. The main question was the highest practicable tube voltage.

> What we require [Rutherford declared, at the opening of a new High Tension Laboratory at Metropolitan-Vickers] is an appara-tus to give us a potential of the order of 10 million volts which can be safely accommodated in a reasonably sized room and operated by a few kilowatts of power. We require too an exhausted tube capable of withstanding this voltage.... I see no reason why such a requirement can not be made practical. [2-57]

Indeed, help was on its way, in a rash of high-voltage projects under way at home and abroad. In England, there was Metro-Vick, with its new laboratory which Allibone would take charge of, and which would continue its close association with the Cavendish. In Berlin, Arno Brasch, Fritz Lange, and Curt Urban of the University of Berlin began tests in 1927 with a lightning catcher strung between two mountains in the Italian Alps 660 meters apart. The potential reached by the antenna was controlled and measured by the air gap between two dangling metal spheres. During thunderstorms, their calculations indicated that the antenna rose to a potential as high as 15 million volts. At least it was high enough for Curt Urban to be killed during the experiments in 1928. Brasch and Lange retreated to the comparative safety of the Allgemeine Elektrizitätsgesellschaft's research laboratory in Berlin; there they began tests of a discharge tube pulsed by a surge generator (a stack of capacitors charged in parallel) capable of 2.4 million volts. In the US, there was the General Electric Company, already noted. In California, Southern California Edison was adapting a million-volt cascade transformer to nuclear physics applications at Caltech. Though hardly meeting Ruther-ford's requirement of convenient size, filling a room 300 square feet in area and 50 feet high, we shall hear more of this particular installation in due course.

In fact, a modest approach much more to Rutherford's liking would prove quite sufficient for atom smashing, in the hands of Cockcroft and Walton. Mainly, however, it was thanks to the insight of a Russian theoretical physicist in August of 1928, just as Cockcroft was finishing his PhD and began casting about for something new to do.

# Chapter 3

# AMERICAN BEGINNINGS

## 3.1. South Dakota pals

If 1932 was the *annus mirabilis* of nuclear physics, 1928 was a good year as well in Berkeley, California. That summer Ernest O Lawrence arrived to begin an assistant professorship in the newly refurbished Physics Department of the University of California in Berkeley. Two years earlier, Merle A Tuve, Lawrence's boyhood friend, had accepted a position with the Department of Terrestrial Magnetism of the Carnegie Institution of Washington in the District of Columbia. The two, from similar third-generation Norwegian backgrounds in Midwestern America, would remain permanently ensconced in their respective institutions for the rest of their careers, and leave their mark on the experimental and administrative conduct of physics in war and in peace. Most important for our purpose, both Tuve and Lawrence established their own distinct 'schools' of accelerator and nuclear physics, characterized by differences in attitudes and motivations, and by similarities in promoting technology-intensive, team-oriented approaches to experimental science that became the *modus operandi* of American physics after 1932. Accordingly, America's two preeminent atom smashers must now claim our attention.

Both physicists were born in 1901 in Canton, South Dakota — a village on the Big Sioux River, not very far from the site of Custer's massacre on the Little Big Horn. Merle Antony Tuve was the senior by all of six weeks, the son of Antony G Tuve, professor and sometime president of Augustana College. After a year in the public high school, Merle Tuve transferred to the secondary-school program at Augustana. While there he took a course in physics, though under the influence of the school's president he decided to major in chemistry when he entered the University of Minnesota in 1919 [3-1]. Ernest Orlando Lawrence was the eldest son of Carl Gustav Lawrence, whom Professor Antony Tuve attracted from Madison, Wisconsin, after Lawrence graduated from the University of Wisconsin, to Canton to be Professor of Latin and History at the

Academy. Ernest's public schooling was partly in the state capital of Pierre, where his father served two terms as State Superintendent of Public Instruction. He entered St Olaf College in Northfield, Minnesota, in 1918, but after a year he transferred to the University of South Dakota in Vermillion.

Both the younger Tuve and Lawrence were of Norwegian descent, as noted. Tuve's grandfather, Gulbrand Tuve, stemmed from Valdres, Norway, while Lawrence's grandfather, Ole Hundale Lavrens, emigrated from Telemark to the Wisconsin Territory in 1843. As boys growing up across the street from each other in Canton, they had something else in common before they were nine: an intense, mutual interest in electricity and electrical devices. They rejuvenated cast-off telephone dry-cell batteries, and later 'monkeyed wireless,' and each even crafted his own vacuum-tube wireless set out of secondhand tubes and old mahogany cabinets. They were less successful in attempting to fly on the golf course, jumping off Tuve's car moving at 30 mph, and flapping wings hinged to a ring that could be fastened around a lithe body. Nevertheless, before long they were the talk of the town [3-2].

Soon after enrolling at the University of Minnesota for the spring quarter in 1919, Tuve switched from chemistry to electrical engineering. The engineering curriculum included a required sequence of introductory physics courses. Those, plus a sequence on theoretical physics taught by John T Tate, clinched his interest in physics. Completing his BSc in electrical engineering in 1922, he stayed on for another year for a master's degree in physics under Tate. For his graduate research project, Tuve wound up studying the ionization of mercury atoms bombarded by positive mercury ions. He also saw much of his old chum Ernest, who had received his BSc at South Dakota in June of 1922. There he had begun his undergraduate career as a premed student, but switched to physics under the guidance of Dean Lewis Akeley. Tuve had written Lawrence, urging him to join him for graduate work at Minnesota that fall. Tuve's letter, and Tuve's insistence at Minnesota that Lawrence was brilliant and would go far in physics, were the deciding factors in shaping Lawrence's career. 'Through Merle's interest in my behalf, I obtained a teaching fellowship in physics at Minnesota' [3-3]. Not only that, 'A duplex team in research would be great,' Lawrence replied to his friend. 'Some phase of the vacuum tube seems to offer infinite possibilities to me' [3-4].

Whether physics in general or wireless in particular, Tuve had no compunction as to pursuing either topic. As a practical matter, physics was taking time that would otherwise be devoted to tinkering with wireless, though he did find time to rig up a continuous wave and phone set. He had initially planned to use vacuum tubes to produce electromagnetic waves shorter than a meter in wavelength for his master's thesis, that limit marking the edge of the wavelength frontier at the time. Though that project fell by the wayside, he would return to radio pulse-technology for his doctorate and subsequent research. Lawrence, too, was diverted from some wireless topic by his mentor in the Minnesota Physics Department, William Francis Gray Swann—a tall English physicist who, among other things, was an accomplished cellist. Originally from London's Royal College of Science, Swann had made his way to Minnesota via the Carnegie Institution's Department of Terrestrial Magnetism (DTM). Designing magnetic instruments had been one of Swann's preoccupations while at DTM. Under his supervision, Lawrence studied the electrical effects produced when a metallic ellipsoid was rotated in a magnetic field for his master's thesis. He built the apparatus, which may still be seen in Berkeley's Lawrence Hall of Science, with the help of the laboratory mechanic and completed the experiment in record time, to everybody's accolade in the Physics Department.

Among other projects in the Department that fascinated young Lawrence, one bears noting in particular. Another of Swann's graduate students, John William Buchta, was working on a device for accelerating charged particles. It was based on the same principle used by Robert Van de Graaff in developing his electrostatic generator a few years later, and which would be exploited so successfully by Lawrence's friend Tuve. Buchta's contraption, made of glass, was too fragile and inefficient to compete with Van de Graaff's subsequent elaboration of the principle, and he and Swann were unable to attain accelerating energies exceeding those of Rutherford's $\alpha$-rays from radium. Nevertheless, the attempt intrigued Lawrence, and he and Tuve discussed it often and at length. Swann was far from discouraged, sensing the theoretical potentials, and encouraged Lawrence to speculate on other accelerating schemes and on the significance of the underlying physics at stake.

Tuve received his MA in 1923—actually a joint degree in physics and music—and went to Princeton on an instructorship; there he hoped to continue his work with positive ions for his doctorate under the direction of Karl T Compton. The same year Swann left Minnesota for the University of Chicago, where Lawrence, with his master's degree in hand as well, followed him to begin his doctoral work on the photoelectric effect in potassium vapor. As it turned out, both Tuve and Lawrence were soon on the road again. Tuve's scholarship at Princeton was not renewed a second year, and Swann, to Lawrence's horror, grew impatient and left Chicago after only a year.

## 3.2. Geophysics, or smash the atom?

After leaving Minnesota, Merle Tuve headed for Princeton University. With a teaching assistantship in hand, he hoped to continue his doctoral work under Karl T Compton, chairman of the Physics Department, in the Palmer Physical Laboratory. Alas, the Department lacked funds to renew his instructorship for a second year. (His teaching post went to Henry Smyth instead.) He spent the summer of 1924 in New York City with Western Electric, the manufacturing subsidiary of American Telephone and Telegraph Company (AT&T). (The engineering arm of Western Electric would be known as Bell Telephone Laboratories from the next year onward.) There he worked under Clinton J Davisson, who had worked his way through the University of Chicago, and obtained his PhD at Princeton under Owen W Richardson. Davisson had married Richardson's sister, and taken up his brother-in-law's field, the emission of electrons from metals [3-5].

The few months spent with Davisson at Western Electric must have left their mark on Tuve. Davisson and his assistant, Charles H Kunsman, were deeply involved in the scattering of electrons, in which they found that a small percentage of the electrons incident on the grids and plates in a positive ion apparatus converted to electron scattering, were scattered back towards the electron gun with virtually no loss in energy. Davisson saw the back-scattered electrons as a tool for probing the extranuclear structure of atoms, much as Geiger and Marsden's scattering of $\alpha$-particles had led to Rutherford's nuclear atom in 1911 (the year Davisson completed his PhD). Three years later, in 1927, Davisson and Lester Germer, still at Bell Laboratories, completed their epochal experiments on electron diffraction that established the wave nature of the electron, first hypothesized by Louis de Broglie in 1923. In these experiments, a narrow pencil of electrons from a gun struck the surface of a single crystal at normal incidence, and were scattered in all directions. A collector for measuring the scattered electrons could be adjusted to any angular position with respect to the crystal. The pattern of the scattered electrons revealed striking maxima and minima, which they were able to explain in terms of diffraction of the electron 'waves.' Davisson would share the Nobel Prize in Physics for 1937 with G P Thomson, the son of J J Thomson, for the discovery of electron diffraction, leading to the quip that J J Thomson had been awarded the prize for showing that the electron was a particle and his son received the award for showing that it was not [3-6].

In the fall of 1924, Tuve resumed his graduate studies at Johns Hopkins University in Baltimore, with Joseph Sweetman Ames as his adviser. A man of great reputation, Ames's motto was 'No matter what it is, everybody who is anybody at Hopkins goes to work on something that deeply interests him' [3-7]. One who did, and happened to be a former student of 'Joe Ames,' was Gregory Breit, whom Tuve first encountered at the University of Minnesota in July of 1923. Born in Nikolaev, Russia, in 1899, Breit arrived on American shores in 1915, and received his PhD under Ames at Johns Hopkins in 1921. Wrote John A Wheeler (another student of Ames) of Breit in 1979, 'Any one in the United States who before World War II contributed more importantly to more fields of physics would be difficult to name' [3-8]. In spite of that, 'insufficiently appreciated in the 1930s, [Breit] is today the most unappreciated physicist in America' [3-9].

We shall appreciate Breit, and Wheeler's remarks, before long. At any rate, Breit and Tuve met just as Lawrence was on the point of leaving Minnesota for Chicago with W F G Swann, and Tuve himself was finishing some piece of work with Jack Tate before reporting to Compton at Princeton. Breit was then an assistant professor at Minnesota, and he, Tuve, and Lawrence hit it off well [3-10]. Nuclear disintegration experiments under $\alpha$-bombardment were then in full swing at Cambridge, as well as in Vienna, and Breit was caught up in the accompanying excitement. He discussed the European work with Tuve and Lawrence, and made 'some crude experiments with Tesla coils.' About all he learned from these experiments was 'how sparks in air were related to potential differences between electrodes' [3-11]. All the same, these brief preoccupations represented an embryonic awakening in the interest in nuclear physics in America. However, the immediate outcome of Breit's encounter with Tuve that midwestern summer was one of steering Tuve in another direction.

> Breit . . . also learned of my 'ham radio' experience and my efforts with 80-cm waves on Lecher wires using Army VT-II tubes with bases removed. I had been trying for highest possible frequencies. They were feeble. [3-12]

Feeble though they may have been, they were good enough to launch Tuve on a short-wavelength transmission project for his doctorate at Johns Hopkins—a project instigated by Breit. While still at Minnesota, Breit had received and accepted an invitation to join the Department of Terrestrial Magnetism of the Carnegie Institution of Washington. No sooner had Tuve arrived in Baltimore, in September of 1924, than Breit looked him up with a joint proposal. Why not build a large parabolic transmitter and direct a beam of very short radio waves at the ionosphere at an angle. Assuming they were reflected back down to Baltimore, the experiment should allow an experimental determination of the height of the 'conducting' or 'Kennelly–Heaviside' layer—the ionospheric layer inferred, nearly simultaneously, by Oliver Heaviside and Arthur Edwin Kennelly in 1902 to account for the long-distance propagation of very long-wavelength radio waves. Tuve's task would be to man a mobile receiver to pin down just where the beam returned to earth.

Tuve, however, was dubious about the critical reflection coefficient— that is, its value near which the radio waves, instead of bouncing off the conducting layer, passes up through the layer instead. He was also skeptical about the attainable sensitivity of a receiver at such short wave-lengths, and proposed instead that they use an approach suggested by Swann at Minnesota while Tuve was still there: broadcasting pulses of conventional-length radio waves and comparing two sets of signals, one arriving directly from the transmitter, and the other scattered via the ionosphere. The amount by which one wave train was offset from the other would be a measure of the time difference between the two different paths, and hence of the height of the conducting layer.

The experiment began in 1924, but clear success was only achieved in the summer of 1925 when Breit and Tuve secured the use of a crystal-controlled transmitter of the Naval Research Laboratory located in Bellevue. Tuve was spending the summer with Breit in Washington, and in June they recorded the pulse 'echoes' on 71 meters (4 MHz). They were received by equipment several miles away from the transmitter, on the roof of the main building of the DTM, and recorded photographically on film.

> It took some weeks to arrange for evening use of the Navy transmitter we used, which then promptly proved that the echoes varied in 'height' and hence were *not* echoes from the Blue Ridge mountains 65 miles west of us! [3-13]

The first conclusive demonstration of the existence of the ionosphere had been made by Edward Victor Appleton and his students at King's College, London, in 1924, by a somewhat different method; however, it was the method of Breit and Tuve that became the standard method for use in ionospheric studies in the following years [3-14].

Joseph Ames accepted Tuve's contribution to the radio echo project as the basis for his doctorate, which Tuve received in 1926. Back at Johns Hopkins, however, he was contacted by John A Fleming, the Assistant Director of the DTM. With the new technique for upper air studies an obvious tool for shedding light of terrestrial magnetism, Fleming and the Department was keen to see it further exploited at the DTM, and he sounded out Tuve, no doubt on Breit's urging, on returning to Washington on a staff appointment with the Carnegie Institution. Tuve hesitated. At Minnesota he had become captivated with Rutherford's $\alpha$-particle scattering experiments of 1911, his nitrogen disintegration in 1919, and subsequent experiments on other light elements. Tuve grew intrigued by the notion that the inverse square electrical force was unable to account for the stability of atomic nuclei, and by the possibility of exploring the presumed short-range nuclear force involved by scattering experiments with high-energy particles. He had considered radioactivity studies for his master's degree (that is, a year before he ran into Breit), but decided against it on second thought. After all, it would be 'hard to compete with Rutherford here alone!' [3-15].

In September of 1924, Tuve heard Rutherford speak in person while in North America to preside over a meeting of the British Association in Toronto. After the meeting Rutherford made several stops on the East Coast of the US; Tuve heard him when he spoke at a meeting of the Engineer's Club at Western Electric, at the end of his summer job there [3-16]. He was sufficiently stimulated to apply for a fellowship from the National Research Council for a year under Rutherford at Cambridge, and so informed Fleming when he came calling in Baltimore. 'I said I was proposing to put a very high voltage on a vacuum tube, to produce a controllable "beam" of high-energy protons for studies of atomic nuclei.' The conversation continued as follows.

'Can you do that in your one year of a Fellowship?' responded Fleming.

'Of course not, but I can start it.'

'Why don't you do it here; then you don't have to stop at the end of the year,' continued Fleming.

'At a Department of Terrestrial Magnetism!' exclaimed Tuve incredulously.

'If you will spend a part of your time on the radio experiments, the rest can be on high-voltage tubes; Carnegie cannot start a new department for every new problem, so we use existing departments to do new things,' promised Fleming [3-17].

With that, Tuve backed off contacting the NRC Fellowship Board, and agreed to join Breit again at the DTM, starting on 1 July 1926. In so doing, not only did the upper air studies receive a new lease on life, but nuclear physics in America would soon enter a new, important era as well.

### 3.3.  Research above all else

We last left Ernest Lawrence at the University of Chicago, where he had enrolled as a PhD candidate in the Ryerson Physical Laboratory. The Ryerson was named after Martin A Ryerson, son of a wealthy lumber tycoon who had largely underwritten the financing of the physics facility; it was completed on New Year's Day, 1894, or two years after the University was formally opened, generously endowed by John D Rockefeller and Marshall Field. The head of the physics department from the start was none other than Albert A Michelson, who became, in 1907, the first American scientist to win the Nobel Prize [3-18]. Having little enthusiasm for administrative chores, Michelson left much of the drudgery of establishing departmental policies with respect to teaching and research in the hands of Robert Millikan from 1896, when the latter arrived with an assistantship and a fresh PhD from Columbia University.

Michelson was still in Chicago, though Millikan was not, when Lawrence arrived in 1923. That year, William Swann was among the new crop of faculty members recruited by Michelson, and Lawrence promptly went to work on a thesis problem, approved by Swann, on the photoelectric effect. That effect, the emission of electricity from a metal plate illuminated by ultraviolet light, was discovered in 1887 by Heinrich Hertz at Karlsruhe. In 1900, Philipp Lenard, Hertz's assistant and successor, demonstrated the identity of the 'photocurrent' emitted from the metallic surface with J J Thomson's cathode ray corpuscles, or electrons. Lenard's subsequent results of 1902, on the dependence of the photocurrent on the intensity of the incident light, and on the velocity of the emitted electrons, remained inexplicable until the publication of Albert Einstein's classic paper in 1905, explaining them in terms of an interchange of energy between light quanta and photoelectrons.

Millikan, too, had tackled aspects of the photoelectric effect, as his first research problem at Chicago during 1905–06, unaware of Einstein's explanation of the phenomenon [3-19]. Whereas Millikan's experiment had involved determining the temperature dependence of the photocurrent from various metals (no dependence was found), Lawrence set out to study the effect in potassium *vapor*. 'Now is the time for all good men to come to the aid,' begins his laboratory notebook on 8 March 1924 [3-20]. Indeed, it took a good man, as his experiment, involving a vapor, was a decided complication, as Professor Arthur J Dempster, in charge of research at the Ryerson and cool towards Swann, the newcomer on the faculty, let Lawrence know in no uncertain terms [3-21]. Light from an iron arc ('Pfund arc') illuminated a monochrometer; the emerging monochromatic beam was collimated by a quartz lens and a series of diaphragms and passed into an ionization chamber exhausted by a mercury pump. It continued through a jet of vapor issuing from potassium heated in a small furnace below the beam tube, and was further collimated and fell upon a photoelectric cell calibrated with a thermopile and galvanometer and attached to an electrometer. Lawrence persisted with jury-rigged apparatus, improvising one thing after another, calibrating and adjusting. 'As he made a final adjustment prior to noting data, he was bumped by a passer-by and the apparatus exploded' [3-22].

Grimly determined, he started over from scratch. In toiling away, he was encouraged by others beside Swann. Michelson himself would pop up in the laboratory on Sundays and offer approving remarks. Even Dempster had to admit he was getting good results at long last. And between data-recording stints at ungodly hours, Lawrence enjoyed company in a laboratory next door. There, Arthur Holly Compton, younger brother of Karl T Compton, who also had worked on the photoeffect, was verifying and extending another effect which he, himself, had discovered at Washington University in St Louis the year before. Nowadays known as the Compton effect [3-23], it would earn him the Nobel Prize in 1927. Watching Compton bending to the apparatus late at night was a fabulous opportunity for young Lawrence to see physics in action at its very best. Professor Compton, in turn, wrote of Lawrence and their time together, munching on sandwiches brought in by Mrs. Compton.

> [Lawrence] had an extraordinary gift of thinking up new ideas that seemed impossible of achievement and making them work. In our conversations in the laboratory our relations had been more those of research colleagues than those of student and teacher. [3-24]

So absorbed had Lawrence been in cooking up new ideas that he had neglected the German and French language requirements for the PhD, though he passed both of Chicago's written and oral physics examinations for the doctorate. His experiment was as good as done, though he had not found time to write it up properly for a thesis. Clearly, it would take another year to complete the PhD. Amidst these considerations, Swann's announcement of his intention to accept a professorship at Yale University came as a shock, but Swann calmed him down right away. 'Come along to New Haven,' urged Swann, suggesting that Lawrence apply for a Sloane Fellowship at Yale. Lawrence did, and won it handily. And so, having passed his first graduate year under Swann at Minnesota in 1923, his second year again under Swann at Chicago in 1924, he was headed for his third and hopefully last year with Swann in New Haven, Connecticut.

Lawrence caught up with Swann on the famous New Haven campus on a rain-swept morning in September 1924. Rain or shine, and first things first, he passed the French and German tests without much effort, and on 29 October passed the Yale physics examinations for the PhD. Without further ado, he resumed his thesis problem — the photoelectric experiment begun at Chicago — with his usual enthusiasm and drive, and with better equipment available for graduate students' research projects in Yale's Sloane Laboratory. Working around the clock as usual, he completed the written dissertation by the end of January 1925. Swann communicated the manuscript to the Royal Society, and, like Lawrence's master's thesis, it was duly published in the *Philosophical Magazine*, dated 2 February 1925 [3-25]. Even before the thesis was finished, however, and with his PhD ensured for the June convocation, Lawrence was off on another experiment, involving systems for collimating and measuring the velocity of monochromatic photoelectrons, and analyzing positive ion rays. The experience gained in manipulating charged particle beams would stand him in good stead a few years later in Berkeley.

In February 1925, Swann nominated Lawrence for a National Research Fellowship, relieving him of the immediate need of having to choose among several academic positions in the offing. As with the Exhibition of 1851 Scholarship, the rules required National Research Fellows to perform their postdoctoral research in an institution other than that which had granted the degree; however, Swann persuaded the Research Council to make an exception for Lawrence, 'who had moved around too much already.' Consequently, the grant allowed him to remain at Yale for the time being, extending his work on the photoeffect to other alkali vapors, and starting an allied investigation of their ionizing potentials. As if not having enough to do, Lawrence also teamed up with Jesse Beams, a visitor from the University of Virginia, who had been engaged in measurements of the time lapse between the absorption of a quantum of light by a metal surface and the ejection of a photoelectron. He and Lawrence set about adopting some of Beam's work with Kerr cells (parallel-plate condensers with a gaseous or liquid dielectric) for a device for producing very short bursts of light. Utilizing a rotating mirror spinning at very high velocity in a jet of air, on one occasion the mirror flew off the optical stand and was nowhere to be found. Months later it turned up, embedded in a corner of the laboratory.

In the spring of 1925, the University of Washington in Seattle upset Lawrence's tranquillity by offering him an associate professorship. Yale, having already tempted him with an instructorship and the promise of promotion to assistant professor the following year, soon countered by offering an immediate appointment as assistant professor — the first time Yale offered a professorship to someone who had never been an instructor. Lawrence had barely made his decision to stay at Yale, the prestige of Yale outweighing Washington's higher offered rank, when a telegram arrived from the Berkeley Physics Department, with an offer for an assistant professorship there. Lawrence had known it was coming; Professor Leonard B Loeb had written that he was being considered for the professorship, and extolling the virtues of the department at Berkeley [3-26]. Lawrence still resisted; he felt at home at Yale. 'I like Yale, the personnel, the laboratory and the facilities for research; perhaps even as much I like the friends acquired in New Haven' [3-27]. However, the announcement by the flighty Swann of his upcoming resignation to assume the directorship of the Bartol Research Foundation of the Franklin Institute, came as a hard blow, both to Yale and to Lawrence. Loeb now hinted that Berkeley might offer an associate professorship instead. Yale countered with better terms. Letters went back and forth [3-28]. Cornell, Columbia, and General Electric came in with attractive offers as well. In the end, Lawrence opted for Berkeley, with their promise of an associate professorship coupled with a lighter teaching load in favor of research. 'I assure you [added Loeb in his letter] that as long as I have any connection with the department, I will do everything in my power to realize your trust in us for your future' [3-29].

Teaching was all good and well, but Lawrence was bent on research, and so were his chief sponsors, Elmer Hall, the Berkeley physics chairman, and Raymond Birge, department second-in-charge. Lawrence hurried west in August 1928, churning dust along US Route 66 in a Reo Flying Cloud coupé.

# Chapter 4

# HOW MANY VOLTS?

### 4.1. New physics, with consequences for smashing atoms

At Cambridge, meanwhile, John Cockcroft was finishing his PhD dissertation in the fall of 1928, while Ernest Walton was still struggling with his problematic linear accelerator. Walton was occasionally assisted with certain engineering details by Cockcroft, who was also in and out of the Mond Laboratory helping Kapitza. Unbeknown to Cockcroft and Walton, however, help of a different nature was on its way, in the hands of another young man attempting to finish his PhD dissertation across Europe at the University of Leningrad just then. Since the 25-year-old Russian theorist, George Gamow, will play a central role in upcoming developments, let us now turn to him.

George Gamow was born in Odessa, Russia, in 1904, the son of Anton Gamow, a teacher in one of Odessa's private schools for boys. As George Gamow tells us, the best student in his father's class was Lev Bronstein, who for some reason tried unsuccessfully to have Gamow dismissed from his job. Years later Anton Gamow ran into Bronstein on one of Odessa's boulevards, and asked what he was doing. 'Just working on the docks,' was the reply. Little did Gamow senior know what Leon Trotsky was really up to [4-1]. Young Gamow's own schooling, during World War I and the subsequent revolution and civil war, was very sporadic. However, when the fighting subsided, he attended Novorossia University in Odessa for a year, and then enrolled at the University of Leningrad. While studying physics there, he earned his keep as lecturer in physics in the Artillery School of the Red October, and later as a technician in the state-owned Optical Institute next to the university's Physics Institute. After Professor Alexander A Friedmann's untimely death aborted his plan to continue in relativistic cosmology for his PhD, he wound up with a problem under Professor Yuri Krutkov on 'adiabatic invariance of a quantized pendulum with finite amplitudes'—an extremely dull thesis topic.

Not surprisingly, Gamow found it hard to concentrate on the problem at hand, especially in light of the excitement caused by the appearance of wave mechanics only two years earlier — a version of quantum mechanics that superseded Bohr's original quantum theory. Fortunately for physics, an adviser suggested to Gamow that he might get a boost by spending a few months abroad, say at Göttingen University, one of the main centers in the development of quantum mechanics. Officialdom at Leningrad University concurred, and in June of 1928 he was on his way.

No sooner had he arrived in Göttingen, than Gamow found himself at a party given by Professor Max Born, the director of the Institute of Theoretical Physics at the University [4-2], for his staff and senior students at a restaurant on Nikolausberg. Max Born was widely known, both for his physics and his textbooks, and physicists flocked to him in Göttingen as they did to Bohr in Copenhagen. In particular, Göttingen University and its charming little city was buzzing with excitement over two major forms of quantum mechanics then making the rounds, which must claim our attention at this point.

It will be recalled that Niels Bohr's early success was mainly with the hydrogen atom, by a brilliant application of Max Planck's 'quantum of action' while spending some time with Ernest Rutherford at Manchester in 1912. Bohr appreciated the inherent instability of Rutherford's nuclear atom. An electron revolving in a closed orbit around a central, positive nucleus would, according to the rules of classical physics, quickly lose energy by radiation, spiral into the nucleus, and be lost. Bohr realized that a fundamental revision of the classical view was called for, and conveniently offered by Planck's universal constant $h$. In effect, what was needed was a new stabilizing quantity involving the dimension of length to define the stable configuration of the orbiting electrons in the atom. In J J Thomson's defunct 'plum pudding,' there occurs naturally such a dimension defining the size of the atom, namely the diameter of the sphere of positive electricity. An atom of this kind cannot collapse to a dimension smaller than this length. Alas, such a length was not an inherent feature of Rutherford's atom. However, Bohr saw after some contemplation, the square of Planck's constant, divided by the product $e^2 m$, where $e$ and $m$ are the charge and mass of the electron, respectively, and with a numerical constant thrown in, yields a value of $0.5 \times 10^{-8}$ cm, which happens to be the classic radius of the atom. A critical dimension, if ever there was one.

Making use of this shrewd gambit as an *ad hoc* stabilizing quantity for fixing the distance and velocity of the electrons in stable orbits (which Bohr took, for simplicity, to be circular), he refined his ideas about the structure of atoms and molecules, first at Manchester and then back in Copenhagen. His ideas would be elaborated in his famous 'trilogy' of 1913, 'On the Constitution of Atoms and Molecules' [4-3]. A famous feature of the work, the centerpiece of the first part of the trilogy, was his use of Johann Jakob Balmer's empirical radiation formula expressing the progression of the frequencies of lines in the hydrogen spectrum, and the spectral regularities it implied. Accordingly, the model incorporated electrons confined to a set of discrete, stationary orbits, unable to radiate energy except by jumping from one orbit to another, and then only in discontinuous quantum packets. The existence of Bohr's stationary states was experimentally verified by James Franck and Gustav Hertz (a nephew of Heinrich Hertz). Their experiment, carried out on the eve of World War I, was not conceived to corroborate Bohr's theory, however; only later did Franck and Hertz realize its importance in this connection.

Modeling the hydrogen atom and its spectrum was one thing; atoms with more than one electron were an entirely different proposition, as Bohr knew full well. The quantum rules had to be generalized, and in this he was aided chiefly by Arnold Sommerfeld, professor of theoretical physics at the University of Munich, who had been an active quantum theorist since 1911 [4-4]. His extension of Bohr's atomic model during 1915–16 involved the introduction of two kinematic generalizations: (a) replacing the circular motion of the electrons with Keplerian elliptical orbits with a precessing perhelion, and (b) introducing Einstein's special relativity in the kinematic treatment. Despite the war, Sommerfeld was able to send his paper to Bohr, then back in Manchester for a longer stint. Bohr wrote back in March 1916, heartily approving of Sommerfeld's initiative.

> I thank you so much for your most interesting and beautiful papers. I do not think that I have ever enjoyed the reading of anything more than I enjoyed the study of them. [4-5]

Sommerfeld summarized Bohr's atomic theory and his own development of it in his masterful text of 1919, *Atombau und Spektrallinien* [4-6]. As the war wound to a halt, in a paper honoring Planck on his sixtieth birthday, Sommerfeld predicted that the greatest scientific upheavals would spring from the connection of the quantum theory with atomism and relativity, if the world were ever again to find peace and quiet for scientific contemplation. In fact, for some time after the ending of hostilities, German physicists remained isolated from colleagues in former enemy countries, and their work was impeded by postwar political and economic dislocations [4-7]. Yet theoretical physics did not noticeably slacken, with Sommerfeld in Munich, Einstein and Planck in Berlin, and Born in Göttingen. In 1922, Planck himself assessed the state of theoretical physics, believing that over the past decade the most important developments in physics were relativity and the quantum theory. Alas, the quantum theory was not yet a 'real theory,' and would not be until its relationship to the wave theory of light was clarified [4-8].

By the time Planck made these remarks, experiments had already lent credence to major features of the Bohr theory and its extension by Sommerfeld, including that of Franck and Hertz, and by Otto Stern and Walter Gerlach in their demonstration of space quantization during 1921–22, dealing with another feature of Bohr's theory. In the latter experiment, a collimated beam of silver atoms passed through a magnetic field, and struck a target screen. If the field was uniform, the atoms, each possessing a magnetic moment, precessed about the direction of the magnetic field, and formed a narrow line on the screen. However, if the field was strongly inhomogeneous, the elementary magnets were deflected from their rectilinear paths, forming two separate lines on the screen. From careful measurements of the separation of the two lines, and of the gradient of the magnetic field, Stern and Gerlach calculated that each silver atom had a magnetic moment in the direction of the field of 1 Bohr Magneton (a unit of the magnetic moment of the orbital electron). The experiment demonstrated the quantization, not only of angular momentum, but also of its component in a direction defined, for instance, by the presence of a magnetic field.

Yet by the early 1920s, quantum theory had been developed about as far as it could go. Dealing properly with atoms of many electrons and complex spectra called for a radical new 'quantum mechanics,' as Born phrased it in 1924. The first to seize the initiative was Werner Heisenberg, a former student of Sommerfeld who, after obtaining his doctorate at Munich, went to Göttingen to study further under Born. Heisenberg grew uneasy over several literal concepts prevalent in quantum theory, such as bound electrons in atomic orbits. Instead, he strove to formulate a new theory that avoided concrete but unobservable representations, retaining only observable quantities, such as the transition probabilities for quantum jumps. His quantum mechanical solution of 1925 was to retain the classical equation for the motion of an electron while rejecting its classical kinematic description in terms of the classic Fourier series, each term containing a single integer. His new 'quantum-theoretical Fourier series' required each term to contain a pair of integers or indexes, corresponding to the initial and final states of the system. The scheme generated a new type of algebra which, upon consulting with Born, he realized was matrix algebra—a mathematical subfield new to Heisenberg. The upshot was that Heisenberg, Born, and Pascual Jordan, another gifted student of Born, joined forces in elaborating a first mathematical scheme of quantum mechanics which became known simply as matrix mechanics.

Heisenberg had spent the period from September 1924 to April 1925 with Bohr in Copenhagen, where he learned to combine the mathematical formalism of Born with the physical insight of Bohr [4-9]. He owed further inspiration to a stay in May 1925 on the small island of Helgoland off the

German coast in the North Sea, where he had repaired to seek relief from a strong bout of hay fever. There he had nothing to do but walk, swim, and think, and into the night he pondered about the subtleties of atomic phenomena, too excited to sleep.

Unbeknownst to the Göttingen physicists, an alternative form of quantum mechanics was being developed about the same time by Erwin Schrödinger in Zürich. Schrödinger was born in Vienna, where he studied under Friedrich Hasenöhrl, who fell in World War I during the battle for Caporetto on the River Isonzo. Schrödinger himself served as an officer in the Austrian artillery, survived, and in 1921 gravitated by way of Jena, Stuttgart, and Breslau to the University of Zürich. Before long he had a distinguished record of investigations behind him, into radioactivity, the kinetic theory of solids, statistical phenomena, and physiological optics (color vision and theory). However, for our purpose the important time in Zürich was late in 1925. At that time he spotted a footnote in a paper by Einstein citing work by the French physicist Louis de Broglie during 1923–24, which suggested that sub-atomic particles might also behave like waves. The upshot of de Broglie's deliberations, generalizing Einstein's own linking, in 1905, of light quanta with wave phenomena, was the celebrated relationship

$$\lambda = h/p$$

which connects the momentum $p$ of a particle to the wavelength $\lambda$ of a wave that is associated with it. Intrigued by de Broglie's ideas, Schrödinger developed them into a formal wave theory, centered on his famous 'Schrödinger equation' relating the wave function for a system to its energy and a potential. Once again Born stepped into the picture; it was shown by him in 1926 that $\Psi$, the rather mysterious field that propagates as a wave, represents the probability amplitude for locating a particle in a given small region of space.

In Schrödinger's approach, the stationary states of an atom are simply standing waves, and transitions between the states take place 'continuously in space and time.' This way of conceiving the atom appealed to Schrödinger as more intuitive, as closer to the ways of classical physics than that of the other quantum mechanics [4-10]. If not closer to the truth, physicists were on the whole more comfortable manipulating partial differential equations than matrices [4-11]. In fact, it was subsequently shown that matrix and wave mechanics, apparently incompatible, are mathematically equivalent.

However, back to Göttingen. Gamow found himself somewhat put off by the feverish excitement over the two forms of quantum mechanics (which were soon shown to be different representations of the same thing). Too many people were caught up in the hoopla, and all were concentrating on atoms and molecules. Why not, Gamow thought, look into what the new theories held in store for the atomic nucleus instead? In poking through recent literature on experimental nuclear physics, he came across an article by Rutherford on the scattering of $\alpha$-particles in uranium, using very fast $\alpha$-rays from RaC′ [4-12]. Even for uranium, Rutherford had found no deviation from his famous scattering formula, suggesting that the repulsive Coulomb forces, which oppose the penetration of $\alpha$-particles into the nucleus, still held sway to very small distances. On the other hand, uranium, being a radioactive atom, *emitted* $\alpha$-particles with only about half the energy of those from RaC′. Rutherford's explanation for why the alphas were so readily expelled from the nucleus was as follows. Starting from the nuclear interior, each $\alpha$-particle carries two electrons which neutralize its positive charge; when the $\alpha$-particle is well beyond the repulsive barrier, the electrons separate and return into the nucleus, 'as two tugs would leave a large ocean liner which they have just pulled out of the harbor' [4-13].

Rutherford's explanation struck Gamow as no good at all, and upon mulling it over he sensed the explanation in the new wave mechanics, which appeared to be valid inside nuclei: $\alpha$-particles obey the Schrödinger equation, just as do electrons. The $\alpha$-particles in the nucleus escape, not by *surmounting* the energy barrier enclosing the nucleus, but by *penetrating* ('tunneling') through it. This explanation also accounted for the old Geiger–Nuttall law, dating from 1911–12, according to which the shorter the half-life of a radioactive nucleus, the greater is the energy of the emitted $\alpha$-particle.

In due course, Gamow learned that his theory of $\alpha$-decay had been conceived independently at Princeton by Ronald W Gurney and Edward U Condon. Their preliminary note appeared in *Nature*, while Gamow's first paper was published in *Zeitschrift für Physik* [4-14].

At the end of the summer of 1928, Gamow was slated to return to Leningrad. However, on the way back he stopped in Copenhagen to pay his respects to Niels Bohr. Bohr asked what he was doing at present, and Gamow told him about his work on $\alpha$-decay. Listening with interest, Bohr then had a suggestion.

> My secretary told me that you have only enough money to stay here for a day. If I arrange for you a Carlsberg Fellowship at the Royal Danish Academy of Sciences, would you stay here for one year? [4-15]

'My, yes, thank you!' stammered Gamow, and that was that. Being free to pursue anything he wished, he took up the potential barrier problem again, but turned it around. Reflecting on why Rutherford's $\alpha$-particles had entered the nitrogen nucleus in 1919, he calculated the probability of an $\alpha$-particle from the outside penetrating *into* the nucleus, even though its own energy was too low to surmount the Coulomb barrier.

Behold! The results agreed equally well with Rutherford's first disintegration experiment a decade earlier.

Bohr and Gamow agreed on the desirability of Gamow going to Cambridge and show Rutherford his calculations in person, though Bohr warned Gamow to be careful in presenting his conclusions, since 'the old man did not like any innovations and used to say that any theory is good only if it is simple enough to be understood by a bar maid' [4-16]. More to the point, Gamow's theory flew in the face of Rutherford's own analysis and easily visualized model. Bohr wrote a diplomatic letter about Gamow, not to Rutherford but to Ralph Fowler; Fowler passed it on to Rutherford, who returned an invitation via Bohr for Gamow to join them in Cambridge for about a month shortly after the New Year of 1929 [4-17]. And so Gamow set off for Cambridge in early January, carrying with him a short manuscript to further placate the Crocodile, as Kapitza was fond of calling the old man. It contained two plots representing Rutherford's latest published data on $\alpha$-bombardment of various light nuclei. One set of points showed that the yield of emitted protons, for a given type of nuclear target, increased rapidly with bombarding energy. The other plot showed that the yield of protons, for a given incident energy, decreased rapidly with the atomic number of the target nuclei. Superimposed on the plots were two theoretical curves by Gamow, based on his wave-mechanical calculations; they fitted the experimental points to a tee.

Apparently Rutherford received Gamow without snapping at him. Cockcroft, for one, got along fine with their personable Russian visitor, who shared with him his little manuscript (Figure 4.1). Impressed,

**Figure 4.1.** *John Cockcroft (left) and George Gamow (right) in Cambridge, January 1929, discussing alpha bombardment of nuclei. Courtesy AIP, Emilio Segrè Visual Archives, Bainbridge Collection.*

Cockcroft penned a memorandum to Rutherford, probably in January 1929, as best he could remember later [4-18]. In it, he summarized calculations of his own, showing that protons with an energy of only 300 kV should be able to penetrate into a boron nucleus with a probability of about 6 in 1000. Rutherford needed no convincing; he himself referred to Gamow's theory of $\alpha$-decay in his opening remarks on a 'Discussion on the Structure of Atomic Nuclei' at a meeting of the Royal Society in London on 7 February 1929 [4-19]. Ralph Fowler, after Gamow spoke on the subject at the meeting, put it as succinctly as anybody. 'Anybody present in this room has a finite chance of leaving it without opening the door, or, of course, without being thrown out through the window' [4-20].

Gamow repeated Cockcroft's calculation before the Kapitza Club, and Allibone and Walton were as impressed as Cockcroft. Why not forget about a million volts, Cockcroft suggested to Rutherford, and concentrate on protons of a few hundred kilovolts? Gamow claims to have repeated the argument before Rutherford during his subsequent, longer stay at the Cavendish under a Rockefeller grant later in 1929. 'We are very happy to have Gamow here,' wrote Fowler to Bohr, '& he is I think making good progress with the nucleus' [4-21]. Gamow, however, may have confused this discussion with his earlier interaction with Rutherford. At any rate, the capital point is that by 1929 Rutherford had accepted Gamow's disintegration argument, if still hedging with regard to $\alpha$-decay. The stage was set for a new approach to atom smashing at the Cavendish Laboratory.

## 4.2. The Three Musketeers of DTM

Merle Tuve rejoined Gregory Breit at the Carnegie Institution's Department of Terrestrial Magnetism in July of 1926, in part to continue their radio pulse-echo technique for studying the conducting layer of the Earth's atmosphere. In these experiments, a crystal-controlled radio transmitter at Bellevue, Anacostia, in the District of Columbia was modulated to emit a series of short pulses that were received and recorded by an oscillograph some miles away at DTM in Chevy Chase. Single or multiple peaks were recorded, corresponding to the ground wave and one or more 'sky waves' arriving at the receiver via the reflecting layer of the upper atmosphere.

In early 1927, Breit and Tuve were joined by the Norwegian Odd Dahl, a curious fellow indeed. Born in 1898 in Drammen, the son of a hardware dealer, Dahl finished middle school in nearby Kristiania, as Oslo was then called. From early childhood he showed great aptitude for mechanical drawing and crafting and improvising mechanical and electrical devices with simple tools such as those found in a hardware store. Like Tuve and Lawrence, he rigged up a radio transmitter and receiver in his spare time.

His first scientific paper appeared in a Norwegian fisheries journal for 1920, entitled 'Short Considerations Concerning Wireless Telephone Stations for Fishing Vessels.'

Following his graduation in Kristiania, Dahl attended technical night-school while holding a succession of odd jobs with various engineering shops and firms, before being admitted to the Norwegian Army Flight School at Kjeller aerodrome in 1920. In spite of 'looping the loop' against strict orders, he graduated head of his class as Flight Sergeant in 1921. Just then, Roald Amundsen was looking for a pilot to join his polar ship *Maud* in a first attempt to reach the North Pole with the assistance of a flying machine. The idea was for the ship to be deliberately frozen into the pack-ice at some high latitude north of the Siberian coast, and drift to a point 'farthest north' with Dahl reconnoitering ice conditions and open leads from the air.

*Maud* left Seattle in June of 1922, under the command of Captain Oscar Wisting, a veteran of Amundsen's march to the South Pole in 1911. Dahl was in charge of a small Curtiss biplane (to be assembled on board from crated parts), as well as of the Marconi station on board, film photographer, and jack-of-all trades. Harald Ulrik Sverdrup, the respected Norwegian oceanographer, was in charge of the geophysics research program during the expedition. The ship was duly frozen in, and drifted ever so slowly with the moving pack ice over the next three years, reaching a maximum latitude of 76°N, north of the New Siberian Islands. The Curtiss was uncrated, assembled, and flown off the ice several times by Dahl, with Wisting as observer in the rear cockpit. Each time they landed with the machine slightly worse for wear, due to rough ice and hummocks. Giving up on what had been a first attempt at using aircraft in the high Arctic, Dahl's technical skills landed him instead the job of instrument maker and technical assistant to Sverdrup; in effect, the ship became Dahl's university campus, and Sverdrup his professor.

When the *Maud* expedition ended in August 1925, Sverdrup returned to the Carnegie Institution of Washington, which had underwritten much of the scientific program conducted during the voyage. On Sverdrup's recommendation, Dahl, who had been his right-hand man in those investigations, was hired by the Carnegie as a technical assistant to Breit and Tuve at the DTM. It was understood, however, that Dahl would only report for work after he got the ice out of his veins by undertaking, with a Norwegian companion, a two-year trek through South America. It took them by mules across the Andes, from Callao on the Peruvian coast to Puerto Ocopa on the headwaters of the Amazon, and thence by Indian canoe to Iquitos and Manao. With the latest adventure behind him, Dahl arrived in Washington to assume his new duties shortly after New Year, 1927. From the outset, his experience with radio equipment and a host of self-made geophysical recording devices under Sverdrup was put to good use, in refining pulsing apparatus with Breit and Tuve; their effort produced fresh determinations of the effective height of the Kennelly–Heaviside conducting layer during 1927–28.

The radio-propagation studies would prove highly successful, and were subsequently pursued by others at DTM, and eventually by Allied forces in World War II. In effect, Breit and Tuve had laid the foundation for the development of radar, though Breit was fond of claiming that 'the correct answer to who invented radar is the bats' [4-22]. Added Tuve,

> [Dahl] was with us at the U.S. Naval Research Laboratory in 1928 when we jointly cursed the planes coming in for landings at adjacent Bolling Field because their reflections of our pulses messed up our recordings of the Kennelly–Heaviside layer reflections, and each incoming plane spoiled our records for several minutes before it landed. These priorities of interest were inverted two years later by Lt. (later Rear Admiral) W. S. Parsons, who had watched our experiments there, and 'radio direction and ranging' (radar) was born in his secret memorandum of 1931 to the Navy Department. [4-23]

Despite the evident importance of the ionospheric studies, even at the time, Tuve had Director Fleming's concurrence that he devote a fraction of his time to high-voltage development and nuclear physics experimentation, building on Rutherford and Chadwick's pathbreaking studies at Cambridge. After all, the structure of the atom was surely the fundamental key to the secrets of terrestrial magnetism! Before long, the latter investigation would occupy most of the working hours of Breit, Tuve, and Dahl. Breit's departure from DTM in 1928 left the nuclear program in Tuve's overall charge. When Lawrence Hafstad joined DTM the same year, the smoothly-working team of Tuve, Hafstad, and Dahl came into being. 'Tough, Hafstad and Dahl,' also known as the Three Musketeers, represented three generations of Norwegians in America, with Lawrence (Larry) Randolph Hafstad born of Norwegian parents who settled in America. His father was originally a cobbler in Steinkjer, Norway, and immigrated to Minneapolis where Hafstad was born in 1904. After graduating from high school in 1920, he took a job with Northwestern Bell Telephone Company. Over the summer of 1922, he and Tuve became friends when they shared the evening shift working on the automatic switchboard equipment. Encouraged by Tuve, he enrolled at the University of Minnesota as an electrical engineering student, and then he, too, switched to physics for his graduate work [4-24]. However, he did not yet have his PhD when he joined DTM.

Breit had sailed for Europe in August of 1928. Arriving in Zürich, he spent his time following up on some theoretical work by Wolfgang

Pauli and Werner Heisenberg. Towards the end of the year he spent a month visiting other scientific centers on the continent. Based on what he learned, he decided to cut his year abroad short, and he returned to the US in January 1929 [4-25].

> In view of the comparative sterility of theoretical physics and its need for new data about the nucleus, it seemed desirable to return to duty at Washington...in order to help obtain this necessary information through our high-voltage developments and investigation. [4-26]

Back at DTM, however, Breit found himself in a situation less happy than it had been before his sojourn abroad. In particular, Tuve now insisted on remaining in charge of the high-voltage project.

> It was during this period that I finally decided to subscribe completely to the idea of a clear division of labor—I would work at experiments and let Gregory and other giants like him work at the necessary calculations and theory.... Frankly, I was pretty much paralyzed by the buffeting of those successive impacts in the period 1925–1929 when theoretical physics made such leaps up the mountainside, but I shall never forget the generous patience with which Gregory tried to keep all of us abreast of these ideas, and indeed did keep me from drowning in discouragement and despair. [4-27]

Breit left DTM in August 1929 to accept a position at New York University.

Dahl's early participation in the high-voltage program was interrupted by his irrepressible wanderlust. This time, however, the Carnegie itself provided the excuse for the next expedition. Dahl somehow convinced Director Fleming of the value of his undertaking terrestrial magnetism observations on behalf of DTM during a yearlong overland trek through remote regions of south central Asia, starting in mid-1928. (Needing a replacement during Dahl's absence, Tuve contacted Hafstad, who was more than happy to join him.) This journey, with his wife and another Norwegian, a boyhood friend from his Kristiania days, took them initially by a Kartusk houseboat down the Euphrates River from Djerabelos on the Turkey–Syria border to Ed-Deir on the Iraqi border. From there they continued in an old Rugby automobile through Persia, Baluchistan, and India, ending in Mandalay. The resulting magnetic observations were not of the highest quality, in Dahl's own opinion; routine, high-precision measurements were not his cup of tea, but designing and building the apparatus certainly was!

Tuve and Breit's original objective, when the high-voltage work was started on a part-time basis at DTM in 1926, was proton–proton scattering as a tool for probing the short-range attractive force between the constituents of atomic nuclei. First, however, they needed a suitable high-voltage source. Although Southern California Edison Company had a cascade transformer producing a million volts (for testing power-line components in connection with Hoover Dam), and General Electric had similar apparatus nearer at hand, such equipment was cumbersome and expensive. 'For scientific purposes,' Tuve argued, 'a high-voltage equipment of small power is more suitable, being less expensive and more easily handled and controlled' [4-28]. The Tesla induction trans-former seemed better suited for their needs, as Brian Goodlet had advised Allibone as well. But was it capable of generating the million volts or more judged necessary to compete with particles from radioactive substances? To find out, Breit and Tuve had a Tesla coil running in DTM's Experiment Building by the fall of 1927, or about the same time that Allibone was similarly engaged at the Cavendish.

The DTM coil, in its final form, consisted of a primary winding made from several turns of copper tubing wrapped into a flat spiral around a secondary winding of many thousands of turns of fine insulated wire wound on a Pyrex tube equipped with spherical metal terminals (Figure 4.2). The primary was driven by the discharge across a spark gap of a very large condenser on loan from the Washington Navy Yard. With the coil in air, the secondary developed a maximum voltage of about 500 kV before sparking to ground. However, by mounting the coil in a wooden box filled with transformer oil, the secondary held 3 million volts, and by pressurizing the oil in a metal tank, they coaxed it to over 5 million volts [4-29].

Generating high voltages was one thing; developing a vacuum tube able to withstand the voltage stress was less easily done. Existing vacuum tubes failed above 300 kV, due to uncontrolled electron emission from the cathode and accumulation of charges on the glass walls, result-ing in flashover and puncture of the tube. The solution to the problem was shown by William Coolidge at General Electric. He recommended cascad-ing two or more tubes, passing the electron beam from one to the next through thin metallic foil windows, as was his own practice with cas-caded X-ray tubes. Coolidge even gave DTM one of their largest tubes, which Breit personally carried back from the GE plant in Schenectady, NY. After some experimentation, the DTM team adopted a modified version of Coolidge's tube: they subdivided the vacuum tube into sections, and omitted the intervening foils. To ensure a uniform distribu-tion of voltage between tube sections, they utilized a potentiometer with taps connected to the successive hollow electrodes mounted on the axis of the acceleration tube [4-30].

All this took time, and as 1928 came to an end, all they had yet to show for their effort was some pedantic alternative work on 'simple types of electrodeless tubes' [4-31], similar to what Rutherford had Ernest

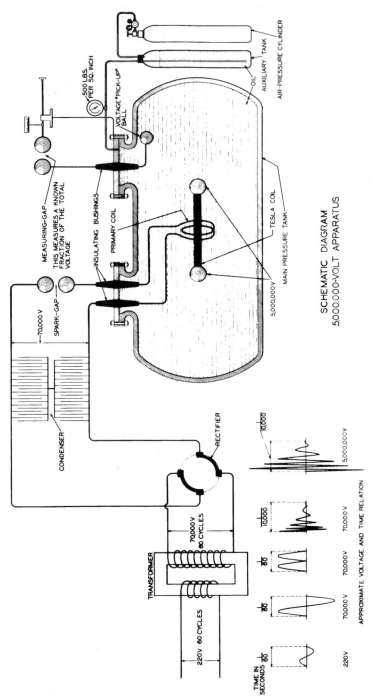

**Figure 4.2.** *Schematic of DTM's Tesla coil in oil under high pressure. Breit, Tuve, and Dahl, 1930, PR, 35, pp. 51–65, on p. 54.*

Walton doing just then, with equally lackluster results. More importantly for the long haul, as the New Year of 1929 came and went, Tuve and company would not have the benefit of sharing first hand George Gamow's latest wisdom about accelerating voltages, though they undoubtedly knew of Gurney and Condon's letter to *Nature* on $\alpha$-decay. Breit certainly did, though seemed curiously indifferent to the simultaneous discoveries in the US and abroad, judging by the tone of his letter to Tuve in October 1928.

> The dope of Gurney and Condon on radioactivity in Nature looks fine though Heisenberg is said to have talked about it last year. I understand that a Russian Gammow (ff?) has a more complete theory which is to appear in the Zeits. [4-32]

On that uncertain note, we leave Tuve and colleagues for the moment, and switch to Ernest Lawrence. The newly appointed professor was just settling in at Berkeley as Breit shared his thoughts with Tuve, though Lawrence was not yet caught up in the mounting excitement over high-energy beams.

### 4.3. Berkeley, and an evening in the library

Ernest Lawrence first paid a visit to his old friend Merle Tuve at the Carnegie Institution in April 1927. Lawrence was riding high just then, having created something of a stir with the results of his and Jesse Beams' experiment on the Kerr Effect; their latest paper, concluding that the lag in the effect is extremely small, attracted letters from Cornell, the University of Washington, and General Electric, all urging Lawrence to reconsider their staff position offers. Tuve, for his part, had his hands full with both the awkward Tesla coil and the sensitive radio receiver, and was perhaps a little preoccupied at the time. At any rate, Tuve advised his friend to quit 'spinning tops' and climb on the high-voltage bandwagon [4-33]. Indeed, fresh from an invited stay and job interviews with GE in Schenectady, Lawrence had viewed high-voltage apparatus at its mightiest. Nevertheless, he allowed that he was impressed with Tuve's effort to reach high voltages, but felt they had yet to find ways of reaching those energy levels cheaply enough to be practical.

If anything, Lawrence was more impressed with what he saw, or with what little he saw, when he, Swann, and Beams visited the Cavendish Laboratory during a European jaunt in June of 1927—a trip he referred to as 'somewhat of a farewell party for Professor Swann and me' [4-34]. That he was impressed with Rutherford himself goes without saying, who introduced them to Kapitza, his brilliant and idiosyncratic assistant. However, their most fruitful discussions were with Cockcroft and Chadwick, who impressed upon them their policy of clean, simple apparatus, inherited from the Old Man; the brass vessel housing Rutherford's disintegration experiment of 1919 was no larger than a shoe box [4-35]!

Leaving Swann in Cambridge, Lawrence and Beams moved on through the Continent, stopping first at Copenhagen, where they missed Bohr, and then in Hamburg, where Otto Stern knew of Lawrence and his work. The Dutch laboratories, Kamerlingh Onnes's cryogenic laboratory at Leiden and the Philips Company's Natuurkundig Laboratorium in Eindhoven, struck them as more luxuriously equipped than any other in Europe, and on par with the very best in the US. In Berlin, at Nernst's Physikalisch-Chemisches Institut, Herr Professor Walter Nernst warmed to them when he learned they were from 'Yawl,' where he had given one of the celebrated Silliman Lectures [4-36]. In Paris they caught up with Swann, who took time from practicing his cello under Pablo Casals to introduce them to Marie Curie. The most humble apparatus of any laboratory was surely that with which Madame had done her famous work, and which she showed her visitors when they came calling at her Institut du Radium.

Tuve made one more attempt at convincing, or haranguing, Lawrence of the importance of nuclear experimentation as the largely unexplored frontier of physics, when Lawrence dropped by the DTM once more, before setting off for Berkeley in late summer of 1928.

> I asked Ernie what research he was aiming to do at Berkeley. He responded, rather vaguely, with some small notions about high-speed rotating mirrors, chopping the tails off quanta and other single-shot ideas. I then talked to him like a Dutch uncle. I said it was high time for him to quit selecting research problems like choosing cookies at a party; it was time for him to pick a field of research that was full of fresh questions to be answered, and sure of rich results after techniques were worked out. I said that any undergraduate could see that nuclear physics using artificial beams of high-energy protons and helium ions was such a field, and that he should stake out a territory there to work and to grow in. [4-37]

Tuve went on to suggest that Lawrence consider not only their 'albatross,' the Tesla coil, but possible indirect acceleration methods, as they too, along with Ernest Walton at Cambridge, had looked into for a starter. Among other things, both Tuve, Breit, and Dahl, and Walton, had fiddled with a toroidal device of a kind that later became known as the betatron; so had Rolf Wideröe in Aachen. Who knows, perhaps Lawrence might do better in Berkeley. 'Ernie was very sober and did not seem to resent the rather harsh way I went after his quantum tails,' adds Tuve in his recollection of their meeting on that sultry afternoon in Chevy Chase. If still not convinced, memories of this discussion with his friend Merle (and perhaps earlier ones in the same vein with Swann) must have stayed with Lawrence when, in April of the following year, he pored over Rolf Wideröe's journal article in Berkeley.

Associate Professor E O Lawrence arrived at the University of California in Berkeley in August, shortly before the opening of the academic year of 1928. Dour Elmer Hall, the physics chairman, had little to say besides welcoming him to the Physics Department in spacious Le Conte Hall, leaving it to his younger side-kick, Raymond Birge, to make their new colleague feel at home. Leonard Loeb, who had worked tirelessly to attract Lawrence to Berkeley, was no less cordial. The department was disappointed that Edward Condon, who had been a student of Birge, had accepted an appointment at Princeton; however, this was offset by the promise that a brilliant young theorist, Julius Robert Oppenheimer, had agreed to join the faculty the next year. Lawrence took up residence at the redwood-clad Men's Faculty Club in a charming glade a hundred yards from Le Conte Hall, where Gilbert Lewis, the influential dean of the College of Chemistry, held forth nightly and encouraged interdisciplinary intercourse between chemistry and physics. Lewis took notice of Lawrence, and would play an interesting role in Lawrence's forthcoming accelerator program, as noted by John Heilbron and Robert Seidel in their tome on Lawrence and his laboratory, and as we shall come to shortly [4-38].

Lawrence threw himself into teaching Electromagnetic Theory with his usual enthusiasm and energy. The key to research, however, lay in graduate students and what research problems their professor could come up with for their mutual benefit. Niels Edlefsen, an assistant in the Physics Department and Lawrence's senior by a few years, requested to do his PhD problem under Lawrence, and Lawrence soon had him working on the photoionization of cesium vapor. Edlefsen, a Mormon, brought in two other students from Utah, to whom Lawrence assigned other photoelectric problems, and before long his laboratory was humming with activity at all hours, to some envy on the part of older colleagues in the department. Not to worry; the most active younger professors viewed his success with admiration and cooperated when they could. Professor Loeb gave up his darkroom for Lawrence and Edlefsen's use, and chairman Hall ensured the superintendent of buildings and grounds that high voltage wires strung along the basement corridor leading to Lawrence's lab in ancient South Hall—the former physics headquarters—would only be powered at night, and always with outside doors locked and marked with warning signs [4–39].

Yet time was passing by, Lawrence was nearing thirty, and Tuve's admonition may have hung over him. Increasingly, he found himself drawn to the nuclear atom and its tantalizing alchemical possibilities looming ahead. Ionization and photoelectric experiments were all good and well, but only brushed the periphery of the atom. Discussions with Thomas H Johnson may have been a factor in Lawrence's new thinking. An old friend from Chicago and Yale, and now with Swann at the Bartol Foundation, Tom Johnson was on leave from Bartol and teaching at Berkeley during the spring semester of 1929. They had much to talk about, reminiscing and looking ahead. Lawrence told him about the ongoing attempts at the Carnegie and the Cavendish to generate high enough voltages to disintegrate nuclei by artificial means and get at the ultimate forces of nature. Despite his close association with Tuve and respect for what he saw of Cockcroft in Cambridge, Lawrence still professed skepticism about the practicality of a brute force approach, say with Tesla coils *à la* Tuve, Hafstad, and Dahl. Johnson agreed; a DC, continuous potential source of high-intensity charged particles seemed more likely to do the trick than the unstable, uncontrollable Tesla coil. But that was easier said than done.

Then one evening, on or about 1 April 1929 [4-40], Lawrence and Johnson were idly perusing scientific journals in the main University library. As it happened, the library had recently opened a subscription to the *Archiv für Elektrotechnik*, a journal catering to engineers and physicists interested in electrotechnology, perhaps on the recommendation of Lawrence himself. So Lawrence picked up the latest issue of *Archiv* and flipped through it. Not much there. But wait! On page 387 began an article by Rolf Wideröe, a Norwegian engineer. Not very good in German, Lawrence scarcely glanced at the text, but the illustrations had caught his immediate attention. Here was clearly something new and important, perhaps the key to smashing atoms in a controlled manner that made sense! Excited, he drew Johnson to a nearby blackboard, grabbing a piece of chalk. In an hour or so, Lawrence knew he had the solution to his problem.

### 4.4.  Wideröe and Van de Graaff: accelerator virtuosos

Just who was Wideröe, and what did he do? We have met him briefly here and there in earlier chapters; however, in light of his pioneering role in the history of particle accelerators, and his importance for E O Lawrence, a little more on his life and work is appropriate in these pages. For a fuller account, there are his own memoirs to fall back upon as well [4-41].

Rolf Wideröe was born in Oslo (then Kristiania) in 1902, the son of an agent for French wines and Cognac and Dutch vegetable oils. Rolf's younger brother by two years, Viggo, became a well-known aviator and founder (with another brother, Arild) of the Norwegian aviation company that still bears the family name. (Viggo Wideröe's daughter Turi became Norway's first female airline pilot.) Rolf was educated in Kristiania, and graduated from Halling Gymnasium in 1920. The year before, while still a student, he read in the local newspapers about Ernest Rutherford's disintegration of nitrogen nuclei with $\alpha$-particles from RaC'. Having speculated on the possibility of 'splitting the atom' at an even younger age, by orbiting electrons in a fanciful process akin to a 'super-Zeeman-effect' [4-42], Wideröe saw that the limited flux and energy of naturally occurring $\alpha$-particles provided an impractical approach to atom smashing; particles electrically accelerated with high-voltage technology appeared much more promising [4-43]. With that in mind, and with his parents' concurrence, he decided to take up electrical engineering at the Technische Hochschule in Karlsruhe, in diffidence to Norway's own Tekniske Høyskole in Trondheim. (He no longer recalls why he chose Karlsruhe in particular. It ranks among the best and oldest German Polytechnic Schools, in contrast, say, with NTH in Trondheim, which was then only 10 years old, and which his parents regarded condescendingly as a 'kindergarten' [4-44]).

At Karlsruhe, Wideröe developed the notion of a 'ray-transformer,' that is, a machine for accelerating particles without the need for dangerously high voltages. The idea was to accelerate electrons, maintained in a circular orbit by a transverse magnetic field, by the electric field induced by the changing magnetic flux linking the orbit — the principle of what later became known as the *betatron*, which is essentially a transformer. The voltage gained per turn would be small, but during the acceleration period the electrons would perform a great number of revolutions and should attain a net gain in energy amounting to millions of volts. He came up with his idea as early as 1922, but only in 1927, with diploma studies safely under his belt, did he broach the concept of his ray transformer to Wolfgang Gaede, professor of physics at Karlsruhe and inventor of the diffusion high-vacuum pump named after him. To his dismay, Gaede gave him a 'cold shower.'

> On the long way the electrons will have to travel before they reach high energy [Gaede retorted] they will be absorbed by the remaining gas molecules. Your idea will not work; forget it. [4-45]

Not to be put off, Wideröe convinced himself, by burning midnight oil in the library, that Gaede's assumptions were wrong. Unfortunately, however, an experiment at Karlsruhe under Gaede was obviously out of the question. Then again, the technology at Karlsruhe's Polytechnic was not entirely up to snuff, Wideröe comforted himself, at least for meeting his own particular demands. After some further mulling it over, he wrote Walter Rogowski, an expert on cathode ray tubes and oscillographs at the Technische Hochschule in Aachen. The upshot was that he wound up outlining his ideas to Professor Rogowski on a train between Karlsruhe and Mannheim. Rogowski was much more receptive than Gaede, and without further ado took Wideröe on as a doctoral candidate.

At Aachen, Wideröe was soon busy assembling his ray- or beam-transformer for his thesis project. As noted earlier, its operating principle went as follows. Current in a primary coil winding induced a changing magnetic flux in the laminated iron core of the transformer. Electrons circulating in a doughnut-shaped glass vacuum tube served as the secondary winding, behaving in the same way (or so Wideröe hoped) as if they were in a copper wire in an ordinary transformer's secondary coil. The electrons were emitted by a cathode and shot into the evacuated doughnut, where they were supposed to revolve on a circular orbit. Acceleration would take place during the quarter cycle of the alternating current while the flux was rising and while the magnetic field at the orbit had the proper strength to guide the particles in a circular path.

The transformer produced a maximum flux density of 14,000 gauss. With a mean orbit radius of 7.25 cm, this would correspond to a maximum increase in voltage of 6 MV. In the event, it proved very difficult to keep the electrons on a stable orbit. Shortly after a single revolution, they invariably struck the wall of the vacuum tube, and no acceleration could be demonstrated. The reason lay in Wideröe's failure to create an optimal arrangement of magnetic fields and electrodes for confining the electrons to the central orbit of the tube, that would have 'a stabilizing effect on the trajectories.'

In fact, Wideröe was not alone in his failed attempt. Ernest Walton tried the scheme at the Cavendish, but he, too, got nowhere. The very same year, Tuve, Breit, and Dahl tried quite independently a similar experiment with only slightly more success at the Carnegie Institution. Their apparatus relied on a spark discharge of their large capacitor, on loan from the Washington Navy Yard, through a coil to produce the magnetic field, and a much more energetic electron gun injector; however, they too were bedeviled by an inadequately-shaped magnetic field unable to stabilize the electron beam. Though claiming an acceleration to 'one and one-half or two million volts,' they admitted that 'the radiations so far ... are very weak and intermittent' [4-46].

A non-working device would not do as a dissertation project, but Wideröe did not give up. Instead, he resurrected a scheme proposed in 1924 by Gustaf Ising of Stockholm's Tekniske Högskola for the acceleration of heavy ions (canal rays) in what amounted to a traveling-wave linear accelerator. In so doing, Wideröe wound up performing the first experimental tests of the principle of *resonance* acceleration. His apparatus became the direct ancestor of all resonance accelerators, both linear and circular, earning its builder the right to be considered the progenitor of all truly high-energy particle accelerators.

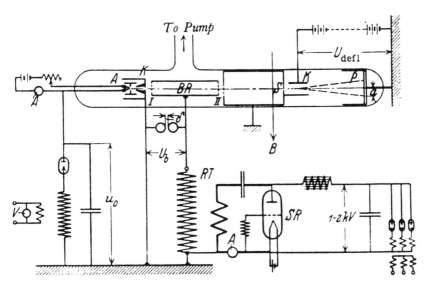

**Figure 4.3.** *Wideröe's linear accelerator and its energizing circuit. Wideröe, 1928, Archiv für Elektrotechnic,* **21**, *pp. 387–406.*

Ising's scheme envisioned a linear array of cylindrical electrodes connected to transmission lines of increasing length [4-47]. A potential pulse generated by a spark gap in an oscillatory circuit was conducted along the wires, arriving sequentially at the electrodes and creating accelerating fields in the gaps between successive electrodes. Inside the hollow electrodes, the beam was shielded and was not accelerated. The gap spacing increased in unison with the increasing ion velocity, ensuring acceleration at each gap.

Ising apparently did not follow up his express intention of conducting practical experiments, but Wideröe did [4-48]. With the goal of simply demonstrating the principle of resonance acceleration (not a feature of the betatron, which does not depend on an r.f. field for acceleration), he chose the simplest arrangement of two accelerating gaps, as shown in Figure 4.3. Sodium and potassium ions were accelerated through a potential of 20 kV across the first gap *I*; thence the ions drifted through an electrostatically shielded tube *BR* and arrived at the second gap *II*. Since the distance traveled by the ions after the first acceleration in one half of the r.f. field was equal to the spacing between the gaps, the ions were accelerated in crossing the second gap, with their energy doubled to 40 kV. Following the second acceleration, the ions drifted through a second shielded region *S*, were electrostatically deflected in a capacitor *K*, and finally struck a photographic plate, where the voltage of the ions could be calculated from the measured deflection $\alpha$. Sure enough, the

55

measurements 'indicated at once that the ions did gain a doubled kinetic voltage, as expected' [4-49].

> Rogowski took hardly any notice of my work, I don't think that he ever as much as looked at my linac. It was expected that my thesis would be published in a periodical and I had no problem getting it into 'Archiv für Elektrotechnik.' ... Rogowski and Professor L. Finzi (physics) were my examiners. I had no problem there either and I finally obtained my title of 'Doktor-Ingenieur' on November 28, 1927. [4-50]

Reproduction of Wideröe's small device (about one meter long), can be seen in a number of places, e.g., Deutsches Museum in Munich, the German Röntgen Museum in Reimscheid, the Norwegian Technical Museum in Oslo, and at the Smithsonian Institution in Washington, DC. They were all, curiously, made at the Radium Hospital in Oslo, to Wideröe's specifications.

Wideröe's senior by half a year, Robert Van de Graaff, an American physicist, also conceived of an accelerator scheme — not for his doctorate, but as a digression while completing his doctorate in 1928, or shortly after Wideröe had his own device in operation. Whereas Wideröe's device became the prototype for all high-energy accelerators in particle physics, based on repeated small boosts in particle energy by a resonance principle, Van de Graaff's scheme involved an elegant method of imparting a single, powerful boost in energy to the charged particles; his electrostatic device became the forerunner for an abundant crop of low-energy accelerators that became the staple of experimental nuclear physics in laboratories and industrial settings around the world. Since it was Tuve and company that first exploited Van de Graaff's accelerator for precision experiments in nuclear research, a little more on the inventor and his classic machine is equally appropriate at this point in our chronicle, to set the stage for the technical turn of events in Tuve's program at the DTM in Washington.

Robert Jemison Van de Graaff was born in Tuscaloosa, Alabama in 1901 and raised in the cotton country around that city [4-51]. He studied mechanical engineering at the University of Alabama, where he earned his BSc in 1922 and MSc in 1923. Next, he held a job with the Alabama Power Company, before studying physics at the Sorbonne, where lectures by Madame Curie wetted his appetite for radioactivity and the atomic nucleus. In 1925 he received a Rhodes Scholarship to Oxford, where he completed his PhD in 1928.

It was while at Oxford that Van de Graaff conceived of his belt-charged electrostatic, high-voltage generator after reading the 1927 anniversary address on St Andrew's Day by Rutherford to the Royal Society on the need for accelerated subatomic particles; he apparently also heard Rutherford in person delivering a lecture on the same theme before the British Association at Glasgow in 1928. Having been trained in the transformation of energy on an industrial scale, he concluded that conventional means of electric-power transformation would provide insufficient energy or beam homogeneity for nuclear experimentation with charged particles. However, the 'electrostatic characteristics of the atomic nucleus' suggested another tack: using a direct-current electro-static method for generating high voltages, and thereby 'meeting the atom on its own terms' [4-52].

From Oxford, Van de Graaff went to Princeton as a National Research Fellow in 1929, where he soon had a tabletop model of his generator run-ning at 80 kV. The principle of the device is rather simple. A motor-driven endless, insulated belt running vertically between two pulleys picks up electric charge from corona points at the grounded, bottom end of the belt, and deposits it, again by corona points, onto the surface of a large insulated, hollow conducting sphere (actually a tin can in the first model) at the top. The accumulated voltage can be discharged as an awe-inspiring spark in a museum, or put to better use for accelerating charged particles through the large potential drop in an evacuated tube.

Karl Compton, still the chairman of the Princeton Physics Department, had nothing but praise for Van de Graaff and his generator, and sup-ported him all along. By late summer of 1931 his latest model could charge a two-foot copper sphere on a Pyrex rod to 750 kV. The next model featured two such spheres, each resting on a 7-foot Pyrex rod. With one sphere charged positively and the other negatively to 750 kV, a potential difference of 1.5 million volts could be maintained between them [4-53], as was demonstrated at the inaugural dinner for the American Institute of Physics at the New York Athletic Club in November 1931. 'Atom Nucleus Yielding to Science,' declared the *New York Times* for Wednesday, 11 November. '1,500,000 Volts Leap From $90 Device in Test Before the American Institute' [4-54].

> The 1,500,000-volt generator was placed in an alcove of the dining room. In appearance it might be taken for two identical, rather large floor lamps of modernistic design, with the copper spheres as the lampshades. A simple lamp cord running to the base of each glass standard bears out the illusion.
>
> The action of the apparatus is similar to that of a bucket conveyor carrying water a little at a time up to a tank. Just as the level of the water in the tank rises with each new bucketful, so the electrical charge or potential rises as the amount of the charge is increased. The electrical charge, however, is said to 'reside on the surface' instead of inside the container. [4-55]

# Chapter 5

# PROTONS, ELECTRONS, AND GAMMA-RAYS

## 5.1. Protons in Cambridge, and γ-rays?

The honor of being the first of our vying laboratories to launch a program on serious disintegration studies with artificially accelerated positive ions fell upon the Cavendish in 1929. In January of that year, John Cockcroft, with George Gamow's guidance and Rutherford's blessing, decided to proceed with protons accelerated to a few hundred kilovolts, assuming he could come up with a vacuum tube able to withstand the voltage stress. Rutherford suggested that Ernest Walton, whom he considered 'an original and able man [who] had tackled a very difficult problem with energy and skill' [5-1], should abandon his frustrated attempts at producing fast electrons and positive ions by indirect means—that is, by emulating Rolf Wideröe in Germany—and team up with Cockcroft in his new undertaking.

Cockcroft decided early on to forego a high-frequency Tesla coil or transient voltages from other induction coils in favor of a steady d.c. voltage for energizing their discharge tube. A Tesla coil, in particular, had two strikes against it. Emissions during the negative half-cycle might well confuse any measurements, and with a damped oscillatory wave the real time available for accelerating particles at the highest crest of the waves was only about one thousandth part of the total time. Electing a d.c. supply, he persuaded his pound-wise boss to ask the University front office for a grant of £1000 for the acquisition of a 300 kV transformer and auxiliary equipment to rectify the alternating voltage and deliver it as a d.c. potential to the accelerator tube [5-2]. Two half-wave thermionic rectifiers (diodes) were designed by Allibone; they were large vacuum tubes of puncture-resistant special 'molybdenum' glass, not unlike bulbous Coolidge X-ray tubes in size and shape. The transformer was specially designed for the job by Brian Goodlet, Allibone's former boss at Metro-Vick, to be assembled inside the designated laboratory space which, despite Rutherford's boast to the contrary, was not sufficiently commodious to allow equipment of decent size to fit through the various doors; standard Metro-Vick testing transformers of 300 kV rating would not get through the entrance arch to the Cavendish from Free School Lane, let alone through the doorway to their laboratory room, even with the doors removed. The rectifiers and the Metro-Vick transformer were first powered up in February of 1930.

The acceleration tube, constructed during the last months of 1929 on Allibone's advice, was not much more than another glass tube with a bulb in the center housing a pair of steel pipes serving as rudimentary electrodes. At its top end sat the ion source, which produced hydrogen ions, or protons. It was based on a design of the canal-ray type first devised by Wilhelm Wien at the Technische Hochschule in Aachen, and further developed at the Cavendish by Mark Oliphant, another of Rutherford's protégés. Oliphant had arrived at the Cavendish from Australia on an 1851 Exhibition Scholarship in October 1927, or at the same time as Walton. Oliphant, too, had a gift for designing and building apparatus, and keeping it running; he was good enough that he was spared the usual training period in the upstairs 'nursery.' He set out to study the properties of positive ions, and devising ways by which beams of positive ions could be accelerated through modest voltages in discharge tubes; true to form, he had his doctorate in two years [5-3]. Oliphant struck it up very well with Cockcroft, and also with Rutherford, who took him under his wing, and for whom he would build a new particle accelerator, less powerful than Cockcroft and Walton's, but one which Rutherford would utilize himself in the waning period of his life.

Cockcroft and Walton's apparatus—acceleration tube with its ion source, transformer with its rectifiers and motor-generator set, plus vacuum pumps and their system—was ready for its maiden run in June 1930. Protons, produced by a glow discharge in hydrogen gas, passed out of the ion source via a fine canal bored through the cathode. They entered the 1-meter long vacuum tube, pumped for all its worth by an oil-diffusion vacuum pump developed by Metro-Vick's C R Burch, exploiting Apiezon oil, the low vapor pressure key to the vacuum technology of the day. After acceleration in the gap between the pair of electrodes, the beam, still reasonably well collimated, reached the experimental chamber—nothing more than a 3-inch diameter glass cup—there to strike a collecting electrode or any target interposed in its path. First, however, the composition of the beam, the relative number of protons and hydrogen molecules, had to be determined in an auxiliary experiment in which the ions were deflected with an 'analyzing' magnet; it showed that approximately half of the beam current was carried by protons, and half by molecular hydrogen ($H_2^+$) ions.

Preliminary experiments were carried out with targets of lead and beryllium salt. Alas, the tube broke down above 280 kV, and up to that energy no convincing evidence was found for any nuclear transformations. In particular, they had expected to observe $\gamma$-rays produced in the target with a gold leaf electroscope as the detector, but saw few if any. Unfortunately, recalled Walton,

> ...we did not get around to looking for scintillations produced by *particles* emitted in a nuclear reaction. We would have had to lie on the floor and peer into a microscope—not a relaxed position for reliable counts. [5-4]

Not finding any $\gamma$-rays, the would-be transmuters concluded that George Gamow was mistaken, and that higher voltages were needed after all [5-5]. In August, adding insult to injury, the transformer failed. With nothing more to show for their year and a half of effort, they dispatched a progress report, such as it were, to the Royal Society on 19 August [5-6]. Though 'very definite indications of a radiation of a non-homogeneous type was found,... the intensity of the radiation was of the order of one ten-thousandth of that produced by an equal electron source....' In fact, had they bothered to get down on the floor and look for scintillations, they might well have seen some, and their overall gloomy conclusion might have been different. Unknown to them, but very probably, they had disintegrated beryllium [5-7].

The reason for their stubborn search for $\gamma$-rays rather than $\alpha$-particles, according to Cockcroft, lay in their 'fixed idea that $\gamma$-rays would be the most likely disintegration products' [5-8]. Their fixed idea probably originated in Chadwick's determined search for spontaneous $\gamma$-rays in hydrogen (section 2.2) and a related experiment by the German physicist Walther Bothe. Bothe, incidentally, had something else in common with Chadwick. Having obtained his doctorate under Max Planck in Berlin in 1914, he, like Chadwick, joined Hans Geiger at the PTR the same year. No sooner had he done so, than World War I broke out, and he, like Geiger, was called up. He was taken prisoner by the Russians in 1915 and sent to Siberia where he was interned until 1920. As resilient as Chadwick at Ruhleben near Berlin, he honed his mathematical skills and studied theoretical physics during the worst of conditions. On his return to Germany, he accepted Geiger's invitation to work at the laboratory for radium research at the Reichsanstalt. Active in both theoretical and experimental work on the scattering of $\alpha$- and $\beta$-particles, he took up nuclear transmutation studies in 1926.

In 1928, just as Chadwick resumed his search for $\gamma$-rays, Bothe and his student Herbert Becker began a systematic study of $\gamma$-radiation from $\alpha$-bombardment of light nuclei. Such radiation seemed not unreasonable, because certain elements did not disintegrate, and hence were not proton emitters. Indeed, beryllium, which did not disintegrate under $\alpha$-bombardment, did, to their great surprise, give off a highly penetrating radiation. The same was soon observed in lithium and boron. Before long both Chadwick and his student H C Webster and Frédéric and Irène Joliot-Curie in Paris were engaged in tracking down Bothe's mysterious 'beryllium radiation' [5-9]. It was Chadwick who cleared up the mystery in 1932, as we shall see. What they were observing were not massless quanta of electromagnetic radiation, but the neutron, which had eluded Chadwick for a full decade!

With the failure of the transformer, and unable to generate more than 280 kV in any case, Cockcroft decided a fresh approach was called for. As it happened, they were also informed that their cramped basement laboratory workspace was being requisitioned by the Physical Chemistry Department, and had to be vacated by May 1931. This piece of bad news proved a blessing in disguise, however. They relocated into a much larger room, with a high ceiling, and in so doing had the opportunity to start over from scratch. And upon mulling it over, Cockcroft knew what he wanted in their new-found laboratory space.

## 5.2.  Protons in Berkeley

We last left Ernest Lawrence in the University of California library, poring over Rolf Wideröe's article in *Archiv für Elektrotechnic* on or around 1 April 1929. Though the principle of resonance acceleration in a linear accelerator had been confirmed to Herr Professor Rogowski's satisfaction, earning Wideröe his Aachen doctorate in 1927, to Wideröe the obtainable beam current seemed disappointingly low for practical applications. Fortunately, unable to read German with any fluency, Lawrence was not put off by Wideröe's pessimism, but focused on the illustrations and numerical formulae. He grasped at once Wideröe's scheme, the multiple acceleration of positive ions by appropriate applications of radiofrequency oscillating voltages to a series of cylindrical electrodes of increasing length in a linear array. Lawrence has told us what happened next.

> This new idea immediately impressed me as the real answer which I had been looking for to the technical problem of accelerating positive ions, and without looking at the article further I then and there made estimates of the general features of a linear accelerator for protons in the energy range above one million volt electrons. Simple calculations showed that the accelerator tube would be some meters in length which at that time seemed rather awkwardly long for laboratory purposes. And accordingly, I asked myself the question, instead of using a large number of cylindrical electrodes in line, might it not be possible to use two electrodes over and over again by bending the

positive ions back and forth through the electrodes by some sort of appropriate magnetic field arrangement. Again a little analysis of the problem showed that a uniform magnetic field had just the right properties — that the angular velocity of the ions circulating in the field would be independent of their energy so that they would circulate back and forth between suitable hollow electrodes in resonance with an oscillating electrical field of a certain frequency which now has come to be known as the cyclotron frequency. [5-10]

The expression for the angular velocity $\omega$ is readily derived. Ions moving in a magnetic field of strength $H$ oriented normal to the plane of their trajectory will follow a curved path whose radius of curvature $r$ is determined by equating the magnetic force on the ions of charge $e$, mass $m$, and velocity $v$ to their resultant centripetal acceleration, or

$$Hev = (mv^2)/r. \tag{5.1}$$

Thus the charge-to-mass ratio $e/m$, or $r$, depends on $v$. However, the angular velocity $\omega$ is *independent* of the velocity and radius, and is given simply by

$$\omega = v/r = He/m. \tag{5.2}$$

This is the all-important 'cyclotron condition'. The angular velocity or frequency of revolution remains constant, independent of the radius or momentum, as long as the magnetic field and the mass does not change. Fast ions describe long circular paths, slow ions short paths, with the frequency of revolution $\omega/2\pi$ the same for all. Ions of mass $m$ and charge $e$ could be made to travel in phase with an r.f. field in a magnetic field $H$, repeatedly crossing a gap between electrodes, spiraling outward and gaining energy each time around.

It seems that Wideröe, and his colleagues at Aachen, narrowly missed inventing the cyclotron, as Lawrence gradually came to call his device. Wideröe recalls a discussion of his resonance accelerator with another of Rogowski's assistants, named Fleger. Fleger asked whether perhaps the ions could be made to circulate, for instance on spiraling orbits in a magnetic field. 'My answer was that this was possible but that it would be difficult to stabilize the orbits and that the ions probably would be lost in collisions with the walls' [5-11].

Another claimant of the notion of the cyclotron was the Hungarian-born Leo Szilard, a man of uncanny intuition about physics, like Gregory Breit.

In maybe 1928 or 1929 I began to think what might be the future development in physics. Disintegration of the atom required higher energies than were available up to that time; there had been no artificial disintegration of the atom. I was thinking of how could one accelerate particles to high speeds. I hit upon the idea of the cyclotron, maybe a few years before Lawrence. I wrote it down in the form of a patent application which was filed in the German patent office. It was not only the general idea of the cyclotron, but even the details of the stability of the electron orbits, and what it would take to keep these orbits stable; all this was worked out on this occasion. [5-12]

Though Szilard speaks of accelerating electrons, it would seem that he must have realized the unsuitability of the cyclotron for that purpose. The light mass of the electron guaranteed a relativistic increase in mass, in clear violation of the cyclotron principle.

Yet another near-inventor of the cyclotron principle was Max Steenbeck, while a doctoral student at the University of Kiel in 1927. He worked out a numerical example involving the acceleration of protons in a magnetic field of 14,000 gauss, but was dissuaded from following up the matter by his professors and subsequent co-workers at Siemens-Schuckert in Berlin-Siemenstadt, where he completed his dissertation [5-13]. At any rate, as Szilard said of Lawrence's success, 'The merit lies in the carrying out and not in the thinking out of the experiment' [5-14].

As already noted, Thomas Johnson was present that historic evening in the library when Lawrence saw the light, though 'it was not until the next day that he realized that the periods would be the same for successive orbits' [5-15]. James J Brady was at his laboratory bench in 216 Le Conte when Lawrence burst in the next morning, drawing him, too, to a blackboard. Resonance would be maintained regardless of the radius, Lawrence demonstrated, scratching out the elementary formulae on the blackboard [5-16]. Yet, Lawrence bided his time in following up the new idea. For one thing, he had a major program going on the photoeffect, involving three doctoral candidates well along on their thesis problems. Then again, practical implementation of the resonance scheme posed a number of technical questions, mostly having to do with keeping the spiraling ions in the median plane, as Szilard had appreciated as well.

In the spring of 1930, at long last, Lawrence decided the time was ripe for proceeding with the cyclotron, encouraged by Otto Stern of Hamburg who was a visiting professor at Berkeley that year, and the world's leading authority on molecular beams. 'Get on with it!' admonished Stern during dinner at Solari's. The next morning Lawrence enlisted the aid of Niels Edlef Edlefsen, who had just completed his thesis on the photoionization of alkali vapors, and grudgingly agreed to lend a hand despite upcoming exams. 'Good!' replied Lawrence. 'Let's go to work. You line up what we need right away' [5-17]. With Lawrence's backing, Edlefsen laid claim to the largest electromagnet in the department, with pole faces about four inches in diameter and capable of generating 5,200 gauss. Next, he fashioned a pair of accelerating electrodes by silvering the interior of a flattened flask. A narrow strip across the center of each inside face was left unsilvered, separating the silvered halves which served as two semi-circular electrodes, much like cutting a pill-box in two equal halves; the two electrodes soon became known as 'dees' from their shape.

The dees were wired to a radio-frequency oscillator, in an arrangement such that ions would be accelerated across the gap between the dees at the proper time in the r.f. cycle. Specifically, the cyclotron was expected to work as follows. An ion liberated from an electron source consisting of a stout tungsten filament near the center between the two dees would be accelerated from one dee to the other and pass into the electric field-free region within the interior of one of the dees. There it was shielded from the r.f. field, but not from the external magnetic field. Under the influence of the magnetic field, the ion would trace out a semi-circular path within the dee. If the time which the ion spent within the dee was equal to half the period of oscillation of the high-frequency circuit, the ion would emerge at the gap again just in time to be accelerated into the electric field-free region within the opposite dee. Around and around the ions would whirl on ever-increasing spiral radii, gaining small increments in energy each time they crossed the gap. The principle is sketched in Figure 5.1.

There remains some doubt as to how far Edlefsen actually got on the project; in particular, whether he achieved any resonance acceleration at all. To be sure, he wound up working virtually around the clock for a few months in early 1930. Giving up on the flask, which had a tendency to crack, he refashioned the dees by cutting a shallow, round copper box in half across its diameter. The two equal electrodes were sealed to plate glass with liberal amounts of red sealing wax, the old ion source re-introduced, and a vacuum system assembled from spare parts. The central chamber with its dees was centered between the pole faces of the electromagnet, and hooked up to the radio-frequency oscillator. Though Edlefsen was preparing to leave Berkeley for an assistant professorship at the University of California campus at Davis, in the middle of it all, in April he wired Lawrence with good news. 'Resonant acceleration achieved with hydrogen ions,' he informed Lawrence, who was in Washington attending the spring meeting of the American Physical Society. Lawrence was delighted, Swann noncommittal, and Tuve skeptical.

Lawrence and Edlefsen's carefully worded report, read by Lawrence before a meeting of the National Academy of Sciences in Berkeley on 19 September 1930, is deliberately vague on just what had been achieved, but ends on a positive note: 'Preliminary experiments indicate that there are probably no serious difficulties in the way of obtaining protons

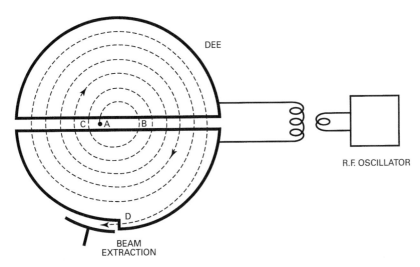

**Figure 5.1.** *Simplified schematic of a cyclotron, showing the hollow 'dees' connected through an inductance, in such a manner that they form a capacity in an oscillatory circuit of high frequency. To this circuit a high-frequency voltage is applied so that an alternating voltage is operative between the dees. A magnetic field is directed normal to the radial plane of the dees; that is, normal to the plane of the paper. Positive ions (protons, deuterons, or α-particles) enter at A, travel in a semicircular arc under the magnetic field, accelerate as they cross the cap at B, accelerate again at C, and so forth. Traversing ever larger spiral paths, the ions arrive at point D where the edge of the dee is cut away to allow them to emerge. They are drawn out of the dee and directed into a target chamber.*

having high enough speeds to be useful for studies of atomic nuclei' [5-18]. What difficulties lay in the way were, in fact, overcome by another graduate student, M Stanley Livingston, who inherited the cyclotron project from Edlefsen when he left for Davis.

Livingston came to the Berkeley Physics Department in the fall of 1929, with an MA from Dartmouth. Born in Broadhead, Wisconsin in 1905, he was the son of a minister in the local church. When Stanley was five the family moved to southern California, where his father became a high school teacher, and later principal. He also acquired an orange grove and a ranch. Apparently it was Stanley's early familiarity with farm tools and machinery that gave him the ability to craft large scientific instruments in later years. After graduating from high school in 1921, he entered Pomona College, where he began as a chemistry major, but switched to physics. He then went east to Dartmouth College, where he obtained a master's degree in physics, and stayed on as an instructor for a year. He chose to continue his graduate work at the University of California, because it was reasonably close to home.

Early in the summer of 1930, he asked Lawrence to propose a topic for an experimental thesis, and Lawrence suggested he undertake a convincing demonstration of the cyclotron resonance principle. After tinkering

**Figure 5.2.** *Lawrence with the first cyclotron, Edlefsen's silvered flask covered with sealing wax. Photograph by Watson Davis, Science Service. Courtesy AIP, Emilio Segrè Visual Archives, Fermi Film.*

fruitlessly with Edlefsen's little device, scraping away gobs of red sealing wax, he discarded it and started over. Livingston's laboratory notebook begins in September 1930, recording progress in 216 Le Conte—shaky at first [5-19]. He began by reassembling and recalibrating the 4-inch magnet used by his predecessor, suspecting an error in Lawrence and Edlefsen's original calibration. Next, he built a glass vacuum chamber with internal electrodes, about the size of Edlefsen's chamber [5-20].

During October and early November, he recorded feeble indications of resonance, but at values of the magnetic field $H$ that depended on the potential $V$ of the oscillator, in clear violation of equation (5.2), rederived on page 3 of Livingston's notebook. Attributing the lack of success, not to hydrogen ions but to residual oxygen and nitrogen ions, he built a new vacuum chamber out of a short section of brass ring with brass end plates sealed with wax. For an accelerating electrode he used a hollow, half pillbox of copper mounted on an insulating stem, with an open grid across the entrance to the single dee, which Livngston wrongly thought necessary for shielding against the electric field. (A slotted 'dummy dee' served in place of a second dee.) An r.f. potential was applied to the electrode from a Hartley-oscillator circuit, while hydrogen gas was admitted to the chamber and ionized by electrons emitted from a tungsten wire cathode mounted near the center of the chamber. The apparatus, with its charging circuit, is shown in Figure 5.3.

The first successful observation of sharp resonance peaks in the $H_2^+$ current collected in a newly designed 'Faraday cup' collector is recorded on 1 December 1930 [5-21]. 'Apparently the peak does not vary with voltage,' reads Livingston's sparse entry on page 19. More importantly, the magnetic field at resonance was closely that calculated from the resonance equation (5.2), which may also be written as

$$H = 2\pi f(m/e), \qquad (5.3)$$

using the measured value of the applied radio frequency $f$ and the $e/m$ ratio for hydrogen molecular ions [5-22].

'We are having good luck with our spinning protons but now are in need of a bigger magnet to get up into interesting energy values,' wrote Lawrence to Tuve [5-23]. And in December Livingston borrowed a stronger magnet capable of 13,000 gauss (13 kG). For $H_2^+$ ions and a maximum orbit radius of 4.8 cm, and with an applied r.f. potential $V$ of about 1 kV, the calculated ion energy, obtained from the relation

$$eV = \tfrac{1}{2}mv^2 = \tfrac{1}{2}mH^2r^2(e/m), \qquad (5.4)$$

was 80 keV, indicating the ions traversed a minimum of 40 turns, or experienced 80 accelerating kicks. These results were good enough to satisfy Livingston's doctoral thesis, completed in May 1931, and were reported by Lawrence at the spring meeting of the American Physical Society [5-24].

It goes without saying that Wideröe's concept of a *linear* array of drift tubes was not lost on Lawrence as an alternative, if less elegant, route to resonance acceleration. Since the length of the electrodes for accelerating ions of different masses are in inverse ratio to the square roots of the ion masses, the acceleration of protons would require an impractically long accelerator; heavy ions would have to suffice with the vacuum tube

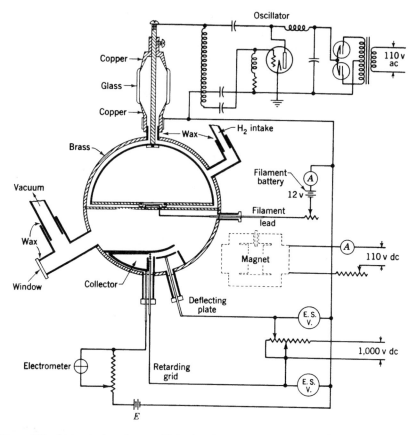

**Figure 5.3.** *Livingston's cyclotron, with its charging circuit. Livingston,* The Production of High Velocity Hydrogen Ions Without the Use of High Voltages, *PhD thesis, University of California, Berkeley, 14 April 1931, figure 3.*

oscillators then available. Another student, David H Sloan, had arrived at Berkeley in 1930 from General Electric on a Coffin Foundation Fellowship, and Lawrence put him to work building a vacuum tube with eight electrodes for accelerating mercury ions. Lawrence and Sloan spent a long weekend carefully inserting eight small nickel cylinders, each 4 mm in diameter and slightly greater in length than its predecessor, in a glass tube blown by Lawrence himself. The cylinders were alternatively connected to the wires from an oscillator, the tube was evacuated, and the whole contraption energized. 'Whoopee!' shouted student and professor in unison [5-25], as almost at once they realized the expected eightfold amplification in energy to 90 keV, with about 11 kV on the oscillator. Lawrence suggested extending the tube with 13 more electrodes 'to jazz up the voltage pr. stage' [5-26]; soon they had 100 keV, and then 200 keV [5-27].

In so doing they out performed Livingston's cyclotron, but only for the time being. And while a subsequent 30-section machine, only 1.14 meters long, exceeded the magical one million volts, heavy ions would prove inefficient in causing nuclear disintegrations compared to protons and light ions from the cyclotron. The proton linac would have to await new developments in radio frequency technology during World War II.

### 5.3. Electrons in Washington, DC

By 1930, then, the Cambridge and Berkeley teams had settled on different tacks in their attempts at achieving nuclear disintegrations, with only marginal success. At the Cavendish Laboratory, Cockcroft and Walton's generator/rectifier approach with moderate voltages had yielded 'indications of radiations of a non-homogeneous type'; at the University of California, Edlefsen's attempt at exploiting resonance acceleration in a rudimentary cyclotron hinted at 'no serious difficulties' in generating protons of respectable energy without the recourse of high voltages.

Very high voltages were the exclusive prerogative of Merle Tuve's team at the Department of Terrestrial Magnetism in Washington, DC, if we ignore industrial laboratories such as General Electric; only DTM had fully 5 million volts at their disposal, albeit with the short duty cycle of the highly unstable Tesla coil as their high-potential source, electrons as their atomic projectiles, and no tube capable of withstanding such high voltages. Which is not to say that the DTM team was unmindful of Gamow's insight into nuclear tunneling at more modest voltages, and the presumed ultimate importance of protons over $\alpha$-particles — and certainly over electrons — in attacking light nuclei. In September 1929, a year ahead of the first publications by Cockcroft and Walton and Lawrence and Edlefsen, Gregory Breit, who had spent that June in the company of Werner Heisenberg, had waxed in print on 'the possibility of nuclear disintegrations by artificial sources.'

> While 'artificial sources operating at potentials [equal to $\alpha$-particles from polonium] will probably require many years of development, it will be readily seen [from Gamow's calculations] that protons accelerated to $2.6 \times 10^6$ volts have a greater probability of penetrating the nucleus than the $5.2 \times 10^6$ "electron-volt-equivalent" [polonium] $\alpha$-particles...'. Indeed, 'even 1500 kV protons should be of interest for the lighter elements.' [5-28]

All this was predicated on the untested assumption that protons would have the same 'disintegration effect' as $\alpha$-particles. And protons on paper were good and well; obtaining an actual beam of well-collimated protons was simpler said than done. 'I have no good notion about producing protons,' admitted ex-team-member Breit to Tuve, 'but you undoubtedly will hit on the right method trying around' [5-29]. While biding their time, Tuve, Hafstad, and Dahl continued their development of multi-section, high-voltage tubes, though as of October 1929 had 'nothing but troubles to report yet on tubes' [5-30].

Tuve was excessively pessimistic; that very summer his group had succeeded in putting 1.4 million volts across a 15-section tube only 7 feet long [5-31]. More recently, however, the tube development work had been hampered by 'a recurrence of tube failures, and quantitative data are still wanting,' as Tuve wrote to K F Herzfeld at Johns Hopkins early in the spring of 1930. 'The problem is at present passing rather slowly through a transitional stage between the tube development work and the "real physics," which is just beginning' [5-32].

'Real physics' called for accelerating electrons and looking for electrons and γ-rays produced in their interaction with matter. Tuve's progress report for May 1930 is more upbeat, as he, Hafstad, and Dahl secured their first photographic plate of the magnetic spectrum of 'artificial beta-rays,' as Tuve called their accelerated electrons [5-33]. At the same time, good progress was being made in understanding and overcoming 'internal shattering' and 'flashing' of their high-voltage tubes by 'heat working' the Pyrex glass during tube production: heating the glass to the softening point, then expanding and contracting it, with the end product free of bubbles. Equally important was 'breaking in' a finished tube during actual operation: bringing the potential up cautiously to the onset of flashing, backing off, bringing the potential back up to a slightly higher level, and so forth.

Having taken up theoretical physics at New York University, Breit still kept in touch with his DTM friends as a consultant.

*Breit to Tuve, 4 August 1930:*
It would seem good to use the Tesla voltages for some scientific problem, imperfect though they are.... The chances of getting money and expanding the work are likely to be best if the Tesla voltages are used for a physical problem other than that of testing tubes. I would be inclined, therefore, to advise you to emphasize the improvement of your present apparatus also because I think you will have satisfaction out of your work that way. An investigation of the efficiency of X-ray production would be of interest.... [5-34]

*Tuve to Breit, 8 September:*
...I have been busy and tried since two weeks ago today, when Dahl and I got the first definite γ-ray results. [5-35]

*Tuve to Breit, 12 September:*
What do you think of Brasch and Lange? After a more calm consideration of their note than was possible at first, I cannot

help wondering if the $2.4 \times 10^6$ [volts] was actually on the tube. They do not state how the voltage was measured, and with a heavily *flashing* tube the inductance (as well as the gap resistance) of the impulse generator may be important. [5-36]

*Breit to Tuve, 15 September:*
Congratulations on the $\gamma$-rays. A letter to the Physical Review is I think fully in order. [5-37]

*Breit to Tuve, 16 September:*
Many thanks for the copy of your letter to the Physical Review. I think it is very good. Your skepticism as to Brasch and Lange is I think justified. Even if they should be temporarily ahead, I am betting on you. [5-38]

To obtain a better assessment of the European competition, in September Tuve dispatched Larry Hafstad abroad, visiting laboratories in Norway, England, and the Continent. At the Norwegian Institute of Technology in Trondheim, Hafstad called on the physicist and geo-physicist Olaf Devik, who was working on Tesla coils in vacuum, in an effort to produce highly penetrating X-rays. Devik's approach to pre-venting corona and sparking was in marked contrast to DTM's practice of operating in oil under pressure as the insulating medium. In fact, his discussion with Devik did not sway Hafstad from the practical advantages of oil over vacuum. Stiffer competition awaited him at Arno Brasch and Fritz Lange's laboratory in Berlin, which had worried Tuve and where Hafstad was shown their thick-walled vacuum tube studded with nickel corona rings of sufficient capacity to circumvent unwanted local buildup of voltage and withstand a surge of at least 1.2 MV and probably twice that at equally impressive transient currents.

Elsewhere in Berlin, at the Physikalisch-Technische Reichsanstalt, Walther Bothe was out of town, but a colleague concurred in viewing with skepticism the much touted Viennese claims to having disintegrated one light element after the other. Sure enough, at the Radium Institute in Vienna, Hafstad found the apparatus of Pettersson and Kirsch to be largely dismantled, with disintegration experiments halted due to diffi-culties in obtaining agreement between their controversial scintillation-counting method of particle detection and the new electric methods (developed to some extent elsewhere as a reaction to the spurious results coming out of Vienna) [5-39].

At Cambridge, his last stop, Hafstad was duly impressed with Patrick Blackett's modified Wilson cloud chamber for photographing $\alpha$-tracks— for many years the ultimate device for studying charged particles and their interactions and one shortly to be put to good use by Dahl and company at DTM. However, at the moment Hafstad was more interested in another development program at the Cavendish, as he stressed in his trip report to John Fleming.

> I also saw the work being done with Tesla coils and tubes by Rutherford's students in an effort to produce an artificial source of high-energy particles. The design is such that the winding of the Tesla is immersed in oil but the high-voltage is in air. Only single-section tubes have been used, and the highest voltage attained is about 500 kilovolts. At present work is being carried on to develop a canal-ray source which will give a large number of particles of this energy. [5-40]

For all its problems, the Tesla coil served Tuve's purpose of facilitating the slow but steady improvement in tube performance, to the point of producing good physics at long last on 'artificial $\beta$- and $\gamma$-rays,' as described by Tuve and his two colleagues at the Cleveland meeting in December 1930 of the American Association for the Advancement of Science. Not only that; shortly afterward, the three were notified that their paper, 'Experiments with High-Voltage Tubes' [5-41], had earned them the Society's $1,000 prize for the best paper presented at the meeting. GIANT X-RAY TUBE WINS SCIENCE AWARD exclaimed the headlines in *The New York Times* for Sunday, 4 January 1931, adding that their two-million-volt tube produced X-rays equivalent to the $\gamma$-rays of 182 million dollars worth of radium [5-42]. Elaborating on the 'death-dealing power' of the $\gamma$-rays that might split atoms and treat cancer, the article by the young Harvard-trained *New York Times* science writer William Laurence went on about the 'friendly race' between DTM, General Electric, Caltech (Lauritsen), and the Berlin scientists, and quoted Dr. Tuve at some length. *Time* was more restrained in its shorter coverage a week later [5-43]. *Science News Letter*, however, even saw a semblance of Tuve and his program in *Wings Over Europe*.

> Some physicists have suggested in the past that if such powerful radiations were allowed to smash into the hearts of atoms, there might be a liberation of the immense internal energy of the atoms with a gigantic explosion that would wipe out the world and this corner of the universe. This idea was the theme of the Broadway success of several years ago 'Wings Over Europe' in which a scientist, not unlike the Dr. Tuve of real life, solved this secret of physics and held the fate of nations and the world in his grasp. [5-44]

Ernest Lawrence lost no time in congratulating them, in a hasty Sunday letter to his friend.

*Lawrence to Tuve, 4 January 1931:*
I have just seen in the paper that you and your co-workers won the $1000 AAAS prize. Congratulations!! This news pleased me tremendously.... Was down in Los Angeles a couple of weeks ago. Lauritsen expressed keen admiration for your work. [5-45]

Tuve himself was far from overjoyed, however, for reasons of his own. Visiting with Breit and the physicist Edward O Salant in Salant's apartment in New York City on 4 January, he and they pored over that morning's Sunday *New York Times*. Tuve's reaction to Laurence's on the whole rather straightforward article in the paper seems curiously excessive.

*Tuve to Hafstad, 4 January, 11 am:*
Have just discovered in Sunday paper what has happened. I am entirely innocent. I gave no interviews in Cleveland except very cautious one to Science Service. Knew nothing whatsoever of prize until five minutes ago. Beware of reporters even if necessary to disappear. Tell Fleming no interviews can be obtained until I return. Several days delay may help us.

Errors now in publicity cannot be corrected except by future modesty. Salant says newspapers will quote and misquote scientists whether they give interviews or not. We cannot avoid what others say about us but it is extremely important that we ourselves should not be arrogantly quoted, especially that no quotations from us about others' work should appear. I shall take the attitude that our publications describe what has been done, that the work in progress is of no further legitimate public interest and that we do not wish to be further misquoted in such a spectacular way. Am with Breit and Salant today. Princeton tomorrow, Washington tomorrow night. [5-46]

Though the prize was awarded to Tuve, Hafstad, and Dahl as authors of the Cleveland paper, Laurence's article notes that 'the work it covered was carried on over several years by the three physicists in conjunction with Dr. Gregory Breit, now at New York University.' Somewhat sheepishly, perhaps, the three recipients assured their former colleague that 'we have wanted you to enjoy with us the accident of this prize which was wished upon us unawares' [5-47]. They had originally planned to share their prize money with him, they explain in their letter to Breit, but concluded that 'for us to offer this would only be inviting you to refuse it.' Instead, they deposited $100 to Breit's credit with a New York book dealer, hoping 'that the books or periodicals which you order will remind you of our constant good wishes.' In the event, Breit declined to accept their kind offer, appreciating their good wishes all the same.

In spite of the apparent good will between Breit and Tuve, Hafstad and Dahl, Breit's extremely abrasive personality was an ever present factor, 'one capable of plunging the most abstract discussion of physics into a heated argument,' according to Louis Brown.

As a consequence he was never able to enter into a relaxed scientific partnership with an equal but continued his dazzling career alone with his students and an occasional collaborating theorist. He bore no grudges, and he and Tuve soon entered the remarkable cooperation of theorist and experimentalist that marked that portion of their lives, but it was carried out through letters and short visits. [5-48]

All in all, and Tuve's excessive standards notwithstanding, 1931 seems to have begun on a positive note, not only in Washington but in Cambridge and Berkeley as well.

# Chapter 6

# PROTONS EAST AND WEST

### 6.1.  One million volt protons

Livingston's 4-inch cyclotron, with a stronger magnet, produced 80-kV protons on 2 January 1931. Exactly one week later, Lawrence set the wheels in motion for building a bigger machine, with one million-volt protons the goal. Its basic parameters were to be those he specified in his and Edlefsen's upbeat report to the National Academy the previous September, namely, a maximum orbit radius of 10 cm and a 15-kG magnet with 10-inch pole diameter. He estimated a thousand dollars would suffice for its construction, and the new President of the University, Robert Gordon Sproul, approved University funds for the project. As with later accelerators in years to come, the actual cost exceeded the allocated funds; however, the National Research Council stepped in with the additional $500 needed to complete the job [6.1].

One million-volt protons was one thing; you also had to have enough of them in a well-collimated beam to do useful physics. However, from the outset Lawrence had emphasized the importance of keeping the hollow space inside the accelerating electrodes free of the electric field, and thought it necessary to shield the interior region with a metallic grid across the entrance to the dees. Unfortunately, in the 4-inch cyclotron too many orbiting ions struck the grid across the single dee employed, if not the top or bottom of the dee, and were lost. Lawrence and Livingston substituted some parallel slits, and the collected current improved somewhat. And when Lawrence was away on a trip to the East Coast in April 1931, Livingston gambled and left the slits out as well. Behold! The collected current rose sharply.

On examining the problem more closely, the reason for the drastic change in performance was soon apparent. In the absence of a grid or similar metallic shield across the dees of a cyclotron, or between the single dee and a slotted 'dummy dee' as in the 4-inch version, there is a slight penetration of the accelerating electric field near the entrance to the dee. Fortuitously, the penetrating field lines, or 'lines of force,' will act like an electric lens: a component of the field normal to the median plane of the vacuum chamber results in a net inward displacement or focusing of ions orbiting off the median plane towards the median plane. This electrostatic focusing, it turned out, is mainly effective for the inner orbits of the spiraling ions, corresponding to the first few accelerations. There is also a *magnetic* focusing at work in the cyclotron, however, due to the unavoidable curvature of the vertical magnetic field near the periphery of the magnet pole faces. This focusing becomes effective for higher-energy ions circling at large radii, again focusing them towards the median plane.

Lawrence's trip east that April had a double purpose. In Washington, as noted earlier, he spoke before the American Physical Society on his and Livingston's success with the 4-inch cyclotron. 'These preliminary experiments,' he concluded his presentation, 'indicate clearly that there are no difficulties in the way of producing one-million-volt ions in this manner' [6-2]. In Cambridge, Massachusetts, he saw a young lady named Mary Blumer, known to everyone as Molly, whom he first met at Yale, where her father was professor of clinical medicine and dean of the Yale Medical School. She had graduated from Vassar with a Phi Beta Kappa, and was now obtaining a Master's in biology at Radcliffe. Slender of figure, shy, mature for her age, Lawrence was drawn to her from the outset, though it had taken her somewhat longer to become fully comfortable with the intense young professor.

Back in Berkeley, construction of a 9-inch magnet for the new cyclotron had been entrusted to the Federal Telegraph Company, who delivered it to the Physics Department in Berkeley on 3 July. The same day Lawrence and his people began setting up the magnet in 329 Le Conte, 'hoping that within two weeks we shall have made our first experiments on the production of high speed protons with its help,' as Lawrence promptly wrote to Frederick G Cottrell [6-3]. Cottrell was an eccentric inventor and former assistant professor in chemistry at Berkeley, who in 1914 had endowed a not-for-profit business, the Research Corporation, headquartered in New York City. A type of 'holding company for academic ideas and inventions, its aim was the protection of inventive rights while furthering development' [6-4], and it would serve accelerator builders well, including Lawrence.

Lawrence had hoped to have the 9-inch running in time for another accelerator meeting of the American Physical Society, this time in Pasadena during 15–20 June 1931, but time ran out while they were still tuning up the machine. At Pasadena, he heard Tuve describing measurements at DTM on high-speed protons in a 'flashing' tube fed by their Tesla coil. The protons were revealed visually by the scintillation method as 'Thomson parabolas' produced by electric and magnetic deflection of protons believed to have arisen from occluded hydrogen on the electrodes of the tube, and from residual gas [6-5]. Lawrence was still unconvinced of his friend's high-voltage approach to atom smashing, but sensed considerable skepticism in the audience over his own 'merry-go-round.' Tuve, for his part, claimed to be quite impressed with all he saw in 329 Le Conte, when he stopped by at Berkeley on his way back to Washington.

Lawrence's steadfast optimism paid off on 17 July, exactly two weeks after delivery of the magnet from Federal Telegraph, as he had guardedly predicted in his letter to Cottrell.

> The proton experiment has succeeded much beyond our fondest hopes!!! [he wrote to Donald Cooksey, an old friend from Yale]. We are producing 900,000 volt protons in tremendous quantities—as much as $10^{-8}$ amperes. The method works beautifully without a hitch.
>
> One reward for this glorious success is a trip to see Molly!! She lands in N.Y. on Aug. $2^{nd}$ and I'll be there! [6-6]

Then in Europe with her mother and younger sister Elsie, Molly had promised to let Lawrence know when she would return so that he could meet her at the pier. 'I am beginning to realize I have two consuming loves—Molly and research,' continued Lawrence's hastily scribbled letter to Cooksey. Cottrell, too, was informed at once of the latest success.

> I am hastening to let you know that the experiments on the production of high speed protons have been successful beyond our expectation. About a week ago we finally got the magnet built by the Federal Telegraph Company set up and during the past week we have been producing high speed protons. The highest speeds so far produced correspond to 750,000 volts. [6-7]

Mainly, however, he sought Cottrell's advice on funding 'to push further developments' 'towards a new goal of 20,000,000 volts.' On a casual note, he added that he was going east again on 25 July 'for unavoidable personal reasons. Possibly you may suggest my calling on someone in New York on the trip.'

In his and Livingston's letter to the *Physical Review*, dispatched to the editor on 20 July, Lawrence viewed the future with confidence.

> These experiments make it evident that with quite ordinary laboratory facilities proton beams having great enough energies for nuclear studies can be readily produced with intensities far exceeding the intensities of beams of alpha-particles from radioactive sources.
>
> Possibly the most interesting consequence of these experiments is that it appears now that the production of 10,000,000-volt

77

protons can be readily accomplished when a suitably larger magnet and high frequency oscillator are available. The importance of
the production of protons of such speeds can hardly be over-
estimated and it is our hope that the necessary equipment for
doing this will be made available to us. [6-8]

The same sentiments are expressed in a letter posted, the same day, to
Cottrell who was still vacationing in Long Beach. Lawrence hoped to
drop by and discuss future funding, on his way east, since 'it appears
that our difficulties are no longer of a physical, but of a financial
nature' [6-9].

The trip to the East Coast was a great success, followed by further
exciting developments touching both his research program and personal
affairs, as Lawrence wrote Tuve back in Berkeley on 27 August [6-10]. In
New York he was well received by the Research Corporation on Lexing-
ton Avenue, where Cottrell introduced him to the Board of Directors and
argued his case before them. While waiting for Molly, he had seen Tuve
on a hasty overnight stay in Washington. Molly arrived, and they had
some time together at her parents' summer place at Haycock Point on
Long Island Sound, where their courting was interrupted on 3 August
by a telegram from Berkeley. Livingston had not been idle in the meanwhile, and had boosted the proton energy another notch. 'Dr. Livingston
has asked me to advise you that he has obtained 1,100,000 volt protons,'
wired the secretary of the Physics Department. 'He also suggested that
I add 'Whoopee'!' [6-11]

Lawrence had more good news for Tuve as of 27 August. The day
before, President Sproul had turned over a whole building, the old
clapboard-sided wooden two-story Civil Engineering Testing Laboratory
located not far from Le Conte Hall, for his exclusive use as a 'high speed
corpuscle laboratory' or what became formally known as the Radiation
Laboratory. Sproul's munificent decree freed Lawrence from direct over-
sight by the Physics Department, to which the 'Rad Lab' nevertheless
would remain formally tied until 1936 [6-12].

26 August 1931 proved an auspicious date in Lawrence's life for more
than one reason. That day, Molly Blumer, along with Sproul, had an
announcement of her own to make during a luncheon at the New
Haven Lawn Club: she and Lawrence were engaged to be married.

In 329 Le Conte, meanwhile, the magnet pole faces of the 9-inch cyclotron were being enlarged to a diameter of 11 inches, with the expectation
of further raising the energy by another notch. However, it remained
equally important to collect a strong current—strong enough at least for
its direct measurement in a Faraday cup by an electrometer suitably
shunted to ground. The partial solution, as earlier, lay in omitting
any grid or other obstruction in front of the dee, and in so doing taking

advantage of the electrostatic focusing thereby provided, as well as of the magnetic focusing provided naturally by the fringing magnetic field. Excellent progress was being made by 10 November.

> The proton merry-go-round is behaving beautifully [wrote Lawrence that day to Tuve]. We have just finished a set of careful experiments and have proved quantitatively that the particles have their expected high speed. We have done this by a system of electro-static deflections of the ion beam. Now I am absolutely certain that we will experience no difficulty in going right up to five million volts or more as soon as the big [27-inch] magnet is installed. [6-13]

Despite the relatively high current, never more than one billionth of an ampere (0.001 µA), the voltage amplification—equivalent ion output voltage divided by voltage on the dee—remained pegged below 100, or 500,000 volt ions for 5,000 volts on the dees. Livingston feared they had reached the relativistic limit of the cyclotron—that is, the point where the increase of particle mass with velocity destroyed the cyclotron condition of equation (5.2). Not yet, countered Lawrence; the problem lay, not in variations in particle mass, but in small but significant imperfections in the guiding magnetic field. On Lawrence's suggestion, this was overcome by inserting small, soft-iron shims here and there in the narrow space between the vacuum chamber and the magnet pole face. With a single shim only 0.025-cm thick, the amplification factor shot up from 75 to about 300. 9 January 1932 was the big day, with the effective amplification 'producing 1,220,000-volts with only 4,000 volts at a wavelength of 14 meters applied to the tube' [6-14].

## 6.2. Protons at DTM

Following the hoopla over their AAAS prize in December 1930, Tuve, Hafstad, and Dahl concentrated on obtaining and photographing tracks of 1000-kV protons from a cascade-type, high-voltage tube excited by their Tesla coil mounted inside a large oil tank. The protons were projected into a Wilson cloud chamber by magnetic analysis. The Wilson chamber, separated from the accelerating tube by a thin mica window, was, like most of the hardware, largely Odd Dahl's handiwork, though it was based on a version of the cloud chamber developed by P M S Blackett in Cambridge.

The cloud chamber, invented by C T R Wilson, is basically a container of moist air, which is allowed to expand rapidly in a cylinder. As it expands, the air cools, and some of the moisture condenses into tiny droplets, provided there are dust particles present acting as condensation nuclei, about which the droplets can form. If a subatomic particle enters the chamber, it forms ions along its path, and these, too, will act as condensation centers; a water droplet will condense on each ion. In this way, not only the subatomic particle, but its pathway as well can be detected and photographed under instantaneous illumination. If the cloud chamber is placed in an electric or magnetic field, the swiftly moving particle will curve in response.

Wilson received the Nobel Prize in 1927 for his invention—the first Nobel Prize awarded for physics hardware, not physics as such. He received his inspiration for the device as a young man volunteering as a meteorological observer in September 1894, at an observatory on lofty Ben Nevis in the Scottish highlands off Loch Linnhe. He began his solitary observations at the crack of dawn, when the sea of cumulus clouds was below the summit and shimmering in the light of the rising sun.

> The wonderful optical phenomena shown when the sun shone on the clouds surrounding the hill-top and especially the coloured rings surrounding the sun (coronas) or surrounding the shadow cast by the hill-top or observer on mist or cloud (glories) greatly excited my interest and made me wish to imitate them in the laboratory. [6-15]

Back in Cambridge, Wilson lost no time in preparing to do just that. His artificial cloud-producing project, pursued over the next several years with singular dedication and unexcelled patience, would result in what has been regarded as the most productive device in the history of nuclear physics, at least up to 1952 when the American physicist Donald A Glaser invented a similar device. In it, instead of a gas out of which liquid drops are ready to form, Glaser used a *liquid* raised to a temperature at which vapor bubbles were about to form. Those bubbles, too, formed along the pathway of an incident, ionizing particle, earning Glaser a Nobel Prize in 1960 for his 'bubble chamber.'

However, for our purpose the important thing was Patrick Blackett's automated version of Wilson's cloud chamber, perfected by him in the course of his long study of the collision of $\alpha$-particles with nuclei, of the resulting ejection of protons, and of the subsequent motion of the residual nucleus and $\alpha$-particle after the collision. In order to photograph large numbers of particle tracks, Blackett developed a modified and automatic form of Wilson's chamber, which made one expansion and took a pair of photographs every ten or fifteen seconds [6-16]. The chamber itself utilized a floating ebonite piston, and the rubber tube used originally to change the volume was replaced by a corrugated metal diaphragm. The reciprocating mechanism in Blackett's version utilized a simple spring and lever action, with the speed and amount of expansion regulated by the tension of a spring and by an adjustable screw, respectively. A special camera, and a system of mirrors, recorded on a single film two photographic images viewed from mutually perpendicular directions. (Later, for the study of cosmic-ray showers, Blackett triggered the chamber expansion by coincident pulses in Geiger–Müller counters located above and below the chamber.)

Blackett, too, received the Nobel Prize for physics in 1948, 'for his development of the Wilson cloud chamber method, and his discoveries therewith in the fields of nuclear physics and cosmic radiations' [6-17].

Two variations on Blackett's automatic chamber were constructed under Dahl's supervision in DTM's instrument shop during 1931 [6-18], assisted by, and with the loan of some parts by Dr. L F Curtiss of the National Bureau of Standards. The first version used a relatively conventional piston with an oil or water seal, and a magnetically-controlled trigger mechanism, both for the chamber and for spark-excitation of the Tesla coil. This chamber is shown in operation in Figure 6.1. Plagued with inevitable leaks of the liquid-sealed pistons, and fogging when introducing air from the room for adjustment of the expansion, they then turned to a new design, consisting of a permanently-sealed chamber using a flexible metal bellows (sylphon)—still magnetically operated. The chamber was illuminated by a pair of mercury spark-lamps, and the tracks were photographed with two 35-mm stereoscopic cameras mounted so as to make an angle of 90° with respect to each other, following Blackett's practice.

The first sylphon-chamber, 6 inches in diameter, eventually withstood 20,000 expansions without difficulty. However, the first experimental run with their 'flashing' tube and Tesla coil, reported by Tuve at the Pasadena meeting in June of 1931, was a dud as far as photographing tracks of high-speed protons. Any proton tracks present in the chamber were masked by dense fogging from the ultraviolet light and X-rays entering the chamber through the mica window. Trouble was also experienced with the stringent requirements for synchronization of the high-voltage impulse from the Tesla coil and the 'sensitive time' of the chamber immediately after the expansion [6-19].

In September, meanwhile, Hafstad began work on his PhD program at Johns Hopkins. In his absence, Tuve cast about for a temporary replacement, and checked with Lawrence as well. The upshot was that he hired Laurence E Loveridge, one of Lawrence's grad students who was just finishing his PhD, with a new type of vacuum tube, GE's FP-54 Pliotron, as his thesis topic. In the event, the FP-54 would be put to good use at DTM, first by Dahl and eventually by Hafstad; however, Loveridge proved disappointing and was soon let go [6-20]. (Hafstad, incidentally, returned to DTM in June of 1932.)

The summer of 1931 passed with little further progress in the Wilson chamber experiment, and with shop work and leave-taking during the heat and humidity of Washington's most oppressive summer weeks. By

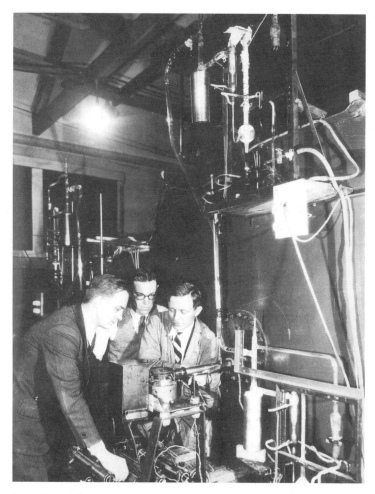

**Figure 6.1.** *Tuve, Hafstad, and Dahl examining the Wilson cloud chamber at the target end of their most powerful Tesla apparatus, 6 January 1931. DTM Archives 2072.*

September, moreover, a new, important development claimed the group's time, as we shall see. However, the ongoing experiment limped along, and by late fall things were looking up. In particular, the troublesome illumination of the Wilson–Blackett apparatus by X-rays from the flashing tube was eliminated by redesigning the slit system in Dahl's magnetic analyzer. Success came in early December.

> The first known tracks in a Wilson chamber of artificially acceler-
> ated high-speed atoms were obtained December 8, when the
> tracks of protons from a 'flashing' tube operated somewhat
> above 1,000,000 volts were seen and photographed. [6-21]

It comes as quite a shock, and testifies to the fragility of the apparatus, that only in December 1931 were the first Wilson 'fog tracks' observed from accelerated protons!

At any rate, the experiment continued around the clock, during the next three weeks, pinning down the range of 1,000-kV protons in air. In so doing, Tuve and colleagues confirmed 'that the law governing the range of protons is nearer to the velocity-cubed law which holds for α-particle ranges' than the corresponding fourth-power law believed applicable for the range of β-particles [6-22]. 'The first known use had been made of the high-voltage technique to obtain new information concerning any law of atomic physics,' added Tuve in his monthly report for December 1931.

Dahl's prowess in designing and building cloud chambers would come to the fore again two years later, in a letter to Tuve by Ernest Lawrence. Lawrence's group had a need for a Wilson chamber, and he thought Dahl might be able to help.

> We are anxious...to get a first-class Wilson camera (with accessory cameras and so forth) — the very best possible (one might say a deluxe model). Robert Oppenheimer has told me how wonderfully fine your arrangement is and the thought has occurred to me that Dahl...might be interested in building a complete outfit for us. Oppenheimer said that he understood that Dahl more or less designed and got your outfit going. At least he seemed to be very much pleased with the apparatus. I know that Dahl is a good instrument man and he might like the opportunity to make a little money on the side. [6-23]

In the event, Tuve decided Dahl could not be spared for the job, and Lawrence understood. One of his graduate students, Milton G White, 'who is rather good,' undertook the job instead [6-24].

Despite December's successful run with the Tesla coil, the difficulties of doing quantitative work with it, due to its oscillating character and brief duty cycle, was apparent to one and all. Providentially, in August 1931, as they were scrambling to get significant results with the equipment on hand, Tuve learned of Robert Van de Graaff's success at Princeton in maintaining a potential difference of 1.5 million volts between his two 2-foot spheres charged to ±750 kV. Though Van de Graaff's preoccupation at the time was mainly with improving the performance of his belt-driven generator by operating it in a vacuum tank, Tuve waxed enthusiastically over prospects for an open-air version that could reliably produce the million volts he needed so badly. As soon as Van de Graaff returned from his vacation in September, Tuve drove up to Princeton. They quickly came to an understanding, and Tuve brought Van de Graaff and his tabletop apparatus, the latter precariously lashed to the back of his car, back to Washington.

At DTM, the little generator was put through its paces hooked up to one of their standard 12-section cascade tubes. These tests immediately showed that there was no difficulty involved in the application of steady (direct-current) potentials to the vacuum tube in the open air, a point hitherto not tested at all since their high-voltage tubes up to that time had been used with fluctuating or alternating potentials in oil under pressure. With a charging current of 40 μA and 600 kV on the spheres connected in parallel, the generator produced the full potential drop across the tube in 20 seconds, as the mercury pump took hold. The tube carried close to a millimicroamp—not much but enough to make tracks in the cloud chamber. And while 600 kV represented the corona limit due to flashover to nearby objects in the limited headroom of the working space, the tube withstood the potential, seemingly indefinitely [6-25].

> This simple device, which Van de Graaff has demonstrated to be capable of such serious possibilities, will undoubtedly alter the whole course of high-potential experiments, here and elsewhere. The important question as to the feasibility of direct current on tubes appears nearly answered, but such experiments will be continued for the next week or more during which we are still allowed to keep Van de Graaff's machine. [6-26]

As he left Washington, Van de Graaff had told Tuve he could hold on to the generator 'as long as it is useful to you' [6-27]. However, once back in Princeton he changed his mind, explaining to Tuve that Compton needed the device 'for a demonstration to some of the benefactors of M.I.T.' [6-28]. Somewhat annoyed, Tuve passed Van de Graaff's letter on to Fleming, with 'Where do we stand?' scribbled in the margin [6-29]. Tuve returned the generator to Princeton on 24 October, fairly sure that 'this voltage source appears to be what we have been wishing for from the beginning [6-30].

> I am satisfied that the machine is well worth trying out [continued Tuve in his December 1931 letter to Lauritsen in Pasadena], and it is cheap enough to build that one can afford to test it. We are just beginning construction of one, a 2 meter Al sphere. I hope to get $1\frac{1}{2} \times 10^6$ volts. [6-31]

The dielectric strength of air at atmospheric pressure, according to a convenient rule of thumb, limits the maximum potential on a sphere to as many million volts as its radius expressed in feet. Accordingly, a sphere two meters in diameter (3.28 feet in radius) should easily hold 1.5 million volts, or half the theoretical limit [6-32]. Tuve's decision to proceed with a full-sized generator was timely. Van de Graaff himself had elected to forego nuclear experimentation for the time being in favor of building an even larger electrostatic generator with generous support from Cottrell and his Research Corporation, and Lawrence was forging ahead without high voltages at all. Only time would tell which approach was best suited for disintegrating atoms and producing other interesting physics. And all three teams realized full well that there were other groups vying for supremacy in nuclear experimentation, at home and abroad.

# Chapter 7

# GIANTS OF ELECTRICITY

## 7.1. X-rays in Pasadena

As if not enough Scandinavians in American accelerator circles, we have yet to meet Charles Christian Lauritsen, a physicist of Danish extraction at the California Institute of Technology in Pasadena, California. Lauritsen's background was as varied as Odd Dahl's. Born in Holstebro, Denmark, in 1892, he graduated in architecture from Odense Tekniske Skole in 1911, and emigrated to the US in 1917. There he became involved, in turn, in the design of naval craft in Boston, professional fishing off the Florida coast, manufacturing ship-to-shore radios for Federal Telegraph in Palo Alto, California, and building radio sets for domestic use with the Kennedy Corporation in St Louis. In 1926, while still in St Louis, he attended a lecture by Robert A Millikan, chief executive officer of Caltech. Inspired by what he heard, Lauritsen moved his family to Pasadena, and at the age of 34 enrolled as a doctoral student in physics at Caltech.

Lauritsen obtained his PhD in 1929, with a thesis on the study of electron emission from metals in intense electric fields [7-1]. He became an assistant professor of physics at Caltech in 1930, full professor in 1935, and remained associated with the Institute for the rest of his life.

In his work on field-emission electrons, Lauritsen had need for an X-ray tube capable of producing highly penetrating rays; indeed, he used cold emission as the source of electrons in producing the rays. He followed Coolidge's suggestion of cascading two or more tubes in such a way that their end window-foils would act as the anode of one tube and the cathode for the next. Lauritsen's tube employed four sections, connected to four rectified transformers also operating in cascade. The tube sections, it might be noted, were assembled from glass cylinders then in common use in pumps in gas stations across the land [7-2]. They were joined at steel rings, which both supported internal metal shields to protect the tube from back streaming electrons, and external concentric metal rings that served as equipotential rings to smooth out the potential gradient along the tube. The finished installation, illustrated in Figure 7.1, measured over 18 feet in height, from the top end of the long central electrode of the tube to the bottom of the massive redwood scaffolding. By August 1928 Lauritsen and Ralph D Bennett, a National Research Fellow, had 750 kV across the tube, generating X-rays capable of penetrating over 2 cm of lead [7-3].

The largest medical X-ray tubes then commercially available were fragile glass tubes rated for 200 kV. Extremely hard X-rays from a tube rated for three-quarters of a million volts promised to be especially effective in the treatment of cancer, as Lauritsen wrote in his patent application of 1930 for his high-voltage apparatus.

I have found that when [1,000,000 volt] potentials are employed, radiations substantially the frequency of the gamma radiation from radium may be obtained from the tube embodying my invention.

Such tubes can therefore be employed as the full equivalent of radium in the treatment of disease, or for therapeutic purposes. [7-4]

Caltech was only nine years old when Lauritsen wrote his patent application. However, its roots went back to 1891, when Amos G Throop founded Throop University, which eventually became Throop College of Technology, located in Pasadena. During World War I, the astronomer George Ellery Hale had raised a large endowment to reshape the small college. To match the endowment with a distinguished faculty, Hale persuaded Arthur A Noyes to head the chemistry division, and turned to Millikan for taking charge of physics. Millikan concurred, and in 1921, the year after the name of the institution was changed to California Institute of Technology, left the University of Chicago for Pasadena.

At Chicago, where Millikan had made his scientific fame by, among other things, doggedly pursuing what he called 'My Oil Drop Venture,' he had grown impatient with complex and cumbersome academic administration. Accordingly, he accepted the call to Pasadena, not to man an administrative post, but to strengthen the physics program in the young but growing institution, as Hale had wished. In the event, Millikan created not only a school of physics, but a whole new institution 'where teaching and research went hand in hand, where a major assignment of resources to research would be achieved, where research would provide the creative atmosphere for stimulating teaching, and where young students would keep the freshness of the research spirit alive' [7-5].

Most of all, however, Millikan attracted outstanding faculty, including R C Tolman (physics and chemistry), Paul S Epstein (physics), Theodore von Karman (aerodynamics), Thomas Hunt Morgan (biology), Harry Bateman (mathematics), Linus Pauling (chemistry), and of course Charles Lauritsen in physics. And with the personalities grew specialized groups,

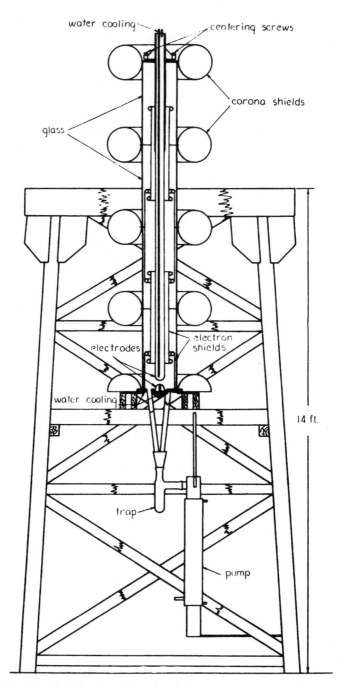

**Figure 7.1.** *Caltech's high-voltage X-ray installation. Lauritsen and Bennett, 1928, PR,* **32,** *pp. 850–857, on p. 852.*

laboratories and divisions, all dependent on Millikan's brilliance as a fundraiser. So it was with the Kellogg Radiation Laboratory, which grew out of his and Lauritsen's High Tension Laboratory, organized originally to assist in importing electric power to the Los Angeles basin. Although named after Will Keith Kellogg of Corn Flakes fame, the initial funding for the laboratory had come from Seeley G Mudd, a Pasadena physician, and his family. Millikan had approached Mudd for advice and money for developing Lauritsen's X-ray tube into a therapeutic facility, and Mudd agreed to a two-pronged program. Half the funds would be spent on a five-year study program involving treatment of cancer patients with 750 kV X-rays, with Mudd himself serving without salary as resident radiologist. The other half of the funds would support a continuing high-voltage R&D program, with Lauritsen in charge.

Preliminary therapy began in 1930, with a somewhat modified X-ray tube [7-6], and with several hundred patients treated with 600-kV radiation at Los Angeles County Hospital. Soon, however, the need for higher potentials and a special building dedicated to cancer therapy demanded more money than the Mudds could reasonably provide. As luck would have it, the 71-year old W K Kellogg arrived from Battle Creek, Michigan, to winter at his Arabian Horse Ranch in Pomona, California, and Millikan prevailed upon him to underwrite the new facility. However, insisted Kellogg, 'for $150,000 I should have some appropriate acknowledgment such as my name on the building' [7-7]. And so it was, with the first patient treated in September of 1932. Four patients could be treated simultaneously in a balcony treatment room pierced by a 30-foot long X-ray tube—a tube consisting roughly of two of the old 1,000-kV tubes joined end-to-end. From the pit, housing two 2 MV General Electric transformers to the ceiling, the X-ray hall measured 68 feet in height. All the shop work was performed in the astrophysical machine shop that would subsequently be used in constructing the metal parts for the 200-inch telescope.

Maintaining the technical facility was the job of physicists and their graduate students, swaying in a bosun's chair while keeping the tube in working condition. While not thus occupied, the 'maintenance crew' could turn to more stimulating matter, namely nuclear physics experimentation, in the adjacent 'High Volts,' as the High Tension Laboratory was usually called [7-8]. While downplayed in Millikan's correspondence eliciting monetary support from physicians and W K Kellogg, Caltech's publicity releases devoted equal time to ongoing basic research, as evidenced by the *Los Angeles Times* for 4 August 1931.

The largest and most powerful instrument ever devised for splitting the atom and combatting cancer was installed today in the California Institute of Technology new radiation laboratory....

The primary object of the institute's X-ray program...was to learn more about the physics of high-speed electron particles. It was known that at about 300,000 volts the electrons about an atomic nucleus are dislodged, and the physicists suspected that with the higher voltage soon to be available the heart of the atom might be broken up. [7-9]

Indeed the *Times* had it right. Cancer research at Caltech was actually withering on the vine, amidst indications that X-ray therapy was not proving very effective; some even argued it did more harm than good. And Lauritsen and his team were not sitting idly. Before long Lauritsen would challenge Merle Tuve on an important point, as we shall see in due course. By the late 1930s the X-ray tube in the large central hall of the Kellogg Laboratory had been replaced with a Van de Graaff generator capable of 2.5 million volts [7-10], and the laboratory's reputation was being made as another active US center for nuclear research. Lauritsen himself, together with his son Thomas Lauritsen, who was born in Copenhagen and obtained his PhD at Caltech in 1939 [7-11], would also assist in the construction of a near-copy of the Pasadena generator at the Niels Bohr Institute in their native land. Pursuing these developments, however, would take us well beyond the time frame of interest here. Instead, let us turn briefly to the largest electrostatic generator of them all.

## 7.2.  Megavolts on Round Hill

In August 1931, as the *Los Angeles Times* hailed Charles Lauritsen's 30-foot, two-million-volt X-ray unit at Caltech, high voltages were being generated at Princeton as well. There, Robert Van de Graaff's seven-foot, tabletop electrostatic generator at Princeton produced 1.5 million volts between the two 2-foot spheres charged to $\pm750\,$kV. Van de Graaff was not unmindful of applications for such high voltages in nuclear research, from his exposure in Paris and Oxford, and subsequent post-doctoral stint at Princeton as a National Research Fellow. Already in March of 1931 he wrote Karl Compton, head of the Princeton Physics department and his own mentor, as follows.

Homogenous beams of protons at voltages that may be expected from the present work could be used for many simple experiments of fundamental nature. Among them would be the investigation of the effect of their impacts on uranium and thorium. These nuclei are already unstable, and it would be interesting to see if an impacting proton of great speed would precipitate immediate disintegration. On the other hand, it might be that the proton would be captured by the nucleus, thus opening up the possibility of creating new elements of atomic number greater than 92.

90

Near the other end of the series of atomic numbers is lithium. Now suppose that a proton is shot into the $^7$Li nucleus, supplying the second component for the second alpha particle group. A consideration of Aston's curve and Einstein's law shows that a nuclear reaction might take place as follows:

$$^7\text{Li} + {}^1\text{H} \longrightarrow 2\,{}^4\text{He} + 16 \text{ million eV energy. [7-12]}$$

Interesting speculation, coming a full year before Cockcroft and Walton were in position to put the Li(p, $\alpha$) reaction to an experimental test, with their new voltage multiplication apparatus finally up and running, and from someone with scant personal experience in nuclear experimentation. Indeed, then and throughout his career, Van de Graaff remained preoccupied with the generation of high unidirectional and steady voltages for such experimentation, in effect sacrificing a career in nuclear research utilizing his own generator. Compton, in any case, was impressed with the 2-foot model generator, built at a cost of under $100; it was the banquet highlight before the American Institute of Physics in November of 1931, and later was put through its paces at Chicago's Century of Progress Exposition in 1933. Full of hope for fundamental and practical applications of electricity in a new voltage range, Compton brought Van de Graaff along as a research associate when he relocated to Cambridge, Massachusetts the same year to assume the presidency of MIT — in effect becoming Millikan's counterpart on the East Coast (on Millikan's recommendation). There the would-be accelerator builder lost no time getting back to work in his chosen line of pursuit, with Compton's steady support and a grant from Cottrell and the Research Corporation. Cottrell, too, was impressed with the up-and-coming machine builder, as he emphasized to Lawrence, his other accelerator client of note.

> Since my return to the East, another matter has come up to the Research Corporation for support in its development, through Compton, at M.I.T. This is the electro-static generator of Dr. Van de Graaff, and matters developing therefrom.... The whole thing looks very interesting and supplements in many ways the approach to this general field which you are already making. Of course, there will be some parts of the general field in which the two lines of approach will in all probability develop some competitive features, but this should only add to the general hilarity and spice of friendly cooperation....
>
> Van de Graaff seems to be quite a fine young fellow with some very constructive ideas, and we certainly could not ask for a better guide and sponsor for him than Dr. Compton and the M.I.T. I am looking forward with real pleasure to the first opportunity when

we can get him and you together around the council table to compare notes and plans in your respective fields.... [7-13]

The maximum charge that can be deposited on a hollow conducting sphere is that which gives rise to an electric field equal to the electrical breakdown strength, or dielectric strength, of the surrounding medium — usually air. Air at normal atmospheric pressure breaks down at a field of about 30 kV per centimeter. However, if the generator is placed in some medium whose breakdown strength is greater than that of air, then the voltage and current (the latter given by the rate at which electricity is carried to and from the sphere by means of belts) both increase proportionally, and the power output increases as the square of the breakdown strength. The two evidently superior media are air or some other gas at high pressure, or a vacuum — that is, eliminating the air by operating the generator in an evacuated steel tank. Van de Graaff initially chose the latter approach, building a small vacuum generator while still at Princeton, and resumed work on another one at MIT [7-14].

In the event, it soon became apparent that a more practical way to make good on higher dielectric strengths was by pressurizing the steel container with air and other insulating gas mixtures. Lauritsen's Pasadena generator used this approach, and virtually all modern electrostatic generators are pressurized as well. However, Van de Graaff's next generator at MIT, begun in early 1932, was designed to operate in air at atmospheric pressure, and was distinguished instead by its sheer size.

Increasing the breakdown strength of air is not the only way to obtain higher sparking voltages in electrostatic generators; another is simply to increase the radius of the spherical terminal, if at the same time the laboratory room housing the generator is correspondingly enlarged to minimize sparking to surrounding walls and objects. Encouraged by the success of the model with 2-foot terminals, Van de Graaff and colleagues next made plans for the construction of a generator capable of the highest performance that could reasonable be expected by operating in the open air. They drew up plans for a generator with a pair of spherical terminals 15 feet in diameter, or a giant machine they judged capable in principle of generating up to ten times the voltage between the 2-foot spheres; with luck, it might yield ten million volts. The largest laboratory space available, it turned out, was a disused airship hangar at the Round Hill estate of Colonel E H R Green near South Dartmouth, Massachusetts.

The Round Hill facility had been available to MIT since 1926 as a Research Station in radio engineering by the generosity of Edward H R Green, an eccentric but generous Boston philanthropist who made his fortune as president of the Texas Midlands Railroad and inherited a second fortune from his mother [7-15]. By 1928 the research at Round Hill had been reorganized, focusing on aerial navigation in fog and fog dissipation, and Edward Bowles of MIT's Electrical Engineering Department, in overall charge of the project, arranged to borrow a dirigible from the Goodyear Zeppelin Corporation for research purposes. When approached by President Karl Compton four years later, Colonel Green, delighted by the spectacular discharges of one of Van de Graaff's prototype generators, gladly gave MIT permission to transform the airship hangar, no longer needed for that purpose, into a high-voltage laboratory.

The airship dock, of structural steel covered by corrugated sheet metal and prominent on the hill, was 140 feet long, 75 feet wide, and 75 feet high. In the back end were located shops and office space, while running lengthwise down the middle of the open floor space was installed a 14-foot gauge railroad track. The track extended through the huge doors at the front of the building, and out into the yard in front for a distance of 160 feet. On this track, each of the two generating units were mounted on separate, movable steel trucks or platforms, so that their separation and the sparking distance between them could be varied.

Each of the two 15-foot spherical aluminum terminals was supported on a hollow 6-foot diameter Textolite cylinder 24-foot high. Within each cylinder ran two charging belts, arranged such as to charge the spheres when going down as well as going up. Each sphere was equipped with trap doors, permitting entrance via a ladder from the ground. The inside of each sphere was essentially a laboratory room, with a floor, a workbench, and various pieces of apparatus, as shown in Figure 7.2. The plan was to mount a discharge tube horizontally between the two terminals, with an ion source in one terminal and a target and a nuclear laboratory for observations in the other.

The concept, illustrated graphically on the front page in the Sunday *New York Times* for 8 May 1932 [7-16], was for

> a giant apparatus with which protons can be shot against atoms at 10,000,000 to 15,000,000 volts. [Dr. R J Van de Graaff's] apparatus consists of two large spheres within which the experimenters sit. Electricity accumulates like water on the outside of a sphere. When there is more of it than the surface of the sphere can hold, it spills over to the other sphere in a blinding flash of artificial lightning. Imprison the flash in a tube and protons are carried to the target — the atom. [7-17]

The Round Hill installation was completed in 1936, and was described by Van de Graaff and his colleagues Lester and Chester van Atta and D L Northrup at MIT in the *Physical Review* that spring [7-18]. The machine developed 2.4 MV on the positive terminal and 2.7 MV on the negative one, giving a total potential of 5.1 MV between the two terminals. In principle, there was 1.1 mA of current available for use with the accelerating

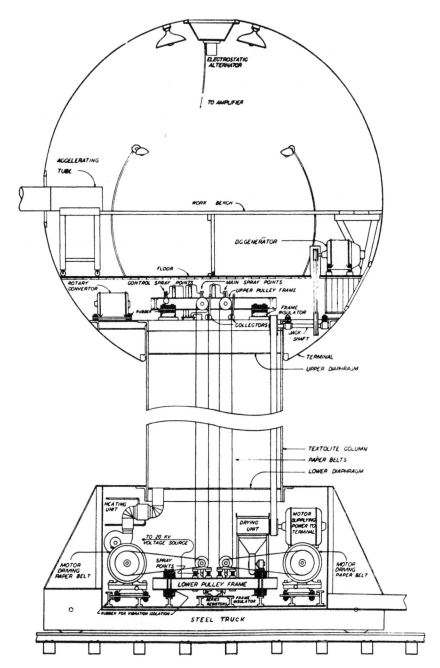

**Figure 7.2.** *Interior of one Van de Graaff generator unit at Round Hill, showing the belt-charging system and equipment in the 15-foot terminal as well as in the truck. Van Atta et al., 1936, PR,* **49**, *pp. 761–776, on p. 762.*

tube at this potential. In the event, however, the difficulties of mounting an evacuated discharge tube between the terminals proved too much, and high humidity near the ocean at Buzzard's Bay and bird droppings on the terminals ruled it out as an accelerator in this configuration and location. It was just as well. Van de Graaff and his little group had coexisted uneasily with Edward Bowles and his own staff, who held the physicists in contempt, considering them inefficient managers and worse engineers, and chafed at their large budgets [7-19].

The high-voltage installation was moved to MIT in 1937, where it was remounted within a metal-domed building with better dust and humidity control. This time the two columns were mounted adjacent to each other with the terminals in contact; one column held the charging belt and the other a vertical discharge tube with an ion source in the terminal. The beam was brought down through the floor to a basement laboratory. Thus reconfigured, desultory experiments were carried out to 2.75 MeV energy in the early 1940s.

Judged obsolete as a scientific instrument, the generator was moved again, this time to the Boston Museum of Science, where it remains on permanent exhibition to this day, occasionally streaming corona and throwing awe-inspiring lightning bolts before the public. To many people, the generator typifies the atom smasher of the nuclear physicist and his power.

# Chapter 8

# DIFFICULT YEARS

## 8.1. Depression and other scientific ailments

Regardless of which accelerator technique was championed by our rival physicists, high voltages by Tuve, Lauritsen and Van de Graaff, medium voltages by Cockcroft, or repeated low-voltage impulses by Lawrence, equipment for nuclear research and facilities for housing it cost money. This simple truth became increasingly evident in the physics community of the early 1930s, as the heady days of scientific growth and prosperity in the 1920s gave way to the Great Depression. Industrial laboratories and their staff bore the brunt early on; by 1933 General Electric had cut its staff by 50% of its strength in 1929, and AT&T had fired about 40% of their laboratory staff. Scientific agencies of the US Government fared no better, with budgets slashed an average of 12.5% for the 1932–33 fiscal year [8-1]. The Bureau of Standards, the largest federal employer of physicists, emerged with a Congressional appropriation in 1932 almost 26% below the previous year's level, and NBS staff were obliged to take a month's furlough without pay.

Academic physicists fared better, but not by much. State allocations fell sharply, beginning in 1932, at the major research universities, among them the Universities of California, Wisconsin, Michigan, and Illinois, and there, too, senior faculty sustained a reduction in salary or 'negative bonus,' while many junior faculty were not re-appointed or simply dropped. Fresh PhDs found it nearly impossible to find jobs. Thus at Caltech, where staff took a 10% salary cut in 1932, and expected more to come, Linus Pauling wrote Samuel Goudsmit at the University of Michigan that 'I haven't the faintest idea as to where [your former student] can get a job. Caltech is filled with our own PhDs and former National Research Fellows hoping for a small stipend' [8-2]. Indeed, a committee of the American Association of University Professors estimated that by December 1933, half-way through what the AAUP judged the worst academic year for American college budgets and staff rosters, 'the number of American scholars dismissed for economic reasons was more than double the number of German scholars dismissed for political or racial reasons' [8-3].

Physical scientists at the prestigious University of Chicago suffered a double blow. The University found itself financially crippled because Standard Oil dividends, which has supplied most of the University's operating funds since its founding by John D Rockefeller, had been drastically cut by the depression. On top of that, however, the University's science departments suffered further insults at the hand of the new president, Robert M Hutchins, the son of a theologian, who took office in 1929, at age thirty, with the announcement that the humanities and social sciences would henceforth receive priority over the natural sciences.

The Depression actually hit the West Coast later than the rest of the country, but with sufficient force to induce California legislators into making serious University budget cuts there as well. At UC-Berkeley in 1932, Raymond T Birge, by then chairman of the Physics Department, projected a 25% decrease in the number of teaching fellows for 1933–34, as well as a cut of more than 30% in the expense and equipment fund for the Department. Robert Oppenheimer had a considerable inheritance from his father's investments, but learned of the crash from Lawrence well after it began in New York's financial district. However, he soon appreciated the gravity of the situation, as he found himself unable to find jobs for his many admiring students. 'Through them,' he testified later, 'I began to understand how deeply political and economic events could affect men's lives' [8-4]. Ernest Lawrence, a modest investor in stocks, hung on, though his first patron, the Research Corporation, suffered badly, starting in 1932, and his colleagues braced themselves for the hard times in sight. As Birge wrote H L Johnston in October 1932,

> ...in regard to your last paragraph I can say that the depression has had relatively little effect on us as yet, but we are expecting to meet it full force next year when a new budget for the University will be in effect. In fact the main result of the depression has been that we have more graduate students than ever before and about a dozen visiting fellows of one kind or another. We are suffering, as it were, from a superfluity of riches. [8-5]

Indeed, Lawrence had half a dozen PhDs working without pay in the Radiation Laboratory, and had to turn down requests from former doctoral students to rejoin him as unpaid research associates [8-6].

As for Lawrence's old pal and friendly rival at the Carnegie Institution of Washington, Merle Tuve, he too felt the pressure at DTM, when the annual income of the richly endowed CIW fell by some $1,000,000 [8-7]. While Lawrence's only budgetary concern of note in early 1932 was funds for procuring a cyclotron magnet massive enough for raising his January energy record by a factor of ten, at an estimated cost between $10,000 and $15,000, Tuve's concern was for housing to accommodate a 2-meter Van de Graaff generator producing a mere million volts or so. But with CIW Director John C Merriam faced with budgetary choices, at a time when Tuve's colleague Robert Van de Graaff was preparing the splendid Round Hill facility for a 15-meter electrostatic generator, Tuve's DTM team didn't have that kind of luxury to count on, as he confided to Van de Graaff.

> The Acting Director [J A Fleming] does not have any critical judgement or appreciation of the relative importance of different things. We have no money for a silo to put the 2-meter generator in, and I can't even get permission to tear out the oil tank and install the 600 keV Van de Graaff generator in our building. [8-8]

Tuve glimpsed the effects of the depression on industry when researchers at General Electric contacted him, inquiring if there was a position at DTM for a technical assistant who had lost his job as a result of the company's recent budget reduction. Moreover, he gathered from his older brother of the effects on academia; in December 1932 George Lewis wrote that he and other faculty members at Case School of Applied Science had recently endured a 10% salary cut [8-9].

The despair of physics group leaders and department chairmen was amplified in public statements by spokesmen for the physics community. Thus Karl Compton, Van de Graaff's champion at MIT and chairman of the board of the newly formed American Institute of Physics, called for an increase in governmental appropriations to offset the decline in private support. At the same time, a mood of public distrust of science was sweeping the nation among legislators, public leaders, and laypersons alike, chastising science and technology for moving too fast and contributing to the collapsing economy. Henry A Wallace, President Roosevelt's secretary of agriculture, argued, perhaps echoing Nicholes and Browne's *Wings over Europe* in London's West End, that scientists and engineers have turned loose upon the world new productive power without regard to the social implications [8-10]. Musing on the complexities of the new physics, a commentator in the *New York Times* wondered whether 'nature needs eight new particles in constructing the cosmos, or it is the physicists who needs them?' [8-11].

Scientists were not unmindful of the need for greater public understanding and support. In his speech dedicating the Hall of Science at the Chicago Century of Progress Exposition, Frank B Jewett, head of Bell Telephone Laboratories, decried the 'senseless fear of science' which seemed to have taken hold in certain quarters, and admonished his colleagues to take serious note of the social problems their work might create [8-12]. Karl Compton, who had served as chairman of the Science Advisory Board, appointed by Roosevelt in 1932 to improve coordination of government scientific work and to develop an emergency scientific program to combat the depression, took a harder line in 1936. Replying to Roosevelt's question whether American engineering schools had introduced economics and social science into their curricula, Compton declared that 'the country does not need engineers and scientists distracted by literature and sociology, but more, better, and costlier science' [8-13].

Of course, the English, too, felt the Great Depression, though Cambridge University and its provincial society were less stirred by the grim economic times than comparable American universities. Ernest Walton's scholarship from the 1851 Exhibition Commissioners expired in June of 1930. Fortunately, the Department of Scientific and Industrial Research agreed to support him for the next four years with a Senior Research Award. He also won a Clerk Maxwell scholarship. Had it not been for such financial aid, it seems likely that neither he or John Cockcroft would have been able to complete their joint experiment, the apparatus of which was at long last ready in early 1932 [8-14]. Employees of Metro-Vick, the company that played such an important role in the background of the Cavendish successes, voluntarily took a 10% salary cut to help the company weather the depression.

For all its ominous harbingers of harder times ahead, 1932 would prove a bountiful year for exciting new results in experimental physics, as we shall see.

# Chapter 9

# 1932

## 9.1. More particles, expected and unexpected

1932 was the *annus mirabilis* of nuclear physics—a year of unprecedented discoveries, perhaps equaled only by 1905, when Albert Einstein in March, May, and June published his three immortal works introducing the light-quantum hypothesis, his theory of Brownian motion (which bears on the reality of atoms), and the special theory of relativity. Two of the four physics breakthroughs announced in early 1932 were anticipated, in a manner of speaking, by that other giant of physics 12 years earlier, Ernest Rutherford. By way of introducing the developments of 1932, let us step back to June of 1920, when Rutherford gave the Bakerian Lecture before the Royal Society on 3 June.

The title of his lecture was the 'Nuclear Constitution of Atoms,' and in it Rutherford spoke of the past and ongoing nuclear disintegration experiments, suggesting that matter consists of two fundamental building blocks, electrons and protons. Referring to the latest round of experiments on scintillations from passing $\alpha$-particles through nitrogen and oxygen, he drew attention to short-range atoms emitted in the process; they appeared not to be the protons emitted in the disintegration of nitrogen, but particles of mass 3 and carrying 2 positive charges. These 'light helium atoms' seemed to be products of nuclear disintegration reactions, just as were protons. They evidently consisted of three H nuclei (protons) bound by one electron, from the analogy with $\alpha$-particles (four H nuclei and two electrons).

> If we are correct in this assumption it seems very likely that one electron can also bind two H nuclei and possibly also one H nucleus. In the one case, this entails the possible existence of an atom of mass nearly 2 carrying one charge, which is to be regarded as an isotope of hydrogen. In the other case, it involves the idea of the possible existence of an atom of mass 1 which has zero nucleus charge. [9-1]

The second possibility, a hypothetical 'neutral doublet,' would have very novel properties, as Rutherford outlined in his lecture and we have cited earlier in Chapter 2 (p. 13). Its prediction sent James Chadwick on a decade-long hunt for the elusive particle, as also discussed earlier. In due course, experiments by Rutherford and Chadwick showed that the 'light helium nuclei' of mass 3 were, in fact, ordinary (mass 4) $\alpha$-particles of 9.3 cm range emitted directly from their RaC source [9-2]. Thus Rutherford had fallen prey to poor statistics and erroneous magnetic deflection experiments, just as L F Bates and J S Rogers had accused Pettersson and Kirsch in their Vienna experiments the year before. However, as Chadwick came to realize, Rutherford's belief in a neutral particle predated the alleged helium-3 results; it sprang from Rutherford's earlier brooding about the difficulty of building up complex nuclei with a large positive charge without invoking a neutral particle [9-3].

Nor were 'light helium nuclei' essential for Rutherford's postulate of a heavy isotope of hydrogen. For many years discrepancies among atomic weight measurements had frustrated scientists; more often than not, the atomic weight of a certain element, as determined by one scientist, did not quite agree with the atomic weight of another sample of the same element. At first blamed on errors in measurements or sample impurities, the differences were gradually traced to the existence of *isotopes* of the element in question [9-4]. Even the atomic weight of hydrogen was suspect, causing Rutherford to suggest the existence of one or more isotopes of that element as well. By 1931, experimental isotopic research had made considerable advances since Rutherford's Bakerian Lecture. That year Raymond Birge at UC-Berkeley, and Donald H Menzel, professor of astrophysics at Lick Observatory on nearby Mount Hamilton, suggested that one part in 4500 of ordinary hydrogen contains a heavy isotope, to which they assigned the symbol $H^2$ [9-5].

Birge and Menzel's letter to the editor of the *Physical Review* caught the attention of Harold C Urey, a physical chemist at Columbia University who had received his PhD in 1923 under the chemical thermodynamist Gilbert N Lewis at Berkeley. America's dean of physical chemistry, Gilbert Lewis was Berkeley's Dean of the College of Chemistry and Chairman of the Department for nearly 30 years. No fewer than five of his doctoral students, including Urey [9-6], would receive the Nobel Prize in Chemistry, and many still feel Lewis deserved the Prize himself on his own merits.

Harold Urey was born on a farm in Indiana in 1893. He graduated from Montana State University with a BSc degree in 1917, worked for a period with a chemical industrial laboratory in Philadelphia, taught chemistry at Montana State, and in 1921 began graduate work in chemistry at UC-Berkeley under Lewis, with a minor in physics. After receiving his doctorate, he joined Johns Hopkins University as an instructor in chemistry, before accepting a professorship at Columbia in 1929.

Birge and Menzel's letter to the editor was highly opportune for the young professor at Columbia. For some time Urey had been pondering certain irregularities in the sequence of atomic nuclei for the lighter members in the periodic table of elements, and Birge and Menzel's $H^2$ proposal promised to help in smoothing out those irregularities [9-7]. Teaming up with George Murphy at Columbia in the fall of 1931, Urey and Murphy soon identified the hydrogen isotope spectroscopically with a 21-foot grating spectrograph. Overly cautious, they decided not to rush into print; better to be quite sure by concentrating the isotopic constituent by fractional distillation of liquid hydrogen. Urey got in touch with Ferdinand G Brickwedde, then at the National Bureau of Standards, and already well versed in low-temperature studies. Samples of residual liquid evaporated from liquid hydrogen were brought to Columbia, and their faint spectral lines examined closely. Sure enough! The discovery of a stable (nonradioactive) isotope of hydrogen was announced at the 1931 Christmas meeting of the American Physical Society at Tulane University [9-8].

If anything, the naming of the new isotope proved more problematic than its isolation. At a special session on heavy hydrogen at the general June meeting in 1933 of the APS in Chicago, organized in conjunction with the Century of Progress Exposition, the ensuing discussion on its naming 'threatened to become acrimonious,' according to Francis Aston of the Cavendish Laboratory — the great authority on atomic weight measurements and a guest speaker at the discussion [9-9]. The argumentation had to do with whether to retain the name 'hydrogen' for the isotope, as Niels Bohr preferred; after all, it was not a new element, and had the atomic number 1. Both Gilbert Lewis and Ernest Lawrence opted for 'dygen' for the $H^2$ isotope and 'dyon' for its nucleus, whereas Rutherford preferred 'diplogen' and 'diplon' instead. In the end, Urey had the last word, as he was entitled to, settling on 'deuterium' for the isotope and 'deuteron' for its nucleus.

Nomenclature notwithstanding, the initial fruits of Urey's discovery were reaped by his old mentor, Gilbert Lewis. Lewis had been engaged, with his assistant Ronald T Macdonald, in the separation of oxygen isotopes when news reached him of a far more interesting isotope of hydrogen. Capitalizing on Urey and Edward W Washburn's finding that the residual water in an old electrolytic cell contained an appreciably larger fraction of deuterium than does ordinary water, Lewis pounced upon a large electrolyzer used next door by William Giauque to liquefy hydrogen. Before long, Lewis and his assistants were routinely producing laboratory quantities of relatively pure heavy water by prolonged electrolysis of ordinary water, studying the chemical and physical properties of the novel substance, and generously shipping vials of heavy water to colleagues here and there — to Rutherford at Cambridge, in particular.

Ernest Rutherford, who had chaired his own in-depth and more civil discussion on heavy hydrogen in a special meeting of the Royal Society in 1933 [9-10], had a very particular interest in the new isotope. Carrying no more repelling charge than protons, but having twice their mass, $H^2$ nuclei might well prove twice as effective as protons as swift projectiles for the transmutation of elements. And so they did, as we shall see.

Rutherford's own opinion about naming the $H^2$ nuclei harked back to the second announced discovery of 1932; it was reported in a letter to *Nature* on 17 February, one day after Urey and colleagues dispatched their first full-length paper on deuterium to the *Physical Review*. The new discovery, vaguely foretold as well in Rutherford's Bakerian Lecture, had its real origin in the purported 'beryllium radiation' reported by Walther Bothe and Herbert Becker in 1930 (Chapter 5, p. 60). News of the penetrating radiation reached Cambridge and Paris almost at once, and James Chadwick and Frédéric Joliot and Irène Curie lost no time in getting experiments going. The French were first.

We will encounter Joliot and Curie in a later development, and they must now claim our attention. Irène Curie, born in 1897, on the eve of the discovery of radium and polonium, was a daughter of Marie Sklodowska and Pierre Curie. She was educated privately and at the Collège Sévigné, and during World War I assisted her mother as a radiologist on the Western Front. At 21 she was named assistant to her mother at the Institut du Radium in Paris. There she was eventually joined by Frédéric Joliot, hired by Marie Curie in 1924 as her *préparateur particulier* (personal demonstrator). Born in 1900 to a well-to-do Parisian tradesman, Joliot received a solid grounding in scientific research under Paul Langevin at the École de Physique et de Chimie Industrielle de la ville de Paris, where Pierre Curie had also taught and been succeeded by Langevin. Joliot and Irène Curie were married in 1926, but continued working on separate problems for the time being. She continued research on the range of $\alpha$-particles in matter, which qualified as her doctoral thesis in 1926. He took up electrochemical studies of polonium, the results of which passed as his own doctoral thesis in 1930. Their first collaboration involved preparing, variously by electrolysis and evaporation, a supply of polonium as a powerful source of $\alpha$-particles. By the end of 1931, after three years of laborious and hazardous effort, they had a polonium source second to none, as well as a large Wilson cloud chamber, augmented by Geiger–Müller counters and ionization chambers for measuring ionizing radiation from the interaction of fast $\alpha$-particles with atomic nuclei.

Joliot and Curie were thus well positioned to tackle forefront problems at the start of 1932. The penetrating radiation observed by Bothe and Becker was just such a problem, and they promptly set about bombarding beryllium with $\alpha$-particles from their powerful polonium source. In so doing, they confirmed the penetrating radiation, as well as the absence of protons, emerging from the target. On 18 January 1932, they reported a result even more surprising than Bothe's radiation. Interposing various absorbing sheets of lead between the target and the detector, they observed a secondary, less penetrating radiation — apparently recoil protons. Not only that; when they interposed a block of paraffin wax or other substance rich in hydrogen, high-speed protons shot into their ionization or cloud chamber. Seemingly, energetic $\gamma$-rays from beryllium had collided head-on with hydrogen nuclei, knocking protons out of the way, in a kind of Compton effect [9-11]. If so, the $\gamma$-rays had to have been ten times as energetic as any previously known $\gamma$-rays.

At the Cavendish, meanwhile, Chadwick had a student, H C Webster, chasing down Bothe's radiation as well, suspecting all along that the radiation actually consisted of neutral particles — quite possibly Rutherford's 'neutral doublet' of 1920 that had eluded him for well over a decade. Then, one morning in February, Chadwick picked up the latest issue of *Comptes Rendus*, and was taken aback by the startling communication by the Joliot-Curies [9-12].

> Not many minutes afterwards [Norman] Feather came to my room to tell me about this report, as astonished as I was. A little later that morning I told Rutherford. It was a custom of long standing that I should visit him about 11 a.m., to tell him any news of interest and to discuss the work in progress in the laboratory. As I told him about the Curie-Joliot observation and their views on it, I saw his growing amazement; and finally he burst out, 'I don't believe it'. Such an impatient remark was utterly out of character, and in all my association with him I recall no similar occasion. I mention it to emphasize the electrifying effect of the Curie-Joliot report. Of course, Rutherford agreed that one must believe the observations; the explanation was quite another matter. [9-13]

The problem was, Chadwick's polonium source was too weak, in part due to the parsimonious attitude of Rutherford. However, by a happy coincidence Norman Feather had just returned from a stay in Baltimore, Maryland, bringing with him a generous supply of old, disused radon ampoules that he came across by chance at Kelly Hospital. Chemical manipulations being second nature to Chadwick, he was soon able to deploy the polonium, a decay product of radium, into an $\alpha$-source on par with the French source, as shown in Figure 9.1. With it, and employing an ionization chamber in conjunction with the new linear amplifier then coming into use at the Cavendish and elsewhere, Webster found that the radiation from beryllium emitted in the forward direction was more penetrating than that emitted backwards. This made no sense,

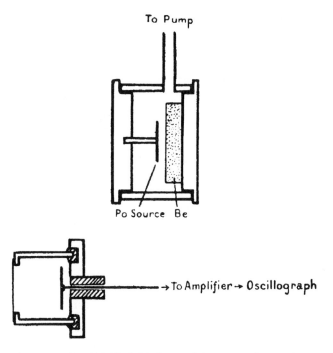

To Pump

Po Source   Be

→ To Amplifier → Oscillograph

**Figure 9.1.** *Apparatus with which Chadwick discovered the neutron. On top, the Po–Be source; on the bottom, the ionization chamber connected to an amplifier and oscillograph. Chadwick, 1932, PRS, 136, pp. 692–708.*

according to the laws of conservation of energy and momentum, if the radiation consisted of $\gamma$-rays—that is, of massless quanta of electromagnetic radiation. However, it made perfect sense if the radiation consisted of neutral particles of mass approximately equal to that of the proton. *Voilà*! Rutherford's neutral particle was at long last in hand.

Cautious as always, Chadwick submitted a letter, 'Possible Existence of a Neutron,' to *Nature* on 17 February [9-14]. Here, then, we also have Rutherford's reason for avoiding the terms 'deuteron' or 'deuton' for the $H^2$ nucleus, wishing to avoid confusion, during casual conversation, with 'neutron.' The latter term, incidentally, first seems to have appeared in print in a paper by one of Rutherford's students, whom Rutherford asked to search for his neutral doublet, soon after his Bakerian Lecture. The paper was forwarded for publication to the *Philosophical Magazine* by Rutherford about a year later, under the title 'Attempts to Detect the Presence of Neutrons in a Discharge Tube' [9-15].

The discovery of the neutron would soon revise Urey's view of his deuteron as consisting of two protons and an electron. However, as late as October 1932, it was still not obvious among physicists that the deuteron consists of 'one proton and one Chadwick neutron' instead of two protons and one electron [9-16].

Just as the deuteron and neutron can be said to have been anticipated well before they were detected experimentally, the same holds true for the third particle announced later in 1932. Indeed, the prehistory of that particle is much more complicated. As may be recalled, the first subatomic particle to be demonstrated as such, the electron, was initially shown to be a material 'corpuscle' in cathode ray experiments by J J Thomson at the Cavendish Laboratory. His classic experiments between 1897 and 1899 demonstrated clearly that the ubiquitous cathode rays, studied in discharge tubes from Faraday's time, were particles carrying a definite, unique charge of negative electricity and with a mass less than one-thousandth of the mass of the hydrogen atom [9-17]. The unambiguous, definitive isolation of the electron was accomplished by Robert Millikan in his oil drop experiments at Chicago, between 1907 and 1916.

The electron carries a negative charge, and for many years it was assumed that there had to be a positive counterpart as well—that is, particles differing only in the sign of their electric charge. However, except for what appeared to have been some erroneous experiments during 1907–19, cathode-ray studies revealed no trace of positive mirror-images of ordinary electrons [9-18], and by 1910 a single negative subcarrier of electricity was generally accepted as an experimental fact. There the matter rested until 1932.

By that year, a burgeoning subfield of physics experimentation was cosmic ray research, with Robert Millikan a controversial head of an active school researching the rays in America. Discovered in 1912 by Victor F Hess, an Austrian physicist and amateur balloonist, early studies of the mysterious radiation showed that it consisted of highly penetrating radiation originating high in the atmosphere, or in extraterrestrial space. The research was discontinued during World War I, but resumed in the 1920s. Millikan, who coined the term 'cosmic rays,' also referred to the rays as 'the birth cries of the atom,' in keeping with his view that much of the radiation was a form of electromagnetic energy accompanying nucleosynthesis in interstellar space. Another American, Arthur Holly Compton at Chicago, found a different explanation, showing from its deflection in the geomagnetic field that the radiation consists mainly of charged particles, not photons.

Which brings us to a student of Millikan, Carl David Anderson. Born in 1905 to Swedish immigrants in New York, Anderson grew up in Los Angeles, attended Caltech and there received his PhD under Millikan in 1930; his thesis involved a study of photoelectrons with a Wilson cloud chamber. Marking time as a postdoctoral fellow at Caltech, on Millikan's suggestion he took up the study of cosmic-ray electrons with a suitably modified cloud chamber in conjunction with a powerful electromagnet. On first operating the chamber with the magnet energized, an unexpected and awkward result was revealed: nearly equal numbers of positive and negative charges. Alternatively, some of the tracks could be interpreted as representing upward-moving negative electrons.

To sort out the confusion, a lead plate was mounted horizontally across the center of the chamber. Now, not only could the sign of the charge be determined from the direction of curvature in the magnetic field, and the energy from the amount of curvature, but also it was possible to distinguish between upward moving and downward moving particles. In traversing the plate, a particle would expend some energy, and its track would be curved more strongly when exiting from the plate than it was upon entry.

Sure enough! Among roughly a thousand photographs scanned, quite a few tracks were produced by upward moving particles, and among them, some by particles with a positive charge and (as indicated by their minimal ionization) a mass small compared to the mass of the proton.

It seems necessary [Anderson cautiously worded his preliminary letter to *Science*] to call upon a positively charged particle having a mass comparable to that of the electron. [9-19]

To Anderson's annoyance, the editor of *Science News Letter*, in which a photograph of one of the renegade tracks was first published, suggested calling the new particle the *positron*, and Anderson left it at that. More importantly, he was vaguely aware that the particle had been predicted in May of 1931 by the Cambridge theorist Paul Adrien Maurice Dirac as one of two energy states of the electron that pops out of his relativistic theory of that particle. Commenting years later on the possible influence of Dirac's work on his own discovery, Anderson allowed that

...yes I knew about the Dirac theory.... But I was not familiar in detail with Dirac's work. I was too busy operating this piece of equipment to have much time to read his papers.... Their highly esoteric character was apparently not in tune with most of the scientific thinking of the day.... The discovery of the positron was wholly accidental. [9-20]

With Anderson's announcement, other physicists went back over their old 'Wilson photos,' and found positron tracks in abundance — tracks too faint to have been noticed, or ignored for one reason or another. Among others who fired up their Wilson chambers anew were Frédéric Joliot and Irène Curie. Still smarting from having handed Chadwick the clue to the discovery of the neutron, they had searched unsuccessfully for Chadwick's particle among cosmic ray debris at the international scientific research station on the Jungfraujoch in Switzerland. During their earlier study of the Bothe–Becker radiation, with the chamber clamped between the poles of an electromagnet, they had occasionally observed something else: tracks of electrons moving 'backwards' from the chamber volume to the polonium source. Now, with Anderson's disclosure, they realized they had missed out on an important discovery once again. However, as if to make up for having missed the positron, they soon came up with a valuable result of their own: the first photograph showing the creation of an electron–positron pair from a hard $\gamma$-ray emitted from their Po–Be source. The creation of the two electrons requires a minimum photon energy given by the sum of the rest masses of the two particles, $2mc^2$ or slightly over 1 MeV.

1932 was, consequently a banner year for the discovery of new particles, with two claimed in the US and one in England. Particles notwithstanding, several teams were poised as well for important progress in nuclear reactions and transformations by artificial means, with particle accelerator technology approaching the requisite stage of maturity. Again the Americans seemed to have the cards in hand, with Tuve, Lauritsen, and Lawrence fielding crack accelerator teams, against Cockcroft and Walton pretty much alone in England. Not too surprisingly, however, the English would even the score in physics achievements worthy of the Nobel Prize for 1932.

## 9.2. Cockcroft and Walton strike

We left Cockcroft and Walton in late 1930, with a failed transformer and apparatus limited, in any case, to an accelerating voltage of 280 kV and with no results to speak of for their effort. Cockcroft decided a new tack was in order. After pondering the possibilities, he came up with a voltage-multiplying scheme for stepping up a modest a.c. voltage to a high d.c. voltage by a circuit based on an intricate set of condensers and rectifier switches. Unknown to Cockcroft, the scheme had been devised by Heinrich Greinacher in 1920. Basically, it relied on rectifier diodes for charging capacitors in parallel during one half of the alternating cycle, and other rectifiers for discharging them in series during the other half-cycle [9-21]. With $N$ capacitors and $N$ rectifiers, one could obtain a voltage multiplication by a factor of $N$, as illustrated in Figure 9.2.

Cockcroft's version of the multiplier circuit called for a vertical stack of four capacitors and four rectifiers, aiming for a fourfold multiplication of the transformer a.c. voltage. This time, instead of Allibone's rectifier bulbs, the rectifiers were enclosed in glass cylinders 14 inches in diameter and 3 feet in length. (Straight glass cylinders were found to withstand high voltages much better than the largest glass bulbs obtainable.) Partly assembled, it all looked promising when another holdup nearly put them out of business once again; their laboratory quarters were

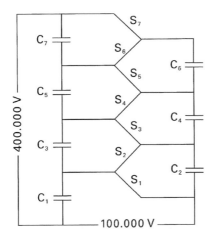

**Figure 9.2.** *Principle of the voltage multiplier scheme adopted by Cockcroft and Walton. In this example, the starting potential is 100 kV, in a circuit employing 7 condensers (C) of equal capacity, and seven switches (S). The action starts with switch $S_1$ allowing capacitor $C_1$ to be charged to 100 kV. Next, switch $S_2$ connects capacitor $C_2$ to capacitor $C_1$; these two capacitors now share a charge of 50 kV each. Switch $S_1$ permits $C_1$ to get charged back up to 100 kV. $S_1$ and $S_3$ permit capacitor $C_3$ to share the charge of capacitor $C_2$, giving each one 25 kV. Switches are now adjusted so that capacitor $C_2$ shares more charge from $C_1$, and $C_4$ shares charges from $C_3$. By alternating the arrangement of the switches, the capacitors are all charged to 100 kV. 40 kV is then available between $C_1$ and $C_7$.*

required by the Physical Chemistry Department, and had to be vacated by May 1931.

As luck would have it, the relocation proved a blessing in the long run. They moved into the disused Balfour Library, emptied of tables and other inventory; as such, the room had much more ceiling height and square footage than their old quarters. Cockcroft and Walton lost no time reassembling the apparatus, as shown in Figure 9.3. With a 200 kV transformer as the a.c. power source and four rectifiers and four condensers, they hoped for a final potential drop of 800 kV. The new acceleration tube, standing six feet tall, came in two stacked glass sections, similar to those used in the rectifier tower; it contained two acceleration gaps between three cylindrical electrodes along the axis of the tube. Protons were generated in a hydrogen discharge in an ion source of the canal-ray type, located at the top of the tube. It was powered by a 60 kV transformer excited by a small alternator. The middle accelerating electrode, maintained at half the full potential drop, carried a little diaphragm to stop backstreaming secondary electrons. The tube was pumped by one of Burch's fast oil diffusion pumps using Apiezon Sealing Compound Q, backed by a Hypervac mechanical fore-pump. At the bottom of the tube, the protons emerged through a thin mica window

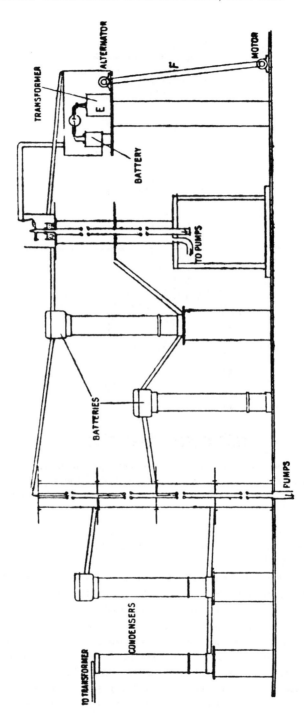

**Figure 9.3.** *Cockcroft and Walton's multiplier apparatus. (The rectifier tower is third from left.) The potential across the discharge tube was obtained from a transformer E, excited by a small alternator driven by a belt F from a motor at ground potential. Cockcroft and Walton, 1932, PRS, 136, pp. 619–630, on p. 627.*

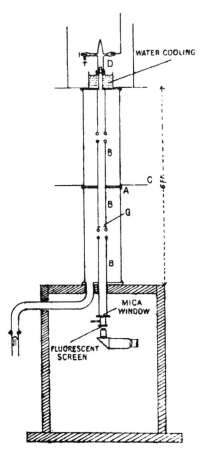

**Figure 9.4.** *Accelerating tube and target arrangement inside 'observer's hut.' The ion source is at D. The two accelerating gaps are seen between the cylindrical electrodes B. The mica window closes the evacuated tube, pumped by a fast oil diffusion pump through a 3-inch pipe. Cockcroft and Walton, 1932, PRS, 136, pp. 619–630, on p. 626.*

into a cramped 'observer's hut' shielded by lead and screened from electrostatic fields (Figure 9.4).

C R Burch, it might be added, had been asked by Metro-Vick Research Director Fleming to look into better ways of impregnating transformer insulation, than the oil hitherto used for such purposes, and in due course he came up with the low-pressure Apiezon oils and greases that played such a crucial role in vacuum systems from 1930 onward. The oil ultimately replaced mercury in diffusion pumps. The new pumps played a vital role in the 40-odd large transmitting tubes in the CH radar transmitters along the British coast during the Battle of Britain (section 13.3, below), and in the hundreds of large oil diffusion pumps in the Oak Ridge uranium isotope separation plant (section 14.2). Burch developed the low viscosity oil, having a vapor pressure less than $10^{-6}$ mm, in a distillation still devised to separate transformer oil fractions. One fraction resembled Vaseline, but had an unmeasurably low vapor pressure, and could be used to seal vacuum joints that fitted tolerably well. According to Allibone, Burch had been a Greek scholar at Caius College, and christened his low vapor pressure products 'Apiezon.' In 1944 he was elected Fellow of the Royal Society, largely for this work [9-22].

Cockcroft and Walton's apparatus was largely completed in the spring of 1931. However, nearly a year was spent breaking it in, with Cockcroft and Walton spending much of their time, along with their laboratory assistant Willie Birtwhistle, perched on wobbly ladders, sealing leaking vacuum joints with the special greases supplied by Burch, or simply rubbing over Apiezon joints with their fingers. Vacuum components were 'outgassed' (cleared of oxide layers and body grease from frequent handling during assembly) and conditioned by repeated spark discharges, and the tube voltage was gradually coaxed upward. It usually required the better part of a day's operation before some maximum voltage could be applied, crudely measured by sparkover between two large aluminum spheres.

In the middle of it all, Cockcroft was diverted by one matter or another, one being a visit to the USSR paid for by Metro-Vick. He sailed for Leningrad on a Russian ship, in a mixed company of mostly British scholars, among them the crystallographer Glenn A Millikan—son of R A Millikan. Service on board being nil, apart from meals, H D Dickinson, a lecturer in economics from Leeds University, proclaimed that 'in a Soviet ship there was no more reason for the passengers to tip the crew than the crew to tip the passengers!' [9-23]. In Leningrad they visited the laboratory of Abram F Ioffe, the prominent Russian physicist and Director of the Leningrad Physicotechnical Institute. They also called on the local Metro-Vick office that was the headquarters of a team of engineers engaged in erecting Metro-Vick turbines in Russia. In Moscow, Cockcroft inspected a well-equipped, high-tension laboratory in—of all places—a disused church said to have been where Pushkin had been married [9-24].

On his return to England, Cockcroft was saddled with the task of organizing celebrations for the centenary of the birth of James Clerk Maxwell, the first Cavendish Professor of Experimental Physics. (The year 1931 was actually the centenary of *three* milestones in British science: of the founding of the British Association for the Advancement of Science, of Faraday's discovery of electromagnetic induction, and of Maxwell's birth.) The celebrations were preceded by the unveiling of a tablet commemorating Michael Faraday and Maxwell in Westminster Abbey on 30 September, and Cockcroft was a member of the Cambridge

University delegation headed by the Vice-Chancellor. At Trinity College, Sir J J Thomson gave a dinner party for 200, including Max Planck, Pieter Zeeman, Niels Bohr, Robert Millikan, and Gugliemo Marconi. Open house at Trinity College and the Cavendish featured public lectures and guided tours, with Cambridge townspeople tramping through manicured lawns and musty corridors.

Niels Bohr was among those who agreed to address the delegates, following Sir Joseph Larmor on the morning of 2 October. Bohr's talk, on Rutherford's invitation, was on 'The Influence of Maxwell's Work on Modern Developments' [9-25]. It fell upon Cockcroft and his secretary to record his talk for publication—mostly mathematical scribbling on the blackboard accompanied by mumbling in broken English. Their combined effort 'did not produce anything coherent,' and Cockcroft 'didn't think anything came of it' [9-26].

At last Cockcroft could rejoin Walton. They brought the apparatus to operational readiness by December 1931, with about 700 kV on the tube, limited by corona losses, and without the rectifier vacuum 'going soft.' Their first priority was determining the range of protons in air, and checking the proton velocity by magnetic deflection. The latter measurements also allowed determining the ratio of protons to molecular ions in the beam, since the deflection of an ion in a magnetic field depends upon its velocity and charge-to-mass ratio. ($H_2^+$ and $H_3^+$ ions, as well as $H_1^+$ ions or protons, are all available from hydrogen ion sources in relative proportions which depend on the particular design of the source.) These various calibration runs, plus desultory searches for the hypothetical Bothe–Becker $\gamma$-radiation from beryllium, occupied them from January through March of 1932. All along, the equipment remained far from stable, and in March they also began work on a new acceleration tube. By the end of March and the first week of April, Rutherford and Chadwick's patience wore thin. As usual, Rutherford wanted results, not refinements in technique. As Walton remembers it,

> Rutherford came in one day and found us doing magnetic deflection experiments and told us that we ought to put in a fluorescent screen and get on with the job, that no-one was interested in exact range measurements of our ions. [9-27]

The crucial day was Friday, 14 April [9-28]. A day or so before, apparently on Rutherford's specific urging, they had mounted a lithium target at the bottom of a brass tube extending from the lower end of the acceleration tube into the small observer's hut. The lithium target was placed at an angle of 45° to the direction of the proton beam. Facing it was a zinc sulphide screen, with a thin sheet of mica interposed between the screen and the target. The screen was viewed with a microscope, by the observer sitting upright on a bench in the little hut.

Early that day, Cockcroft was in the Mond Laboratory, lending Kapitza a hand, as he often did. While waiting for him, Walton turned on the accelerator. The accelerating voltage was controlled by varying the field of the alternator exciting the main high-voltage transformer. The secondary voltage of the transformer, related in a well-defined manner to the steady potential produced by the rectifier system, was read by a microammeter on the control table. When the voltage reached 'a fairly high value,' Walton left the control table while the apparatus was still running, and crawled (to avoid danger from the high voltages) across the room to the little hut under the acceleration tube [9-29], as we see in Figure 9.5. On peering into the microscope, he immediately spotted scintillation flashes on the screen. He went back to the control table and switched off the power to the ion source. On returning to the hut, no scintillations could be seen. After a few more repetitions of this, he was convinced the effect was genuine.

To Walton, the flashes seemed quite like what he had read about $\alpha$-scintillations but never actually seen before. He phoned Cockcroft, who came right away and confirmed the observations. Cockcroft then rang up Rutherford, who arrived shortly afterward. With some difficulty they maneuvered him into the little hut, and he, too, had a look at the scintillations.

> He shouted out such instructions as 'Switch off the proton current! Increase the accelerating voltage!,' etc., but he said little or nothing about what he saw. He ultimately came out of the hut, sat down on a stool, and said something like this: 'Those scintillations look mighty like alpha-particle ones. I should know an alpha-particle scintillation when I see one for I was in at the birth of the alpha-particle and I have been observing them ever since.' I have since thought that he might have added that he was also in on the christening! He did not stay very long but came back next day and did some scintillation counting. He did about a dozen counts of which the details are in my notebook. [9-30]

Walton himself made no attempt to count scintillations in his first observations, and recorded neither the proton current nor the bombarding voltage. Later in the day, he and Cockcroft counted scintillations commencing at 252 kV and diminishing in steps to a lowest value of 126 kV. Clearly they could have done so at 280 kV two years earlier, had they bothered to look for scintillations.

The next day, during his second visit to the high-tension hall, Rutherford did a most unusual thing, quite against his principles and the normal ethos of the Cavendish. He swore both Cockcroft and Walton to strict secrecy, ostensibly so that they could analyze the effects of the

**Figure 9.5.** *Ernest Walton observing scintillations in the lead-shielded target hut at the bottom of their accelerating tube. At the left, rear are battery columns plus a spark gap between two aluminum spheres. At the right, rear is the transformer providing the accelerating potential across the discharge tube, excited by an alternator at high potential, driven by an insulating belt from a motor at earth potential. Courtesy The Cavendish Laboratory.*

scintillations without interruption from visitors. By working late into the evening—another break with Rutherford's normal practice—they had all the essential information in hand, including photographs of Wilson chamber tracks, by 16 April. Late that evening they went round to Rutherford's home at Newnham Cottage and reported their results. Then and there they drew up a letter to *Nature* which was published on 30 April [9-31]. In it, they claimed that

> the lithium isotope of mass 7 occasionally captures a proton and the resulting nucleus of mass 8 breaks into two alpha-particles, each of mass four and each with an energy of about eight million electronvolts. The evolution of energy on this view is about sixteen million electron volts per disintegration.

The equation for mankind's first nuclear disintegration with artificially accelerated projectiles in laboratory apparatus may be written as

$$_3\text{Li}^7 + {_1}\text{H}^1 \longrightarrow {_2}\text{He}^4 + {_2}\text{He}^4 + 17\,\text{MeV}.$$

The energy released, 17 MeV, is in good agreement with that calculated from Einstein's equation $E = mc^2$, and equals the difference between the masses of the original atoms of lithium and hydrogen, and the masses of the two product atoms of helium.

Rutherford was elated, as was everybody at the Cavendish. On 21 April, in writing to Niels Bohr about the recently discovered neutron by Chadwick, Rutherford expressed himself 'very pleased that you regard the Neutron with favor.' He then turned to the latest cause for excitement in the laboratory.

> It never rains but it pours, and I have another interesting develop-ment to tell you about of which a short account should appear in *Nature* next week. You know that we have a High Tension Labora-tory where steady D.C. voltages can be readily obtained up to 600,000 volts or more. They have recently been examining the effects of a bombardment of light elements by protons.... [9-32]

Describing Cockcroft and Walton's experiment, he noted that

> In the case of lithium brilliant scintillations are observed, begin-ning at about 125,000 volts and mounting up very rapidly with voltage when many hundreds per minute can be obtained with a protonic current of a few milliamperes....In experiments in the last few days similar effects have been observed in boron and fluorine but the ranges of the particles are smaller although they look like $\alpha$ particles. It may be boron 11 captures a proton and breaks up into three $\alpha$ particles, while fluorine [atomic weight 19] breaks up into oxygen and an alpha....It is clear that the $\alpha$-particle, neutron and proton will probably give rise to different types of disintegrations and it may be significant that so far results have only been observed in $4n + 3$ elements. It looks as if the addition of the 4th proton leads at once to the formation of an $\alpha$-particle and the consequent disintegration.

116

In closing, he expressed himself

> very pleased that the energy and expense in getting high poten-
> tials has been rewarded by definite and interesting results.... You
> can easily appreciate that these results may open up a wide line of
> research in transmutation generally.

Rutherford was equally pleased to bring Cockcroft and Walton to the
Royal Society on 28 April, where the discovery was formally announced.
Later that day all three dined as Rutherford's guests at the Royal Society
Dining Club. According to Allibone, Rutherford invited Cockcroft,
Fowler invited Walton, Aston invited Chadwick, John Cunningham
McLennan invited Dirac, Jeans invited Mott, R Robertson invited C D
Ellis, and a total of 24 sat down for dinner at the Royal Society Dining
Club [9-33].

Bohr's reply to Rutherford's letter on 2 May was enthusiastic and
appreciative of the significance of the latest development at Cambridge.

> By your kind letter with the information about the wonderful new
> results arrived at in your laboratory you made me a very great
> pleasure indeed. Progress in the field of nuclear constitution is at
> the moment really so rapid, that one wonders what the next post
> will bring, and the enthusiasm of which every line in your letter
> tells will surely be common to all physicists. One sees a broad
> new avenue opened, and it should soon be possible to predict the
> behavior of any nucleus under given circumstances. [9-34]

The world press waxed even more enthusiastically over what must
have been uplifting news during somber economic times [9-35]. Thus,
the *New York Times* carried consecutive articles with the following head-
lines: 1 May 1932, 'Atom Torn Apart with Energy Rise;' 2 May, 'Atom
Torn Apart, Yielding 60% More Energy than Used;' 3 May, 'Hail New
Approach to Energy of Atom,' and 'Value Put in Energy Gain;' 4 May,
'Atomic Energy.' As for the reaction of other fellow physicists abroad,
we shall come to that in the next chapter.

As emphasized in the news releases and popular accounts, at the
lowest effective bombarding energy, 120 kV, only one hit was scored
per ten million proton shots. At 400 kV, the marksmanship was somewhat
better — up by a factor of ten, though way below Gamow's optimistic
predictions. More interesting, however, was the amount of energy
released: 16 million volts for a maximum energy expenditure of a few
hundred kilovolts.

Following their initial experiment with lithium, boron, and fluorine,
Cockcroft and Walton took a host of other elements under attack, from
beryllium to uranium, using both the scintillation method and the
new electronic counters (ionization chamber plus linear amplifier and

oscillograph) provided by Wynn Williams; oscillograph kicks from the counters confirmed the ejection of $\alpha$-particles with varying degrees of success. As usual, Rutherford, who participated personally in some of the experimental runs, kept Bohr informed of their progress. On 26 May he wrote that

> We have so far only had time to examine a limited number of elements, but everything so far tested shows some result; although the magnitude varies over a wide range. So far pure iron gives the lowest result, a very minute fraction of that from lithium, and small compared with nickel and cobalt.
>
> I may tell you also, privately, that uranium and lead also give a positive effect. I made the initial observations with uranium myself and we were able to increase the natural activity several times. I think, however, this is not a case of stimulating the natural transformation for it looks as if the range of the particles is greater than the normal. I am inclined to believe that in these heavy elements the effect is likely to be due to capture of the proton through one or more resonance levels and this is a point that I propose to examine using comparatively low voltages. Of course the results are so far in a preliminary stage but I think the results are quite clear about lithium which give a prodigious effect compared with any other element. [9-36]

As for the two alphas from lithium, using a pair of ZnS screens and two tapping keys for independent recording, Cockcroft and Walton showed that the $\alpha$-particles were emitted in coincident pairs. Their results were reported in a full-length paper communicated by Rutherford to the Royal Society on 15 June [9-37]. Subsequently, as we shall see, they traced the $\alpha$-particles from most of the other elements besides lithium and boron to boron impurities in the targets, perhaps from the glass cylinders.

# Chapter 10

# RUNNERS UP

## 10.1 Disintegrations at Berkeley, if with difficulty

Ernest Lawrence appears to have been in the habit of receiving timely communications from Berkeley while residing with the Blumer family in Connecticut. May 1932 was no exception. He learned that month of Cockcroft and Walton having anticipated him, while preparing for his wedding with Molly Blumer in New Haven. 'We have just gotten the Tues. [3 May 1932] N.Y. Times with the account of the new Cambridge work,' wrote Raymond Birge. 'Very exciting!' [10-1] Lawrence immediately wired James Brady in Berkeley. Brady had recently completed his PhD thesis under Lawrence on some aspect of the photoelectric effect, and was staying on to learn something of the art of cyclotroneering before assuming a fellowship at St. Louis University [10-2].

> COCKCROFT AND WALTON HAVE DISINTEGRATED THE LITHIUM NUCLEUS STOP GET LITHIUM FROM CHEMISTRY DEPARTMENT AND START PREPARATIONS TO REPEAT WITH CYCLOTRON STOP WILL BE BACK SHORTLY

Turning the beam from the 11-inch cyclotron in Le Conte Hall onto a LiF crystal, Brady found only hints of disintegrations; the beam current was too weak, and the detector too insensitive for the job [10-3]. Things looked more hopeful with the arrival that summer of two friends from Yale, Franz Kurie, a new PhD, and Donald Cooksey, his mentor; they were scheduled to stay with Lawrence's group from early August through mid-September. Cooksey was a 40-year old research fellow as well as curator of instruments at Yale's Sloan Laboratory. He was also Best Man and organized the bachelor's luncheon at New Haven's Lawn Club prior to Mary Blumer and Lawrence's wedding. The party, naturally, had included Merle Tuve and Jesse Beams, as well as Linus Pauling, among others.

Kurie set about building a cloud chamber, and Cooksey, who became the associate director of the Radiation Laboratory in 1936, looked into better instrumentation in general. They were assisted by an interesting, non-salaried volunteer on campus, Commander Telesio Lucci, retired Italian naval officer and diplomat. He was 'a most obliging pair of hands, if told what to do,' and a popular member of the group [10-4].

Alas, as of mid-August they still had nothing more than weak evidence to show for their effort, as we learn from Lawrence's letter to Cockcroft on 20 August.

> At the present time we are attempting to corroborate your experiments using protons accelerated to high speeds by our method of multiple acceleration. We have some evidence already of disintegration, though as yet we can not be certain. Unfortunately our beam of protons is not nearly as intense as yours although of higher voltage. Whenever we obtain some reliable results of course we will let you know promptly. [10-5]

Why, it may be asked, had Lawrence and his capable assistants been so slow in developing detectors of the requisite sensitivity to seriously exploit their accelerators for nuclear physics instrumentation—even with the beam currents on hand? The answer, paradoxically, according to Heilbron and Seidel, seems to have been the weak beam current, $10^{-9}$ amp, or one milli-microampere (nanoampere). Undoubtedly the current could be improved [10-6]; in the meanwhile they had been too busy improving the accelerator itself, leapfrogging from the 11-inch to a 27-inch cyclotron. The next cyclotron would raise the energy by a factor of ten or more, to 10 million volts or perhaps even twice as much.

The new machine was on the drawing board long before Livingston had completed the 11-inch cyclotron. As usual, the first technical hurdle was the magnet. By a stroke of luck Lawrence came across a suitable magnet yoke in a derelict Poulsen-arc magnet in Palo Alto, 50 miles south of Berkeley. It belonged to the Federal Telegraph Company, built years ago for use in wireless transmission in China, as an alternative to Marconi's system. The system had been rendered obsolete by the vacuum tube. Through the good offices of Leonard T Fuller, professor of electrical engineering at UC-Berkeley and a vice president of Federal Telegraph, the 80-ton white elephant was donated free of charge to the University for Lawrence's use. However, to transport it to Berkeley, house it and convert it to an 18-kilogauss, cyclotron magnet required funds. Again Lawrence turned to Cottrell and the Research Corporation, assuring them that 'with the aid of the big Federal Telegraph magnet we can go right on up to 20,000,000 volts' [10-7]. In the end, the Research Corporation provided $5000, and its president, Howard Poillon, secured $2500 from the Chemical Foundation—a foundation organized by the government in 1920 to administer German chemical patents seized during World War I. In addition, Robert Gordon Sproul, president of the University of California, came up with the balance of the $12,000 Lawrence deemed necessary to complete the job [10-8].

The big magnet first went to the Pelton Waterwheel Company in San Francisco, where the poles were re-machined to specifications provided by Professor Fuller. It gave Lawrence as well a breathing spell for renovating the old Civil Engineering Testing Laboratory that President Sproul had turned over to him in late August 1931, and allowed Livingston and Sloan to get on with the 11-inch cyclotron—the last to be built in Room 329 of Le Conte Hall. The reworked Federal Telegraph magnet was installed in Room 114, reinforced with concrete piers, in the refurbished Rad Lab building at the turn of the year. The energizing coils encircling the 27.5-inch pole pieces were layer-wound from copper strip and immersed in oil tanks for cooling. 'The oil tanks leaked,' recalled Livingston. 'We all wore paper hats when working between the coils to keep oil out of our hair' [10-9].

Early in 1932 Livingston turned the 11-inch cyclotron over to Milton White to use for his thesis problem, so Livingston himself could spend more of his time on the larger machine. The vacuum chamber for the 27-inch cyclotron was a 28-inch brass ring fitted with iron plates for top and bottom lids. As in the 11-inch cyclotron, initially only one D-shaped electrode was installed, facing a slotted bar at ground potential ('dummy dee'). In the semi-circular space behind the bar the Faraday cup beam collector could be mounted at any chosen radius. The beam was first observed at a small radius, and as the oscillator circuit, using a Federal Telegraph 20-kW water-cooled tube, was tuned and the magnet repeatedly 'shimmed' the collector was gradually moved out to some practical maximum radius (10 inches). At that point in time two symmetric dees were installed, allowing somewhat higher beam energies and intensities to be reached [10-10].

Lawrence had hoped to have the 27-inch cyclotron running by the end of February 1932 [10-11]. However, March came and went with him and Livingston still fighting leaks in the vacuum tank from the enormous electromagnetic forces generated by the big Federal magnet. He vented his frustrations on Curtis Raymond Haupt, an ex-graduate student of his who received his PhD on ionization phenomena in mercury vapor in 1930.

> ...I have been working night and day with Livingston and Brady on the big magnet and associated apparatus. I have neglected everything else—even my fiancée has suffered.... [10-12]

In May Lawrence made up for his neglect of Molly by heading East for their wedding, leaving Livingston to carry on. Success at last! On 13 June, with Lawrence still with Molly at Haycock Point, the Blumer summer place on Long Island Sound, a first beam of 1.24 MeV hydrogen

**Figure 10.1.** *Livingston and Lawrence beside the 27-inch cyclotron, about 1934. Courtesy AIP, Emilio Segrè Visual Archives.*

molecular ions was finally recorded in the electrometer. Not much, but by August the $H_2^+$ beam energy was raised to 3.6 MeV—the equivalent of 1.8 MeV protons. (Acceleration of protons was impractical at that stage, as it would have required an oscillator with impractically high frequency.) By that time Lawrence was back home and in charge once again, though preoccupied with installing Molly in their new home on Keith Avenue in Berkeley and with hopes of disintegrating lithium early on. First, however, a letter to the editor of the *Physical Review* was in order, announcing successful operation of the new machine. It should be signed by Livingston alone, insisted Lawrence, as the 'credit is yours and my name is not to go on the paper' [10-13].

Figure 10.1 shows Livingston (left) and Lawrence besides the completed 27-inch cyclotron, as it appeared in 1934, and Figure 10.2 shows Lawrence at the controls.

The electromagnet for the 27-inch cyclotron can still be seen at the entrance to the plaza in front of the Lawrence Hall of Science on the hill above what is nowadays the Lawrence Berkeley National Laboratory. Actually, it is a reconfigured version of the magnet dating from 1937, when the pole-pieces were replaced with poles of larger diameter—in effect converting the accelerator to a 37-inch cyclotron. The oil tanks

**Figure 10.2.** *Lawrence at the controls of the 27-inch cyclotron. Courtesy AIP, Emilio Segrè Visual Archives.*

housing the coils are actually devoid of the coils, and therein hangs a little tale. Following its decommissioning after World War II, the cyclotron passed on to the Physics Department at the Los Angeles campus of the University of California. At some point during the construction of the Hall (1965–68) Harvey E White, Berkeley professor of physics and the first director of the Hall, heard that UCLA had no further use for the cyclotron. He wrote, asking if he might bring it back to Berkeley to serve as an exhibit near the entrance to the Hall. UCLA answered in the affirmative. However, upon contacting a large draying company, White found it would cost $4000 to haul the machine from Westwood to Berkeley. Not having money for such an operation, White asked the UCLA officials if they might sell the copper windings of the magnet. They did, and thus obtained the necessary funds to pay for the hauling [10-14].

Part of the plaque attached to the cyclotron exhibit reads as follows:

'This cyclotron was a leviathan of science in its time. It led the scientific world in particle energies from 1932 to 1939, opening new frontiers in nuclear research.'

However that may be, it was the 11-inch cyclotron that was the center of attention at the start of the 1932 fall term in Berkeley. Brady left for St

123

Louis on or about 31 August, and Cooksey and Kurie returned to New Haven two weeks later. Just before the short-lived team broke up, however, they obtained the first unambiguous counts with Cooksey's point Geiger counter. A beam of 700 kV protons, from a filament source of Cooksey's in the little cyclotron, produced, upon striking a lithium fluoride crystal, what appeared to be a hefty number of $\alpha$-particles — about 100 counts per minute per milli-microampere. Unfortunately, time ran out before they could analyze the results with any degree of confidence. In September, the experiment was continued by Lawrence, Livingston, and Milton White. They were joined by Malcolm Henderson, a research fellow at Yale who had received his PhD at Cambridge, who would spend the fall semester at the Rad Lab as a research associate without stipend. Like his Yale colleagues, Henderson was skilled with Geiger counters, and lent the current project an invaluable hand in keeping a counter and pulse amplifier going.

With Henderson's counter, and the intensity of the proton beam boosted somewhat with a new filament source, they soon verified a prodigious yield of $\alpha$-particles from the LiF crystal at bombarding energies of 360, 510, and 710 kV. When plotted in a curve of alpha-particle counts per unit time per incident proton versus bombarding energy, they overlapped nicely with Cockcroft and Walton's data extending up to 500 kV [10-15]. Their results were rushed posthaste by letter to the editor of the *Physical Review* [10-16], with a copy dispatched as well to John Cockcroft, among others.

In their letter to the editor, Lawrence and his teammates note an apparent disagreement with Cambridge over the observed number of counts per unit current and per unit solid angle, the Berkeley value being somewhat higher than Cockcroft and Walton's. Neither Lawrence nor Cockcroft found the slight discrepancy to be of concern, easily accounted for by several experimental uncertainties in the measurements. However, in his return letter to Lawrence, Cockcroft drew attention to a rather puzzling, internal numerical discrepancy in the Berkeley report. If corrected for, he noted, the Berkeley results appeared in close agreement with the Cambridge results [10-17]. In his reply, Lawrence sheepishly allowed that he 'made the silly mistake of neglecting to change alpha-particle counts per minute to counts per second' [10-18].

A less silly mistake would befuddle Lawrence and his team in the year to come, as we shall see. In the meanwhile, a nagging discrepancy came to light almost at once — one alluded to earlier and not between experiments, but between experiment and theory. In their initial report to the Royal Society on 15 June 1932, Cockcroft and Walton had observed what appeared to be radiation from proton-irradiated elements as heavy as uranium. They regarded these results with skepticism from the outset.

In view of the very small probability of a proton of 500 kilovolts energy penetrating the potential barrier of the heavier nuclei by any process other than a resonance process, it would appear most unlikely that such processes are responsible for the effects observed with the heavier elements. [10-19]

Nor could George Gamow, who had been kept informed of their progress by Cockcroft, account for these seemingly anomalous results.

.... The bad thing is I cannot understand the theoretical possibility of those [results] for the heavier elements. Recently I have done some calculations based on the usual formula for penetrating through the barriers. Can the resonance hypothesis help us? It is very pity [sic] but it can't. [10-20]

Cockcroft, for his part, chided Gamow by taking advantage of the latter's unparalleled sense of humor.

I am most disappointed in you. I always believed it possible for a really good theoretical physicist to explain any experimental result and now you fail me at the first test. [10-21]

Clearly more work on the heavy elements was in order, as Cockcroft wrote Lawrence as well [10-22]. In October he put it more bluntly. 'We feel...that the Gamow theory [as verified for the Li(p, $\alpha$) reaction by Robert Oppenheimer on Lawrence's urging] is far from accurate for the heavier elements' [10-23]. For the next word on the matter, we turn to Merle Tuve in Washington.

## 10.2. Disintegrations in Washington as well

What about the three musketeers of the Carnegie Institution's DTM in all of this? Following Tuve and company's satisfactory demonstration test of Robert Van de Graaff's double, 2-foot electrostatic generator at DTM in September 1931, Tuve resolved to abandon the fickle and erratic Tesla coil in favor of a larger, practical electrostatic generator. In order to fairly test the prediction that the attainable voltage with such a generator was proportional to the diameter of the conducting sphere, a generator with a sphere of 2-meter diameter was drawn up by Odd Dahl. It was intended to erect it on the lawn behind the Experiment Building, since no suitable housing was available on the DTM campus for accommodating a generator of that size. The design was completed in December 1931 [10-24], and the generator was erected during the following months. The first tests were under way in May 1932, with a calibrated generating voltmeter and with results about as predicted [10-25].

Because of flying bugs, lint and other debris electrostatically attracted to the sphere, difficulty was encountered in obtaining a

> trustworthy measure of the peak voltage and voltage steadiness, but during short periods after cleaning the sphere... well in excess of 2000 kilovolts were indicated by the generating voltmeter. [10-26]

The maximum voltage was limited by heavy sparks down the 20-inch diameter Textolite cylinder which supported the sphere and protected the silk and rayon belt carrying the charging current from the elements. These lightning bolts 'splintered the redwood base parts in impressive fashion.' Because of the unstable conditions it was impractical to undertake proton bombardment of targets with this outdoor equipment; however, 1000 kV was applied without internal discharge or breakdown to a 34-section vacuum tube 9 feet long.

These tests corroborated 'the wantfullness of spending more time endeavoring to use for further serious work such an uncontrollable source as our Tesla-coil set-up,' declared Tuve [10-27], and his resolve was underscored in a special report, dated 27 May 1932: 'On Replacing the Tesla Coil by the Van de Graaff Electrostatic Generator in Air' [10-28]. However, in the absence of a suitable 'silo' for housing the 2-meter generator, Dahl, assisted by C F Brown, went ahead with the design and construction of a smaller 1-meter generator that would fit in their existing indoor laboratory space in the Experiment Building, once the oil tanks for the Tesla coil experiments were removed. The tanks were hauled away in October 1932, and the 1-meter generator was installed without delay.

The smaller generator consisted of two hemispherical aluminum shells, joined by a short cylindrical section containing a belt pulley, comb and spray wire for collecting the charging current, a.c. and d.c. generators, transformers, controls, a hollow-anode low-voltage ion source and the high-voltage end of a segmented, six-section discharge tube of Pyrex [10-29]. A 12-inch wide silk and rayon belt, running 4000 feet per minute, deposited 180 to 200 μA onto the outside of the terminal, and drove the ion source generators as well. The drive motor and the charging apparatus were placed directly overhead, some 12 feet above the concrete floor. A 12-inch diameter Textolite cylinder supported the generator 5.4 feet above the floor. The discharge tube, mounted at an angle with respect to the vertical, was 5.5 feet long. It was provided with focusing electrodes in the form of brass cylinders, across which the total potential drop was divided into many smaller potentials. They were fitted with toroidal copper rings on the outside of the tube. The focusing action of such a tube with coaxial cylindrical electrodes had, incidentally, been predicted by Ernest Lawrence more than a year earlier, on the basis of his work with David Sloan on the acceleration of mercury ions; however, Tuve and company had not been able to properly confirm the focusing with their fluctuating Tesla-coil voltages.

126

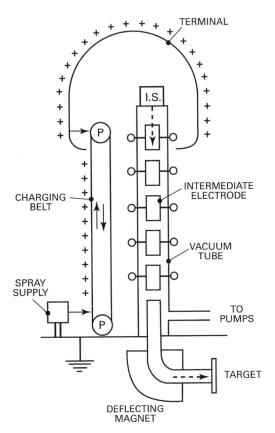

**Figure 10.3.** *Schematic diagram of a Van de Graaff generator. A high-voltage terminal is supported on an insulating column (not shown), and positive charge is conveyed to it by an insulating belt running between the lower and upper pulleys, P; the lower pulley is at earth potential and motor driven. Charge from a rectifier unit is sprayed onto the belt by means of corona points, and is removed from the belt at the top by corona discharge inside the terminal. (Sometimes additional spray points are provided in the terminal, maintained at a potential different from the first set, so that the belt leaves the terminal not merely discharged, but carrying charge of sign opposite to that on the rising belt, thereby doubling the current output of the machine.) Positive ions are produced in the ion source and are accelerated in the evacuated accelerating tube. The beam emerging from the tube is usually deflected by a magnet onto the target. Inside the vacuum tube are focusing electrodes that also give a uniform potential drop from section to section along the tube.*

The accelerating tube with its focusing electrodes is shown schematically in Figure 10.3, which indicates the principal features of a Van de Graaff generator.

Under most weather conditions, the 1-meter terminal held approximately 500 kV, constant to perhaps 3%; the voltage varied between 400 kV during high humidity and a maximum of 600 kV in dry weather. (Later work would show that the useful maximum voltage was closer to 400 kV.) When the humidity was high, as it often was in the District of Columbia, the attainable voltage was limited by heavy sparks up the belt or down the Textolite column.

> Although no parts were outgassed [the] tube withstood full voltage ... with target currents of 0.1 microampere within one-half hour of the first application of voltage immediately after being pumped down. Target currents up to 10 microamperes were obtainable at full voltage within several hours. [10-30]

The target arrangement was similar to that used by Cockcroft and Walton, and is shown in Figure 10.4. Soft X-rays from the tube with 10 μA on the

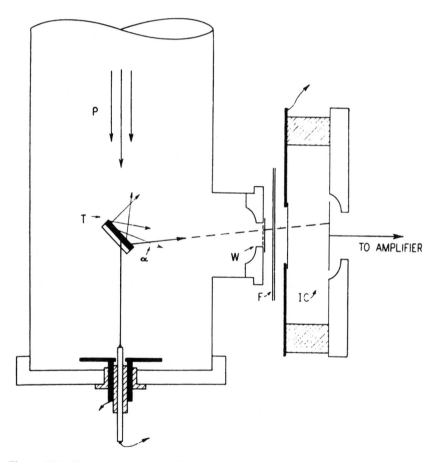

**Figure 10.4.** *Target arrangement with the one-meter generator, as used for the study of the disintegration of lithium and boron. P is the incident proton beam, W is a mica window, and F are absorbing foils used to measure the range of the α-particles produced by the disintegrating nuclei. IC is an ionization chamber. Tuve, 1933, JFI, 216, pp. 1–38, on p. 32.*

target required operation of the generator by remote control from a small observer's hut outside the concrete wall of the laboratory.

The low voltage, 500 kV or so, was really not so worrisome, reflected Tuve in planning for the even larger generator, in view of 'Cockcroft and Walton's success with 120- to 400-kilovolt [having shown nuclear disintegrations] to be possible at these lower voltages.'

> This important verification of the wave-mechanics prediction that the penetration of the potential wall of the nucleus does not become impossible as the energy of the bombarding particle is reduced below a certain limit, but only that it becomes rapidly (exponentially) less probable, in a sense yields the alternative of more current and less voltage for higher voltages and less current. It is unknown to what extent this 'exchange' is valid, however, and the importance of the investigations in the region 1,500–2,000 kilovolts for which we have laid our plans and made our preparations (since even on a classical basis the nucleus can be attacked at these voltages) is by no means minimized by the astonishing and valuable results at lower voltages. [10-31]

That August Tuve took time off to attend a symposium on theoretical physics at Ann Arbor, sponsored by the University of Michigan. (The Michigan Physics Department had held these conferences every summer since 1923.) Breit was scheduled to deliver a series of lectures at the 1932 symposium, in place of Gamow, who was unable to attend. Breit suggested to Tuve that he attend as well, mainly to hear Heisenberg, who was also to give a series of lectures at Ann Arbor. Heisenberg had recently developed a theory of nuclear structure in which the neutron and proton replaced the electron and proton as the basic nuclear constituents. 'Heisenberg may have some new experiments to suggest,' wrote Breit, 'and some may occur to you when you hear about the theory' [10-32]. Once there, Tuve listened attentively to Heisenberg, who emphasized the importance of the DTM physicists to proceed with their nuclear studies using voltages already available, rather than pressing forward towards ever-higher bombarding voltages [10-33].

The first piece of physics research with the one-meter generator, shortly after the New Year of 1933, was obvious enough: confirming Cockcroft and Walton's disintegration of lithium, as well as Chadwick's neutron from beryllium (not with the generator but with a polonium $\alpha$-source). And so they did, with the results deemed sufficiently important that they first appeared in print under the name of John Fleming, the DTM Acting Director. A fuller account of the preliminary experiments, as well as their earlier work, was given by Tuve in a lecture on 2 February before the Franklin Institute on 'The Atomic Nucleus and High Voltages' [10-34]. In that lecture Tuve also discussed atomic theory in terms of wave mechanics, including the nuclear potential barrier and energy levels, Heinz Pose's first observation in Halle of resonance disintegration, and subsequent work on that subject by others, including Hafstad with his FP-54. (Replacing the standard Hoffmann duant electrometer, the FP-54 was a specially designed triode by General Electric, and developed by Hafstad to record currents from ionization chambers reliably and automatically for observing cosmic rays.)

There was, however, considerable pressure, especially from Gregory Breit in New York, for them to move on into new territory; in particular, to pursue disintegration attempts on heavier elements, for which the English results were less clear-cut.

> I agree with you about the relative importance of Li and neutrons from Be [wrote Breit to Tuve in January]. Am not sure that it is wise to be leaving heavy elements alone. Oppenheimer disposes of these results of Cockcroft and Walton by blaming it on impurities. I am not so sure only some impurities will do and these could be tested for say spectroscopically. If Cockcroft and Walton are right about the heavy elements the ordinary disintegration there needs considerable revision. [10-35].

Tuve, Hafstad, and Dahl chose targets of Al, Ni, and Ag as representative of the heavier elements studied at Cambridge. In all three cases, the DTM measurements indicated far fewer $\alpha$-particle counts per microampere per minute in a linear pulse-amplifier, with 600 kV hydrogen ions, than the number of scintillation counts observed at Cambridge at 300 kV. However, the *range* of what few particles they observed agreed, within experimental errors, with that of the particles observed from a boron target in the same apparatus. Ergo,

> ...the observed yields from Al, Ni and Ag can thus be explained by boron impurities of amounts 1/33,000, 1/9000 and 1/100,000 in these targets. Contamination to this extent by boron is certain to be expected unless severe precautions are taken. [10-36]

The reason why Tuve considered boron a likely contaminant, it might be added, was the widespread use of borax soap in cleaning operations.

Tuve also read their preliminary report before a special session on nuclear disintegration of elements at the joint meeting of the APS and of the AAAS in Chicago in June, in connection with the Century of Progress Exposition noted earlier. The special session, on 23 June 1933, was chaired by none other than Niels Bohr, who, along with Frances W Aston, Enrico Fermi, and a few other notables, were invited for this historic occasion. Unfortunately, Bohr, in his opening address, rambled unintelligibly and inaudibly for about an hour, squashing any chance for a useful discussion, as John Cockcroft complained to his wife

[10-37]. Added *Time*, 'the blare of the loudspeaker as Bohr fiddled with the microphone was almost a relief' [10-38]. Cockcroft, too, was an invited speaker, and Lawrence, naturally present as well, was impressed by his talk. *Time*, on the other hand, was mainly impressed with Lawrence, who reported on disintegrating everything in sight with his cyclotron. In contrast to Bohr, wrote the reporter for the newsmagazine, 'it was much easier, and more pleasant, to understand round-faced young Professor...Lawrence...tell how he transmuted elements with deuton bullets' [10-39]. Cockcroft, for his part, thought Tuve's paper was 'not at all convincing and gave the impression of great enthusiasm but too much hurry' [10-40].

Apropos 'deuton bullets,' the Chicago meeting was also the setting of the aforesaid session on heavy hydrogen, where names for the hydrogen isotope and its nucleus were hotly debated. Lawrence, too, was in the thick of that discussion, as of course was Harold Urey—someone equally opinionated. Wrote Lawrence to Gilbert Lewis after the lively session:

> I suppose you had eye-witness accounts of Urey's temper as displayed at Chicago—stimulated by 'deuton' [at the time Lawrence and Lewis's preferred term for the $H^2$ nucleus]. He was certainly 'sore' and I believe the general impression was that he made an ass of himself. [10-41]

From Chicago Cockcroft went to Washington, where he called on Tuve, Hafstad, and Dahl at DTM. There he was more impressed by Dahl's characteristic Norwegian use of linen fishing line for remote control of the ion source situated at the top of the high-voltage discharge tube. By that time, however, Cockcroft was also in full agreement with Tuve's conclusion about boron impurities, as he had suspected when he wrote Lawrence earlier in the year.

> We are beginning to suspect that the alpha-particles we observe from the heavier elements may very likely have been due to boron impurities, since we find that in the process of outgassing a tube it is possible to contaminate a target and get results which are quite absent with a perfectly clean target. [10-42]

Among those attending the Chicago disintegration symposium had been Charlie Lauritsen of Caltech's Kellogg Laboratory. He, too, would eventually turn to Van de Graaff's electrostatic generator for quantitative nuclear physics research. However, in the spring of 1933, while Tuve and company were reaping the early rewards of abandoning their Tesla coils in favor of the generator, Lauritsen was still not convinced of the unsuitability of the Tesla coil for serious work.

*Lauritsen to Tuve, 1 March 1933:*

I have often wondered if you got completely disgusted with the Tesla coil. If so, I wish you would tell me the reason. We have been playing a little with an arrangement as shown [in the attachment], and it looks to me like it might be pretty good. [10-43]

*Tuve to Lauritsen, 30 March 1933:*

We were not exactly disgusted with the Tesla coil but became convinced that for any quantitative results it was an extremely difficult device to work with.... Our objection to the Tesla coil was primarily the 'on time' of only a few microseconds per primary spark during which the really high voltage is available, and the fluctuating value of that voltage.

I do not wish to discourage you in any ideas you may have regarding better Tesla-coil arrangements than we had but for the purpose we had in mind, namely scattering experiments and dis-integration experiments, a more continuous voltage source became practically imperative. Continuous excitation of a Tesla coil runs into money for tubes and equipment, but the Van de Graaff generator appears to be a very satisfactory device considering its extremely low cost. [10-44]

Midsummer 1933 was as hot and muggy as Washington tends to be at that time of year. Tuve and his teammates suffered as much as their generators. While the trio appear in photographs neatly attired in suit and tie, in point of fact they often toiled at odd hours of the day or night clad in pajamas (there being no air conditioning or even humidity control to speak of in their laboratory), and the Carnegie administration didn't mind if they reported for work at unconventional hours. Tuve bore the brunt of their casual life style, being referred to as 'the late Dr. Tuve.'

In addition to following up on the $(p, \alpha)$ reactions investigated in Cambridge, the strong, steady and homogenous beam from the Van de Graaff generator would prove excellent for detecting narrow resonances by 'resonance excitation' in a multitude of light and medium weight nuclei, and in so doing establish the basis for a system of isotopic masses comparable in accuracy to those from mass spectrometers at Cambridge and elsewhere.

Meanwhile, as the one-meter generator was put through its early paces in DTM's Experiment Building, the Executive Committee of CIW's Board of Trustees met in New York to consider a 'Memorandum on the Emergency Necessity for Immediate Housing of the Two-Meter High-Voltage Generator,' dated 9 January 1933.

The outstanding opportunity for the Department of Terrestrial Magnetism to make a highly significant and original contribution to nuclear physics by its high-voltage technique...is in immediate danger of being lost unless housing can be obtained in the next two or three months for the 2-meter Van de Graaff generator built at the Department nearly a year ago. The significance of the results which can be definitely predicted, the completeness with which we are prepared to meet the opportunity, and the necessity for immediate action if the Institution is to obtain due recognition for its contributions in the high-voltage field are briefly summarized in the following outline. [10-45]

The Memorandum then lists a strong source of neutrons, and disintegration and scattering experiments above 1000 kV, as essential ingredients for a significant program in fundamental physics. Apparently the Trustees were sufficiently impressed, since they approved a special allotment of $4500 for a building to house the generator [10-46]. 'We will be...much slowed up by the construction [wrote Tuve in his letter to Breit], Dahl especially, since it is to be done by the Department and not by an outside contractor.' And so it fell on Dahl to design the new high-voltage building, actually a large extension to the Experiment Building, as well as a permanent mounting for the big generator within it. The building was completed by August 1933, and installation of the generator began the same month [10-47].

The high-tension hall had a floor space measuring 32 by 36 feet, or not so different from Cockcroft and Walton's stripped Balfour Library and larger than Room 114 in Lawrence's reworked barracks building; however, it bore no comparison with Lauritsen's X-ray hall. Tuve's hall had a slanted ceiling with a maximum height of 28 feet above the floor and minimum clearance from the generator shell to the nearest ceiling rafter of about 8 feet. This time, to gain the many advantages of a vertically oriented high-voltage vacuum tube, a tripod support was adopted for the generator terminal, with an extension of the tube passing through the floor into a target room in the basement (Figure 10.5).

The generator itself featured two concentric aluminum shells: an inner one-meter shell and an outer two-meter shell resting on the tripod legs (Figures 10.6 and 10.7). The ion source was placed within the one-meter shell, with two sections of the acceleration tube bridging the gap between the shells. The inner shell served primarily as a potential divider. That is, the two shells in unison gave a total voltage for accelerating the ion beam which exceeded the maximum voltage of the two-meter shell by whatever additional voltage was obtainable between the outer and inner shells [10-48]. In the event, with the acceleration tube installed, the generator

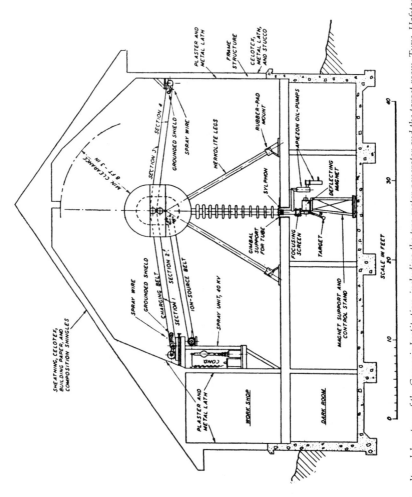

**Figure 10.5.** *The high-voltage laboratory of the Carnegie Institution, including the two-meter generator and the target room. Tuve, Hafstad, and Dahl, 1935, PR, 48, pp. 315–337, on p. 323.*

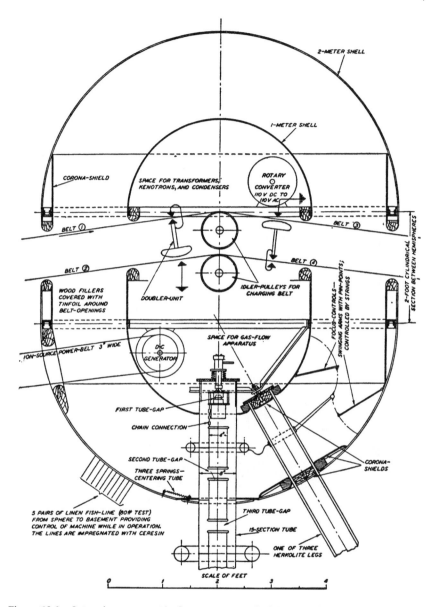

**Figure 10.6.** *Internal arrangement in the two-meter terminal. Tuve, Hafstad, and Dahl, 1935, PR, 48, pp. 315–337, on p. 323.*

held a maximum voltage of about 1200 kV. As opposed to the earlier one-meter generator, the new machine had a separate belt driving the ion source generators. Ions accelerated down the multi-section tube passed through the floor into the target room below. There the beam, a mixture

**Figure 10.7.** *The completed two-meter generator, 18 May 1935. Dahl is on the ladder; on the floor (left to right), Meyer, Hafstad, and Tuve. DTM Archives 3662; Tuve, Hafstad, and Dahl, PR,* **48** *(1935), pp. 315–337, on p. 322.*

of protons, hydrogen molecular ions, and deuterons, was deflected by a magnet into a target chamber where the ion species struck different spots on the target, depending on their charge-to-mass ratio. Thus, the 'mass-1 spot' represented protons alone; the mass-2 spot registered the combined impact of deuterium nuclei and $H_2^+$ molecular ions, and so forth. The pumping system used a pair of Apiezon oil diffusion pumps backed by a mechanical Hypervac. The design of the diffusion pumps

136

was thanks to Ernest Lawrence, in whose laboratory such pumps were designed and served the many accelerator projects under way.

Lawrence would play a larger, albeit indirect, role in the first use of the 2-meter generator. It would prove its worth in verifying additional anomalous results—not at Cambridge but in Berkeley.

# Chapter 11

# DEUTERIUM

## 11.1. Deutons on target

Gilbert Lewis's heavy-water still was in full swing by March of 1933, with a cubic centimeter of the precious liquid soon on hand. Rendered into gaseous form, it was more than enough for Ernest Lawrence to feed the ion source of the 27-inch cyclotron across the street. 'Heavy protons,' Lawrence expected, should prove better atom-smashing projectiles than ordinary protons, and Lewis agreed; early on he had his own notion about the structure of atoms, at considerable variance with the opinions of contemporary physicists. However, Lewis was equally serious about determining the effects of heavy water on living organisms, and fed his entire stock in doses to a mouse; it survived, though showed 'marked signs of intoxication' [11-1]. Commented Lawrence, rather annoyed, 'this was the most expensive cocktail that I think mouse or man ever had' [11-2]. However, Lewis brewed another batch in his electrolyzer, and on 20 March Lawrence accelerated deuterons (to use Urey's term) for the first time in the big cyclotron. Bombarding lithium fluoride, he and Livingston found the yield of $\alpha$-particles was ten times as great as observed with protons on the target. Moreover, as he wrote Merle Tuve on 3 May,

> There are two ranges, 8.3 cm and 14.8 cm. About one-third of the particles have the longer range. These correspond to nearly 13 million volts and are the most energetic alpha particles so far obtained either from radio-active sources or artificially. Their range checks with the assumption that lithium 7 splits into two alpha particles and a neutron. The heavier proton disintegrates abundantly nitrogen, magnesium, and aluminum, also. The nitrogen particles have a single range of 6.3 cm, magnesium 6.8 cm, and aluminum around 6 cm. The disintegration yield for the substances is comparable to that of lithium when using ordinary

protons.... You will be glad to know that these disintegration processes do not require particularly high voltage. 600 kV is enough for a good yield. We are working with 1,300,000 volts $H^2$ protons. [11-3]

Writing to Cockcroft the very next day, Lawrence was less sanguine about low voltages sufficing with his 'Hemi-alpha' particles on medium-weight targets of aluminum and magnesium: reducing the ion voltage a mere 100 kV below 1300 kV reduced the $\alpha$-particle yield by about a factor of ten. And while the swifter $\alpha$-particles appeared to come from $Li^7$ according to

$$_3Li^7 + {}_1H^2 \longrightarrow 2\,{}_2He^4 + {}_0n^1,$$

the slower $\alpha$-particles seemingly came from the breakup of $Li^6$:

$$_3Li^6 + {}_1H^2 \longrightarrow 2\,{}_2He^4.$$

At any rate, 'this interpretation [was] in excellent agreement with the mass ratio from band spectra of $Li^7$ and $Li^6$ recently determined in the laboratory....' [11-4].

A news release went out in May, claiming lithium, beryllium, boron, nitrogen, fluorine, aluminum, and sodium all underwent disintegrations under deuterium bombardment. However, in their first letter to the editor of the *Physical Review*, thoughtfully with Gilbert Lewis as the first author and dated 10 June 1933, results are less clear-cut, except for lithium, beryllium, and nitrogen. It was even less obvious which isotope of lithium gave rise to the fast $\alpha$-particles [11-5].

A second letter to the editor from Lawrence and his companions, also dated 10 June, deals with the emission of *protons* from targets bombarded with deuterons. Its conclusions would prove the source of a major *faux pas*, on Lawrence's behalf, having to do with the neutron and the deuteron. We postpone dealing with this celebrated controversy, and how it was resolved, until the next section.

Lawrence was not the only one to benefit from Lewis's largesse. So was Rutherford, among several others. Rutherford, the conceptual father of deuterium, had been equally keen on procuring heavy water for accelerating deuterons in low-voltage transmutation experiments in the company of Mark Oliphant with equipment of their own. Only three days after Lawrence first accelerated deuterons, Rutherford had spoken on 'The New Hydrogen' before the Royal Institution in London. 'If we assume [he argued], as seems not unlikely, that the D-nucleus consists of a close combination of a proton and a neutron,' it should produce some very interesting results when used in accelerators either as projectile or target [11-6]. And, of course, the potential results loomed interesting, indeed, in light of Cockcroft and Walton's ongoing experiments.

Rutherford had not only followed Cockcroft and Walton's work with great interest, but had been sufficiently impressed to have a go at it himself. In keeping with his penchant for simple, inexpensive equipment, and his curiously strong faith in George Gamow, he opted for a maximum bombarding energy far *lower*, not greater, than Cockcroft and Walton's modest 800 kV. A mere 200 kV would do, he reckoned, providing the ion *current* was well above Cockcroft and Walton's microampere or so. And he had just the man for producing high-current beams: Oliphant had shown himself very adept at designing intense positive ion sources, and would be an excellent partner in any case. Persuading Oliphant to join him in what would prove to be Rutherford's last experimental venture, he had Cockcroft contact Allibone at Metro-Vick about procuring a suitable transformer-rectifier set. Allibone obliged with a 100 kV transformer and a simple voltage-doubling circuit. Because of the low ceiling in the room where the equipment was to be deployed, it was necessary to use a horizontal accelerating tube. When Rutherford grumbled about all of £100 for the apparatus, Allibone replied that it was only a farthing per volt, whereas a flash-lamp battery cost two or three pence per volt! [11-7]. Rutherford saw the humor and his point, and accepted the Metro-Vick invoice without further complaint.

By February 1933, Rutherford and Oliphant's simple accelerator, shown in Figure 11.1, was up and running at 200 kV with a beam current of 100 µA or more of protons, thanks in part to the helpful advice of Cockcroft at every stage as well as the patient assistance of George Crowe, Rutherford's laboratory steward. Wasting no time, they found that it took protons of no more than 100 kV to disintegrate lithium. They also soon found a lithium threshold at 20 kV, and a boron threshold at 60 kV. For elements heavier than boron, even 200 kV protons gave no sign of disintegrations, except for a trace with fluorine [11-8]. Rutherford's own participation in the experiments was mostly limited to discussions about what to do next, to interpreting the results, and keeping Oliphant and B B Kinsey on their toes [11-9]. By that time, well into his sixties, Rutherford's hands shook rather badly. Every now and then, when he did lend a hand in running the apparatus, he might puncture Crowe's thin mica absorbers with a pencil stump, with air rushing into the tube with a loud report.

In May Gilbert Lewis shipped to Cambridge a vial of nearly pure heavy water, as we learn from his correspondence with Rutherford.

*Lewis to Rutherford, 15 May 1933:*
While Fowler was here — and we enjoyed his visit enormously — I promised him that I would send you some of our concentrated hydrogen isotope in case some one in the Cavendish Laboratory would like to experiment on the new projectile. I have been

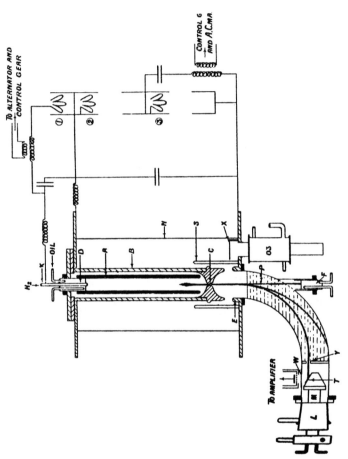

**Figure 11.1.** *Oliphant and Rutherford's accelerator. C is a perforated steel cathode, screwed tightly into the steel tube, A. Hydrogen is fed in through a small tube K, and the whole discharge tube is mounted in a tall glass cylinder N. The anode is a second steel tube B. The proton beam ('canal ray') emerging from the perforated cathode is accelerated by up to 200 kV applied between C and E. The beam can be collected in the Faraday cylinder F, or deflected by the magnetic field between the shaped pole pieces P. A slit Y defines the deflected beam, which strikes the steel target T, shaped like a truncated pyramid. The metal to be bombarded is pressed or hammered into recesses in each face of the truncated target. Above the bombarded face of the target is a mica window W, and above it sits the chamber of an ionization counter connected to a Wynn-Williams amplifier. Oliphant and Rutherford, 1933, PRS,* **A141**, *pp. 259–281, on p. 261.*

working with Lawrence and Livingston and we already have found that it is going to be a remarkable aid in the study of the nucleus. With the new particle, which we have decided, with some compunction, to call the deuton, we have disintegrated lithium ($Li^6$?), ...

Altogether we have used about the same quantity of the heavy water which I am sending to you. Ours contain fifty per cent $H^2$, while the sample that I am sending you contains ninety three per cent. We are now succeeding in making small quantities of pure isotope and later on, if you desire it, we could send you a sample.... [11-10]

*Rutherford to Lewis, 30 May 1933:*
I was delighted to receive your concentrated sample of the new hydrogen isotope in good shape, and we shall certainly take an early opportunity of examining its effects in our low voltage apparatus which Dr. Oliphant and I have been using the past year. [11-11]

And so, in June Oliphant and Rutherford, assisted by B B Kinsey, whom Oliphant regarded in voice and manner as straight from Wodehouse [11-12], fired up the apparatus once more. About 20 cc of hydrogen gas, prepared from Lewis's sample, was mixed with purified helium; as a result, 'the magnetic bending of the accelerated beam was able to affect a complete separation of the $(H^2)^+$ ions from ions of ordinary hydrogen, and the effects of the heavy isotope as a disintegrating agent [could] be studied in the absence of any extraneous effects' [11-13]. In addition to $\alpha$-particles of 8.4-cm range, ascribed to lithium under proton bombardment, their measured absorption curve for $H^2$ ions on the lithium target at 150 kV revealed a highly energetic group of $\alpha$-particles with a mean range of 13 cm — presumably in qualitative agreement with the 14.8-cm particles reported by Lawrence and company and attributed to the $Li^7(d,n)2\alpha$ reaction [11-14]. Wrote Rutherford to Lewis,

You will be interested to hear that Dr. Oliphant and myself have made a few experiments with the 'diplogen'! you so kindly sent us and, as far as we have gone, have in general confirmed your results with lithium. We are working under 200,000 volts and get plenty of effect even when we dilute the diplogen with 99% of helium.

I was very interested to hear, through Lawrence's letter to Cockcroft, of the extraordinary interesting results he has obtained with diplon bombardment. Please convey to him my warmest congratulations on such a fine reward for his labour in developing his accelerated [sic] system. The whole subject is opening up in great style. [11-15]

Oliphant and Rutherford's report to the Royal Society, submitted on 1 August, was accompanied by one from Philip Dee and Ernest Walton reporting similar observations in a Wilson cloud chamber (instead of Oliphant's purely electronic counting system), allowing them to 'see the events,' with the original Cockcroft–Walton accelerator at 400 kV [11-16]. They used a new discharge tube similar to Oliphant and Rutherford's, giving ten times greater positive ion current on the target, and fed the ion source gas extracted from some of Rutherford's precious heavy water. Dee and Walton claimed definite confirmation of the two principal modes of disintegration,

$$_3\mathrm{Li}^7 + {}_1\mathrm{H}^1 \longrightarrow {}_2\mathrm{He}^4 + {}_2\mathrm{He}^4$$

and

$$_3\mathrm{Li}^6 + {}_1\mathrm{H}^2 \longrightarrow {}_2\mathrm{He}^4 + {}_2\mathrm{He}^4.$$

Cockcroft did not partake in Dee and Walton's experiment, as he was still in the US, by then watching Livingston blasting lithium and other elements with their own $\mathrm{H}^2$ projectiles in Berkeley. (Lawrence was still back East, but had left strict instructions with everybody to take good care of their important visitor.) While being shown how to operate the cyclotron, Cockcroft received an urgent telegram from Oliphant, asking that he bring back a fresh batch of heavy water. It seems that what was left of the original material sent by Lewis to Rutherford had been contaminated by accident.

> We have had an unfortunate accident with the last half of the sample of water you so kindly sent me [wrote Rutherford to Lewis]. The tube for conversion of the water into hydrogen cracked at the critical moment and I am afraid what little we have managed to save is badly contaminated with water vapour from the atmosphere. [11-17]

Unfortunately, a drought in the Berkeley area prevented speedy replacement of fresh material, though Lewis did his best by mailing off some older stock—not quite as pure as the original batch, but good enough for Rutherford's use [11-18]. Providentially, by that time a Viennese physical chemist, Paul Harteck, had arrived in Cambridge for a one-year postdoctoral Rockefeller Fellowship under Rutherford to learn a bit about the exciting new field of nuclear physics and artificial disintegrating with fast protons with Oliphant and the Old Man himself. So he would; however, his first contribution was not as a novice physicist but as the good physical chemist he was.

'Fortunately for me and the Cavendish,' recalled Harteck of this memorable year in his early career, 'within a week of my arrival at the laboratory, there arose a need for heavy water in many experiments—just the right task for a physical chemist' [11-19]. Harteck set about implementing their meager heavy-water stock with a series of small electrolytic cells, based on Lewis's procedure and with his helpful advice [11-20]. As feed water for his electrolyzer he used two gallons of water with 2% deuterium concentration, obtained by Cockcroft from Lewis and hand carried by him back to Cambridge. Only with great difficulty did he get it past the British customs officials, who could not understand why a liquid resembling water should be brought into the UK. On top of that, Rutherford scolded him for having spent $10 on the heavy water without authorization.

## 11.2.  A controversy erupts

We left Lawrence, Livingston, and Gilbert Lewis firing deuterons at a multitude of target nuclei in June 1933. Alpha-particles shot out from the targets in great profusion, and so did protons, but with an important difference. Whereas both slow and fast $\alpha$-particles were ejected from the targets in varying amounts, every target gave off protons under deuteron bombardment, and all with practically the same range, 18 cm, irrespective of the target composition. How could every target, from easily smashed lithium to gold, surely immune to nuclear disintegration, yield protons of virtually the same range? 'I am almost bewildered by the results,' confessed Lawrence to Cockcroft in early June [11-21]. Only one explanation made sense: the deuteron itself was breaking up, presumably into a proton and a neutron [11-22].

The problem with this 'hypothesis of the instability of the deuton' was that it implied a lower value for the mass of the neutron than Chadwick's value. Chadwick's mass was based on the transformation of boron according to the $B^{11}(\alpha, n)N^{14}$ reaction, with no accompanying $\gamma$-radiation. From the known masses of $B^{11}$ and $N^{14}$, and the known kinetic energy of $\alpha$-particles from polonium, the mass of the neutron came out as 1.0067 mass units (on the scale of $O = 16$), with a probable error of 0.001 [11-23]. This value, as Chadwick explained in his Bakerian Lecture to the Royal Society in May 1933 [11-24], was evidence for the neutron as a *composite* particle, a proton plus an electron, because it was less than the sum of the masses of the proton and the electron. The *mass defect* equaled the *binding energy* of the composite nucleus formed, explained by Einstein's equation $E = mc^2$ and was equal to the energy needed in pulling the two particles apart again.

Be that as it may, and Chadwick remained rather open on the question of the neutron being a composite or elementary particle, Lawrence's argument in June went as follows. Assume, for simplicity, that a deuteron of energy 1.33 MeV strikes a gold nucleus, presumably too massive to recoil significantly under the impact. The deuteron splits into a proton of 18-cm range and a neutron of unknown energy; Lawrence and Livingston had weak evidence that the two constituents flew apart with equal and opposite momentum. Since an 18-cm proton has a kinetic energy of 3.6 MeV, the energy released in the explosion was from $2 \times 3.6$ MeV to 1.33 MeV, which equals 0.0063 mass units. This value, with the mass of the proton and neutron, must equal the mass of the deuteron, determined to be 2.0136 by Kenneth T Bainbridge from his analysis of a sample of heavy water supplied by Gilbert Lewis [11-25]. Hence, $m_n \approx 1.000$. A subsequent, somewhat improved calculation based on new experiments by Malcolm Henderson and Livingston [11-26], raised $m_n$ slightly, to 1.0006. As for the underlying physical picture of the deuteron breakup, Lawrence viewed it as the disruption of the deuteron in the Coulomb field of the atomic nucleus, with the charged proton flung out by the electrostatic repulsion.

The light neutron was the centerpiece of a symposium at Caltech in May of 1933, in honor of Niels Bohr, who was on his way to the Century of Progress Exposition in Chicago that June. Bohr welcomed Lawrence's news as a 'marvelous advancement' and Robert Millikan, self-centered to a tee, chimed in with his own congratulations. Following the symposium, Professor and Mrs. Bohr were houseguests of the Lawrences in Berkeley, while feted at the Rad Lab and by the UC-Berkeley community. Bohr remarked to Herbert Evans, the prominent Berkeley anatomist, 'I've seen and heard the first person I could compare with Rutherford' [11-27].

In Cambridge, however, Oliphant and Rutherford were unable to excite 18-cm protons in their low-energy accelerator with deuterons, and Walton and Dee saw only a few 18-cm Wilson tracks at best.

> I noticed Lawrence's views about the nature of the tracks [wrote Rutherford to Lewis in late June], but we are at the moment not inclined to view with favor the conversion of a deuton into a neutron of mass about 1. However, it is too early to take definite views. [11-28]

Only in October did Lewis reply to Rutherford's skepticism, still firmly confident in Lawrence's light neutron.

> Experiments made during the last two days in the Radiation laboratory, and which Lawrence will report at the Solvay Congress, show that an enormous number of neutrons are produced when deutons strike a great variety of targets. These experiments strongly confirm, although they do not demonstrate, our theory of the instability of the deuton. [11-29]

The Solvay Congress of 1933 was shaping up as the ultimate forum for Lawrence's big news. Ranking as the most prestigious of physics gatherings, the Solvay Conferences were named after the Belgian industrial chemist Ernest Solvay, who had made a fortune on a process for manufacturing sodium carbonate. Solvay used a portion of his fortune in support of philanthropic causes. He also happened to be an amateur dabbler in physics. In 1910 he had encountered the German physical chemist Walther Nernst, who got him fired up over the radical new quantum concepts of Max Planck and Albert Einstein. The upshot was that Nernst and Solvay organized an international conference to 'review current questions in connection with the kinetic theory of matter and the quantum theory of radiation.' It was held at the Hotel Métropole in Brussels during October–November of 1911, and attended by Planck, Einstein, Rutherford, and a score of other leading lights of physics. Thus was born an important tradition within the physics community, with the Conference held every few years, always in Brussels and each time with a different underlying theme.

The seventh Solvay Conference was scheduled to convene in Brussels during 22–29 October 1933, with the general theme the 'structure and properties of atomic nuclei.' John Cockcroft, one of eight attendees from Cambridge, would be responsible for reviewing particle accelerators at the conference, and James Chadwick for the state of knowledge concerning the neutron. Lawrence, too, received an invitation at the eleventh hour — a singular honor, since he was the only American participant that year, having been invited on Bohr's suggestion. Reports were to be read by Rutherford, Cockcroft, Chadwick, Bohr, Heisenberg, Gamow, and the Joliot-Curies. Prior to the conference, Lawrence wrote Paul Langevin of the Collège de France, President of the Scientific Committee, about extensive responses he intended to make concerning two points raised in the pre-conference release of Cockcroft's paper, and perhaps comments on the papers by Chadwick, Joliot, and Gamow [11-30]. (The Congress would be the first international meeting he had ever addressed.)

At the conference itself, Cockcroft's report, 'Disintegration of Elements by Accelerated Protons,' was first on the agenda. He began with Rutherford and Chadwick's scattering experiments as a background to his own project with Walton, and ended with a somewhat bland review of the Berkeley program. In it, he drew attention to Lawrence's $\alpha$-particle yields from lithium, but merely noted their 18-cm proton data without further comment. As for the cyclotron, he stated that its weak current, one-thousandth the flux from the high-potential Cambridge approach, might well render moot the advantage of higher bombarding energy. Lawrence's rebuttal of Cockcroft's claim that only small currents are possible from the cyclotron, and his irritation at Cockcroft having quite ignored his deuteron-breakup hypothesis, is underscored by marginal notes scribbled on his own copy of Cockcroft's report [11-31]. Lawrence was also emphatic in sharing credit with his colleagues, Henderson, White, Sloan, Lewis, and Livingston where Cockcroft's paper mentioned only Lawrence.

Chadwick's report came next, covering 'Anomalous Scattering of $\alpha$-Particles; Transmutation of Elements by $\alpha$-Particles; The Neutron.' Among other things, he expressed doubts over the dissociation of the deuteron on energetic grounds. From Lawrence's subsequent objections to Chadwick's remarks it appears that Lawrence may not have fully appreciated Chadwick's calculation of the neutron mass [11-32]. In arguing for his own calculations, giving a neutron mass close to unity, Lawrence came under fire from Werner Heisenberg, who countered that if the deuteron disintegrated in the Coulomb field of the atomic nucleus, the proton yield should be less from heavier targets, since the deuteron penetration should decrease with increasing atomic number; for sufficiently high Z, one ought to see no disintegrations at all. If that were so, injected Bohr, one might expect the deuteron to disintegrate after entering a nucleus, in which case the speed of the ejected protons should increase with increasing Z. Alas, neither suggestion was borne out in Lawrence's results [11-33].

Chadwick was followed by the Joliot-Curies, who reported on the 'Penetrating Radiation from Atoms under the Action of $\alpha$-Rays,' and Lawrence followed their presentation with interest. As we shall see, the positive electron, or positron, occupied an important role in their report. Yet, curiously, the discovery of the positron by Carl Anderson of Pasadena, another milestone in the 1932 *annus mirablis* of particle physics, was related by P M S Blackett almost as an afterthought in the discussion following the French report. (Anderson himself was not among the invited participants, nor was his prominent mentor, Robert Millikan.)

The Joliot-Curies marshaled the positron in support of a neutron even heavier than Chadwick's. Observing positrons as well as neutrons emitted under $\alpha$-bombardment of boron and other light elements, they tacitly assumed that both decay particles were emitted simultaneously, according to the reaction $B^{10}(\alpha, ne^+)C^{13}$. If so, they argued, instead of Chadwick's composite neutron, $n = p + e^-$, it is the proton that is a composite of $n + e^+$. Moreover, calculations based on the disintegration of $B^{10}$ gave $m_n = 1.012$. In fact, it would turn out that $B^{10}(\alpha, n)$ and $B^{11}(\alpha, n)$ are competing reactions. What the Joliot-Curies had actually observed was *artificial* or *induced* radioactivity. The neutron and positron were *not* emitted simultaneously; the positron was slightly delayed. In this particular nuclear reaction involving $B^{10} + \alpha$, the reaction proceeded via the formation of a hitherto unknown nucleus, $N^{13}$, according to $B^{10}(\alpha, n)N^{13}$, followed by the radioactive decay of $N^{13}$, with a half-period of about 10 minutes, according to $N^{13} \longrightarrow e^+ + C^{13}$ [11-34].

For our purposes, it suffices to note that Frédéric Joliot's report of neutrons from $B^{10}$ was strongly criticized from the audience, by, among others, Lise Meitner of the Kaiser Wilhelm Institut in Berlin; she stated flatly that she had been unable to spot a single neutron in a similar experiment. After the session, however, Bohr took Frédéric and Irène aside, assuring them that he found their results very important. Not knowing what to think, Joliot and his wife returned to the Radium Institute in Paris, determined to re-check their observations and settle the matter. They were right, and Meitner wrong, as it turned out. Two months of exhaustive work culminated in January of 1934, with the announcement by the Joliot-Curies of a new type of radioactivity involving delayed positron emission [11-35].

Lawrence, though challenging the Joliot-Curies on the mass of the neutron, actually got along well with Joliot at Brussels; Joliot was eager to learn more about the work at Berkeley, and to have a cyclotron installed in Paris. (Construction of a cyclotron, in a sub-basement under the Collège de France, began in 1936, though the first beam was only circulated in 1939.) Equally important, Lawrence met Rutherford for the first time at the Solvay conference, and Rutherford was favorably impressed with the young man, despite his misgivings over some of Lawrence's purported results. After the lively session following Cockcroft's report, Rutherford nudged Chadwick. 'He's just as I was at his age!' [11-36]. Back in Cambridge, Rutherford voiced similar sentiments in a letter to Lewis.

> I have just got back from Brussels where I had the pleasure of meeting Lawrence and hearing all about his latest results. He is a broth of a boy, and has the enthusiasm which I remember from my own youth. We are expecting him at the Laboratory tomorrow. [11-37]

P.S., added Rutherford, Lewis's latest shipment of heavy water arrived safely. Fortunately, Paul Harteck had done better than anticipated by Rutherford in concentrating heavy water; 'from now on we must be responsible for our own supply.'

As noted by Rutherford, Lawrence stopped in Cambridge on his way home, staying with the Cockcrofts where he and John went over the Berkeley data again, while looked after by Elizabeth Cockcroft. Cockcroft was of the opinion that Lawrence's supposed unstable neutron was more likely the result of the transformation of some common target contaminant under deuteron bombardment. At any rate, Lawrence's two days in the 'Lion's den' was on the whole a friendly experience, with Ralph Fowler treating him like an old friend, and Rutherford personally showing him around the laboratory, apologizing that nowadays he functioned chiefly as a sort of father figure for his younger colleagues. Admiring the comparatively simple apparatus and careful experimental attitude at the Cavendish, Lawrence nevertheless suggested they might profit from the higher bombarding energy of a cyclotron. He would be only too glad to supply technical data and advice. 'We only use what we can make here,' replied Rutherford. 'But don't you use a spectroscope?' asked Lawrence tongue in cheek, pretty sure that even the Cavendish didn't make them any more [11-38]. Cockcroft, listening to the good-natured exchange and despite his remarks at Brussels, was inclined to agree with Lawrence; the cyclotron was surely the most promising accelerator in sight.

Lawrence also stopped briefly in Washington, where he paid a visit to Tuve and company, who were engaged in fine-tuning the focusing action on the tube of the new two-meter generator. They were also finding that the voltage measurements made by the generating voltmeter during October were systematically too high. The true voltage on the tube, as measured by magnetic deflection and proton range, was about 1,200 kV maximum, with a proton spot of about 2.5 μA on the target [11-39].

> I persuaded Tuve to investigate the origin of the 18 cm protons and the hypothesis of the disintegration of the deuton right away [wrote Lawrence to Cockcroft on his return to Berkeley, in expressing thanks for Cockcroft's hospitality]. I want to get the matter cleared up as soon as possible and it will be a great help if Tuve, with his independent set-up, will investigate the problem. [11-40]

Back in Berkeley, Lawrence himself did just that, enlisting Livingston and Lewis in going over the problem once more. Using sets of carefully cleaned targets, they found the yield of protons from various substances was quite independent of their cleaning procedures and agreed within a factor of two with their earlier measurements. To further eliminate the possibility of any organic film settling on the target when placed in the cyclotron, Lewis prepared pairs of targets, one containing light hydrogen and the other heavy hydrogen. These they bombarded with protons – the idea being that the protons would disintegrate the deuterium in the targets, giving rise to long-range protons. They did indeed find long-range protons from both targets, with protons from the light target accounted for by the small concentration of deuterons in their molecular ion beam, and the more numerous protons from the heavy hydrogen target accounted for by proton bombardment.

> I think that now even Chadwick will have to admit the evidence is preponderably in favor of the instability of the deuton [wrote Lawrence to Tuve]. . . . Soon now we shall write this work up in detail for publication and I sincerely hope that Chadwick will see the point. [11-41]

Regardless of Chadwick's views, and more importantly for Lawrence, Tuve himself did not see the point.

## 11.3. The controversy is resolved

Still smarting from his interaction with Chadwick in Brussels, yet confident in their results, Lawrence dispatched their 'unambiguous proof of deuton disintegration and the instability of the deuton' in two installments to the *Physical Review* — a letter to the editor with Livingston, dated 3 January 1934, and a fuller paper with Lewis and Henderson as well, received by the journal on the same day [11-42]. Lawrence also mailed their latest results to Cambridge. At the Cavendish, Cockcroft, as well as Oliphant and Rutherford, were hard at work on the problem, thanks to Paul Harteck's now functioning electrolyzer. The correspondence between Berkeley and Cambridge speaks for itself.

> *Cockcroft to Lawrence, 21 December 1933:*
> We have so far not worked beyond 600 kV, and it may well be that some groups [of long-range protons] appear at higher voltages. I feel myself, however, that the evidence so far is against your interpretation of the break up of $H^2$. [11-43]

Despite his respect for Cockcroft, Lawrence replied in a note of annoyance.

> *Lawrence to Cockcroft, 12 January 1934:*
> It seems to me that you are hardly justified in feeling that the evidence obtained by you so far is against the interpretation of the break-up of the deuton, since you have not worked at voltages above 600 kV.... It seemed pretty evident from our first preliminary observations that the yield of the group of protons which we ascribe to deuton disintegration is in all cases very small below eight or nine hundred thousand volts. [11-44]

February's mail brought more serious evidence against Lawrence's claims. Note in particular Cockcroft's allusion to Oliphant's 'queer results' — results that would settle the matter before long.

> *Cockcroft to Lawrence, 28 February 1934:*
> We have been investigating copper, copper oxide, iron, iron oxide, tungsten and silver, with stronger heavy hydrogen, and we find from all of these we get three groups of particles of identically the same range.... The third is a proton group of about 13 cm. This latter group is the one which you ascribe to the break up of the deuton. It seems ... clear that these three groups cannot all be due to this break up, and we therefore feel strongly that the alpha particle group and the 7 cm proton group are at any rate due to an impurity which is probably oxygen. We are not yet certain about the 13 cm group, but are carrying out experiments with white hot tungsten targets which I hope may finally dispose of this possibility.

*Added in a handwritten postscript:*
We have now found that on boiling in caustic and cleaning thoroughly the 13 cm group is reduced by a factor of 10.... Oliphant is getting queer results with $H^2 + H^2$. [11-45]

There was little doubt, then, that the 13-cm group was due to contamination.

*Lawrence to Cockcroft, 14 March 1934:*
I think it is quite possible that the effects we observe when bombarding targets of heavy hydrogen with hydrogen molecular beams were due, as Lauritsen suggested, to an increase in deuton contamination resulting from partial decomposition of the target. I cannot understand my stupidity in not recognizing this possibility when the experiments were in progress. Needless to say, I feel there is now little evidence in support of the hypothesis of deuton instability. [11-46]

The very same day, Ralph Fowler wrote Lawrence, expanding on the Cambridge experiments.

*Fowler to Lawrence, 14 March 1934:*
Things have been very exciting, and for a long time Rutherford and Chadwick were nearly convinced that your explanation was right and there might really be the light neutron arising from the disintegration of $H^2$, but now the matter has been really cleared up, entirely satisfactorily I think, in terms of the heavy neutron, and in entire agreement, I think, with all the facts you have recorded so far as I understand them. [11-47]

We will return to Fowler's fine summary of the situation at Cambridge shortly. However, let us first, at this stage of the various ongoing experiments, note that the Cambridge physicists were not the only ones at work on the stubborn problem. So were Charlie Lauritsen and Richard Crane, a collaborator of Lauritsen at Caltech, and Tuve, Hafstad, and Dahl at the DTM. At Caltech, as Lawrence himself indicated, 'beautifully clean' experiments seemed to clinch arguments against his deuton hypothesis by showing that calcium deuteroxide bombarded by protons 'lose hydrogen two which joins the beam and contaminates it' [11-48]. Tuve and company, meanwhile, had been both handicapped and aided by a forthcoming demonstration of the two-meter Van de Graaff generator to the Trustees on 15 December. 'This scheduled visit affected considerably the items which could be attempted previous to their visit because the machine had to be in commission and operating with certainty on that date' [11-49]. Alas, the charging belt of paper for the generator broke even as CIW President John Merriam and a group of Carnegie investigators were on their way to the laboratory on 12 December. Happily a new belt was installed and running successfully by the time the trustees arrived three days later.

Experiments had not come fully to a crashing halt because of the important visitors at DTM, however. As of 6 December, several targets had been tested for disintegration by 1000-kV protons, and on 11 December preliminary observations were made on the disintegration of the same targets by deuterons, the team having introduced into the tube the vapor of heavy water presented to them by Edward W Washburn of the Bureau of Standards the previous spring. 'The emission of neutrons by beryllium bombarded by deutons was very definitely confirmed,' wrote Tuve in December's Report.

Only in January 1934 was the new apparatus running satisfactorily, albeit with only 1 µA on the target, pending Dahl's installation of a new high-voltage ion source. The beam current proved adequate for their initial purpose, in any case. For a starter, they were 'unable to verify the emission of alpha-particles from either beryllium or aluminum in the ranges and voltage region which Lawrence specifies even though we work with currents one hundred times as strong as he used' [11-50]. At the end of the month, further observations of disintegrations produced by deuterons were made using the balance of Washburn's heavy water. 'No evidence was found for the break-up of the deuton as reported by Lawrence, but a careful study of the matter is in progress using the supply of deuterium kindly furnished us by Dr. Urey and his associates at Columbia.'

Tuve's report for February is devoted largely to investigating the fact,

> ...as we have suspected since our first work with heavy water in December, that even in the extensive data obtained during February we are having difficulty in finding almost anything which agrees with the data published by Lawrence and his colleagues in support of their hypothesis of the break-up of the deuterium nucleus with the emission of additional energy of 4.8 MeV.... The outstanding claim of Lawrence and his colleagues was the emission of an 18-cm proton group by all targets investigated. There is certainly no 18-cm group present with any of the six targets so far investigated by us.... A careful search for the neutrons emitted by these targets supposedly due to the break-up of the deuton into a proton and a neutron, and which Lawrence and his colleagues have reported that they found in several of their targets, yielded negative results. [11-51]

As for the underlying cause of Lawrence's claim, 'the similarity of his results for all targets,' Tuve's group had no ready explanation yet to offer. Only in early April were definitive experiments on the bombardment of gases carried out at DTM for the express purpose of identifying the elements responsible for the various purported effects. Since Tuve's report for March was actually written on 20 April, the March report alludes to April's interesting finding, namely that 'the suspected contamination responsible for most of Professor Lawrence's erroneous reports was deuterium itself occluded on the targets' [11-52].

Naturally, Tuve had kept Lawrence informed of their ongoing investigation of his claims — an investigation encouraged, after all, by Lawrence himself. In fact, Tuve's January letter is rather cautious in tone.

*Tuve to Lawrence, 6 January 1934:*
We regard all our observations to date as preliminary and will report them to you in detail as soon as we are confident of them ourselves. Until then do not worry about the fact that we as yet have not observed as many disintegrations as you have. [11-53]

Tuve's letter on 28 February must have come as a shock, however; curiously, it was written the very day that Cockcroft, too, had bad news for Lawrence.

*Tuve to Lawrence, 28 February 1934:*
I know you have been anxiously awaiting this letter. We have just as anxiously been awaiting the results which you wanted to see in it.... In general, I must say we have been having a great deal of difficulty in correlating our observations with those you have published.... In the first place there must be an error somewhere regarding the 18-centimeter group although this may be simply an error in stopping-power determinations....We find no sign of neutrons from any of the targets except Be, beyond a suspicion of neutrons from $CaF_2$ which may be fairly definitive.... As far as we can tell we check none of your alpha-particle results as published....

I have never encountered quite such a situation as this before. Recognizing its importance, especially to you, we have tried to be as careful as possible in our work, which is still very much unfinished. We might suggest that you check over your apparatus very carefully, since at present, as well as from the type of results you have reported, there appears to be basis for suspicion that at least part of your observations are due to some factor common to all of your targets.... [11-54]

By now Lawrence had also heard from Lauritsen, through Oppenheimer, of the latest experiments at Caltech, and of course also from Cockcroft.

*Lawrence to Tuve, 14 March 1934:*
We are suspicious that some of the alpha particles observed were due to contamination of deutons in our hydrogen molecular

beam as well as contamination in the target. Cockcroft writes me that they find $3\frac{1}{2}$ cm alpha particles from copper, copper oxide, tungsten and silver bombarded by deutons. I have the impression that he thinks the effects are due to a common impurity.... [By boiling tungsten and thereby reducing the number of long-range protons] Cockcroft has thereby clearly disposed of the matter of universality of the proton groups.... The neutrons are accounted for by the reaction resulting in the radioactive isotopes [stimulated by deuton bombardment]. Thereby there is now little evidence for the hypothesis of instability of the deuton. [11-55]

The next day, Lewis, Livingston, Henderson, and Lawrence dispatched a letter to the editor of the *Physical Review*, voicing similar misgivings about their hypothesis, leading them to place further, intended experiments with protons on hold [11-56]. Alas, this was not good enough for Tuve, who was by now more than a little annoyed with his old friend. Even so, he wished to avoid publicly exposing Lawrence's sloppiness by giving him the opportunity of retracting his erroneous observations.

*Tuve to Lawrence, 17 April 1934:*
I wrote you at the end of February warning of the direction which our results were undoubtedly taking. After working up all of our data, we reached the astounding conclusion that we were unable to check a single one of the observations which you have reported.... We had for some time been suspecting that the most obvious feature, namely the proton groups between 10 and 20 cm, was due to a contamination probably connected with the bombardment of the targets. The correctness of this assumption was shown by the work described at the end of the enclosed Letter to the Editor [of the *Physical Review*].

In the face of the very general interest which has been aroused by your publications and the many people who have asked us whether we verify your conclusions (we have been very non-committal up to the present time except for the abstracts submitted to the Washington meeting), we have decided that the only way to handle the situation was to make a bald statement of the extent to which we have endeavored to check your results and failed. There is no way of evading the question much longer and it is preferable to us to make a signed statement in the *Physical Review* rather than let the fact that we have been unable to check your work be disseminated by the grape-vine route. I must say that we here have certainly not enjoyed the position in which we have been placed. Once in a lifetime is once too often. [11-57]

Lawrence replied on 20 April, conceding his errors on a slightly peevish note [11-58]. And there the matter rested, with protons and neutrons from deuterons bombarding deuterons as the explanation. Credit for unraveling the puzzle first and most clearly goes to Oliphant and Rutherford (and Harteck as coauthor of their final paper on the subject). As Fowler continued in his letter of 14 March to Lawrence, collisions between protons and deuterons were found to be 'practically unimportant,' while collisions between deuterons and deuterons were 'supremely important' [11-59]. Added Rutherford in his own letter to Lawrence on 13 March, 'Oliphant and I have been particularly interested in the bombardment of D with D ions, and I am enclosing a note from Oliphant giving an account of our results' [11-60].

Two new reactions had been discovered. In one, the target and projectile deuterons fused to produce another isotope of hydrogen (tritium or 'triterium' as they first called it) consisting of one proton and two neutrons, with the simultaneous ejection of a proton and the release of 4.0 MeV of energy:

$$_1H^2 + {}_1H^2 \longrightarrow {}_1H^3 + {}_1p^1 \quad \text{or} \quad H^2(d, p)H^3$$

for short. In the second d–d reaction, the two deuterons fused to produce a helium-3 nucleus (two protons and a neutron) plus a neutron and 3.2 MeV. That is,

$$_1H^2 + {}_1H^2 \longrightarrow {}_2He^3 + {}_0n^1 \quad \text{or} \quad H^2(d, n)He^3.$$

'I personally believe that there can be little doubt of the reaction in which the hydrogen isotope of mass 3 is produced [continued Rutherford], for the evidence from all sides is in accord with it. The evidence for the helium isotope of mass 3 is...at present somewhat uncertain but it looks to me not unlikely.... We suggest, very tentatively, that your results may be explained as due to the bombardment of films of D and of D compounds.' Indeed, Rutherford's confidence in their results was understandable; he had predicted both mass-3 nuclei in his famous Bakerian Lecture of 1920. As for Lawrence's experiments, it appeared that whenever deuterons were used in the bombardment of targets of *other elements*, the targets were invariably contaminated with deuterium. Oliphant and Rutherford's results with carbon, beryllium, or most any element could all be accounted for by deuterium films less than one monomolecular layer in thickness.

Lawrence replied to Rutherford in a thank-you letter on 10 May, of necessity, perhaps, in a rather more gracious tone than in his letter to Tuve on 20 April. Referring also to Cockcroft and Walton's more recent work, he assured Rutherford that 'I certainly appreciate the manner in which this complexity of nuclear phenomena already brought to light makes it clear that it is easy to fall into error, and that a good deal of cautious work must be done for trustworthy conclusions' [11-61]. Tuve, writing to Cockcroft on 18 April, stressed the importance of 'judgment and point of view' over the 'errors in technique which can give rise to such a situation' [11-62]. Responding, Cockcroft focused on the errors themselves—their own over impurity effects in heavy elements, now Lawrence, and the mess in Vienna [11-63]. 'There is a real danger of the subject getting into a mess, and I feel that the only thing to do is to delay publication until we are reasonably sure.' Tuve's opinion as well, to be sure.

Lawrence had learned his lesson. 'From that time onwards,' opined Oliphant, 'the contributions made to nuclear physics in the Radiation Laboratory were above reproach and of rapidly increasing importance, as the energy and intensity of beams available from the cyclotron increased' [11-64]. Partly, we must add, the upswing in quality of experimental nuclear physics at Berkeley was a consequence of an influx of new talent in the Rad Lab. Much of the new talent, notes Heilbron and Seidel, would look with disfavor on the 'lust after machinery' that had dominated Lawrence's program in the past, leaving 'little time for physics' [11-65]. In contrast, sadly, to the new mood in Berkeley, by the mid-1930s the Cavendish Laboratory had seen its better days as a premier nuclear physics laboratory, as Rutherford grew older and more staid, and his disciples left to pursue their own interests—circumstances which we touch upon in the next chapter.

## 11.4. Mopping up the heavy water

Two additional developments in connection with the heavy-water experiments are worth noting briefly. The first concerns the mass-3 isotopes. With the identification of the two d–d reactions at the Cavendish, a rush began on both side of the Atlantic to verify and concentrate the two mass-3 isotopes of hydrogen and helium. While the tritium and helium-3 nuclei 'appeared to be stable for the short time required for their detection, the question of their permanence [required] further consideration' noted *The Times* of London in March 1935 [11-66]. Merle Tuve, for one, had analyzed their heavy-water sample from Urey, and concluded it contained a stable isotope of mass 3 [11-67]. However, nobody else could reproduce their results, which were in due course recognized as simply erroneous.

In the spring of 1935, in any event, Rutherford wrote to Norsk Hydro, by then the world's leading producer of heavy water by prolonged electrolysis of ordinary water, suggesting an experiment designed to isolate the isotope $H^3$ in the form of tritium oxide ($T_2O$) by electrolysis, in much the same way that deuterium oxide, $D_2O$ or heavy water, was concentrated [11-68]. 'As the discoverers of this isotope,' he and his

Cambridge colleagues were keen on obtaining a sufficient quantity of the material to determine its mass accurately in Aston's mass spectrograph, and generally explore the effects of $H^3$-ions in their ongoing transmutation experiments. The importance of the tritium concentration experiment was underscored by Urey in his own letter to Hydro's Director Sverre Brænne, soon after returning from a visit with Rutherford at Cambridge [11-69].

Norsk Hydro General Director Axel Aubert agreed not only to Hydro undertaking the experiment, but generously offered any forthcoming sample of $H^3$-enriched material free of charge [11-70]. Rutherford and Oliphant also discussed the experiment with the Norwegian physical chemist Leif Tronstad during the latter's visit to Cambridge in May 1935. Tronstad was then a newly appointed professor in technical inorganic chemistry at the Norwegian Technical University in Trondheim. It was on his urging that Norsk Hydro had initiated heavy-water production on an industrial scale in early 1934, using electrolyzers in charge of Jomar Brun, head of Hydro's hydrogen electrolysis plant at Vemork in Telemark. Trondstad remained an active consultant to Hydro and, like Gilbert Lewis, was soon a prominent authority on the physical, chemical, and biological properties of deuterium and deuterium oxide.

The painfully slow reconcentration experiment was duly performed at Vemork by Jomar Brun, using as starting material 43 kg of 99% pure heavy water concentrated electrolytically from 13,000 tons of ordinary water. Between April 1935 and March 1936, this sample was electrolyzed to a final volume of 11 cc, and this material was delivered to Cambridge for analysis by Aston. Behold! No trace of tritium was observed — 'a disappointment for both of us,' as Rutherford put it in his report to Norsk Hydro [11-71].

The negative result is not surprising; the natural abundance of tritium is less than $10^{-10}$% (compared to 0.02% for deuterium). In retrospect, however, the effort at Norsk Hydro may not have been in vain. Visiting Norway in 1948, Urey reported that the sample had been forgotten on a laboratory shelf during the war, but was subsequently re-analyzed with a newer, more accurate method. It was found to be radioactive — a telltale indication of tritium, which is indeed radioactive with a half-life of 12 years. This finding also vindicated work at Princeton in 1935. There a group of scientists, including H S Taylor and Walker Bleakney, had concentrated 0.5 cc of heavy water and claimed to detect spectroscopic traces of tritium in it. Doubts were subsequently cast on that finding, when tritium was found to be radioactive. Rutherford, with his own heavy-water sample seemingly devoid of tritium, had lectured Taylor for having arrived at a 'foolish conclusion.' Learning much later that Rutherford's own sample was highly radioactive, Taylor is said to have been very happy 'to learn that Rutherford's sample was lousy with tritium' [11-72].

The other, albeit more positive development in the aftermath of the deuterium-bombardment controversy, soon provided the missing, definitive mass of the neutron. One day in May 1933, Maurice Goldhaber, a young graduate student in physics at the University of Berlin, retreated to the library during Nazi rioting in the city. There he came across a popular account of Gilbert Lewis's separation of a full cc of heavy water, discovered not long before by Urey and his colleagues. Goldhaber asked himself to what use such a large amount of heavy water might be put [11-73]. Among other things, he thought the heavy hydrogen nucleus might be split in two parts by very energetic $\gamma$-rays, rather than by a bombarding particle, in a weak analogy with the spectroscopy of the hydrogen atom. If so, the photodisintegration of the deuteron, as the process later became known, might put to an explicit test the question whether the deuteron was indeed a composite of a proton and a neutron.

On account of the mounting political tension in Germany, Goldhaber wrote to Rutherford for advice, who promptly responded by offering him a research studentship at the Cavendish. Goldhaber arrived in Cambridge in May 1933, and as his first research problem under R H Fowler that fall he chose a theoretical analysis of the lithium reactions studied earlier by Cockcroft and Walton. In writing up his results in April 1934, he needed to know the masses of the various isotopes involved. Nevill Mott told him to see Chadwick, 'who knew all about masses.' Goldhaber did, and while having Chadwick's attention decided to inform him about his idea for the photodisintegration of the 'deuton,' as the deuteron was still called by Rutherford.

> Chadwick, gentleman that he was, listened politely, but seemed to catch fire only when the point was brought out that the mass of the neutron could be determined rather accurately from a measurement of the 'photoproton' energy. Perhaps his interest was aroused because the question of the correct neutron mass was hotly debated at that time. It was then not even certain whether the neutron was heavier or lighter than the proton. Our conversation ended with no explicit commitment from Chadwick. [11-74]

About six weeks later, Goldhaber had occasion to see Chadwick on some other matter, and Chadwick interrupted him. 'Were you the one who suggested the photodisintegration of the diplon to me?' he asked. When Goldhaber answered yes, Chadwick said: 'Well it works—for the first time last night. Would you like to work on it with me?' Goldhaber went to clear it with Ralph Fowler, who readily agreed. And so he joined Chadwick, switching from theory to experiment. Since Chadwick was pretty busy as assistant director of research under Rutherford, he allowed Goldhaber considerable leeway in the experiment, ably assisted by H Nutt, Chadwick's laboratory assistant.

For the $\gamma$-ray source they used thorium C″, plentiful at the Cavendish. It emits a $\gamma$-ray of 2.62 MeV, the highest photon energy then available and, they estimated, somewhat above the binding energy of the emitted particles [11-75]. They filled an ionization chamber with deuterium gas, and observed the photoprotons from the following reaction:

$$_1H^2 + \gamma \longrightarrow {}_1p^1 + {}_0n^1.$$

Since the masses of the proton and neutron were known to be approximately equal, the available energy from the incident radiation ($h\nu - W$, where $W$ is the binding energy of the two particles) could be assumed to be equally shared between the photoproton and photoneutron; hence the mass of the neutron could be calculated from

$$M_n = m_D - m_p + E_\gamma - 2E_p.$$

With the known values for $m_D$ and $m_p$, the equation gave a neutron mass of $1.008 \pm 0.0005$ [11-76], or close to the modern value of 1.0085. Since the known mass of the proton was 1.0078, the neutron was heavier than the proton and hence an elementary particle, not a composite of a proton and an electron as had long been assumed by most everybody, at least until the Solvay Conference of 1933 [11-77]. It was also, Goldhaber reflected somewhat taken aback, radioactive, with a half-life for decay by $\beta$-emission which he estimated might be about half an hour [11-78].

Finally, according to Heilbron and Seidel, though the discovery of the d–d reactions had been precipitated by Lawrence's cyclotron experiments, the above and other ($\gamma$, n) reactions, which did not depend on accelerated particles, would provide an inexpensive substitute for neutron beams from energetic accelerators, and supported Rutherford's stubborn view that a cyclotron had no place at Cambridge [11-79].

# Chapter 12

# THE AMERICANS FORGE AHEAD

## 12.1. Slowdown at the Cavendish; Rutherford's death

Nuclear physics dominated the agenda of an International Conference on Physics, that began in London at the Royal Institution on 2 October 1934. Rutherford opened the gathering with a survey titled simply 'Opening Survey' [12-1]. He was followed, somewhat later, by papers read by Joliot, reporting on their work in Paris on artificial radioactivity via $(\alpha, n)$ reactions followed by delayed positron emission, by Enrico Fermi on their own program on radioactivity induced by neutron bombardment, and finally by Robert Millikan, who dwelt on cosmic rays. Fermi had correctly supposed that neutrons, carrying no electric charge, were more effective than $\alpha$-particles in initiating nuclear reactions. His appearance in London, it might be noted, was less than three weeks before he and his colleagues at the University of Rome made a singularly important discovery, of the effectiveness of *slow* neutrons in provoking nuclear reactions and artificial radioactivity by $(n, \gamma)$ reactions. By a curious twist, while he was still in London his partners Edorado Amaldi and Emilio Segrè had searched with contradictory results for signs of the $(n, \gamma)$ irradiation of aluminum, and so cabled Fermi, who was 'angered and embarrassed at having communicated an erroneous result.' He need not have fretted, as it turned out [12-2].

From London the Conference relocated to Cambridge on 4 October, on Rutherford's invitation, for VIPs to behold the fine facilities comprising the Cavendish Laboratory. There, also, Chadwick, Cockcroft, Oliphant, Norman Feather, Max Born (by then a refugee from Nazi Germany), Gamow, Lauritsen, and Crane participated in a three-day meeting on the 'Artificial Transformation of Atoms and on the Constitution of Atomic Nuclei.' The international character of the meeting, notes Edward Andrade, was illustrated by the fact that an Italian, speaking in French, gave to the predominantly English-speaking audience an account of recent work carried out by a German who was unable to attend [12-3].

The London/Cambridge meeting heralded the disintegration of Rutherford's coterie of brilliant collaborators and disciples [12-4]. They left, one by one, to pursue their own careers and interests in other institutions, at home or abroad. In fact, the first of Rutherford's intimate associates to depart did so not on his own volition. Peter Kapitza had been in the habit of visiting the Soviet Union during the long vacations at Cambridge, and did so again in July 1934. As he was preparing to return to Cambridge, he was detained by the Soviet authorities, ostensibly to undertake work of an unspecified nature for the greater good of the Soviet Union. Rutherford first learned of his fate from his wife Anna, when she returned to Cambridge alone that October. Having warned Kapitza before of the risk he took with his habitual home visits, half-jesting that 'they'll getcha,' Rutherford was not entirely surprised. Nevertheless, he lodged strong protests with the Soviet Embassy in London, to no avail. He also mobilized the international scientific community on Kapitza's behalf, including Bohr, Langevin, and other heavyweights, though it became increasingly clear that his brilliant associate would be unlikely to return to Cambridge.

In April 1935 a statement was issued by the Soviet Embassy, explaining that Soviet scientists who hitherto had worked abroad were urgently needed to help bolster the national economy of the Soviet Union.

> Dr. Kapitza...has been appointed director of a new institute for physical research under the Academy of Sciences. This institute has been specially founded for him by the Soviet Government and large sums have been set aside for the building and its equipment under the directorship of Dr. Kapitza, and in accordance with his desires and requirements. As far as his personal life is concerned, he is very comfortably situated and receives a good remuneration. [12-5]

Rumors nevertheless flew, ranging from Kapitza having orchestrated the whole affair with some ulterior motive, to his having been 'taken by Bolsheviks as a hostage for Professor Gamow.' Many of those who knew him well, including Rutherford, believed Kapitza's often brusque and superior attitude while in Soviet circles, had contributed to sealing his fate. More to the point, Rutherford didn't deny the Soviet Union's right to enlist Kapitza in work within his homeland, and early on set about assisting him in his newfound situation. For their part, the Soviets made overtures to Rutherford about purchasing Kapitza's equipment from the Mond Laboratory, in order for him to continue his pioneering work in the USSR.

In August 1935, as a result of talking with English colleagues visiting Moscow, Kapitza wrote to Rutherford, confirming that the Soviet Government was prepared to purchase the Mond equipment. By that time Rutherford had appointed Cockcroft, who had been so closely associated with Kapitza and his equipment at Cambridge, as Assistant Director of the Mond Laboratory. A committee of Cavendish scientists, chaired by the Vice-Chancellor, went over the whole matter. It concluded that, since the Mond was not in position to continue the high-field and low-temperature studies Kapitza had initiated, any equipment that was not needed for nuclear research or could be readily duplicated, could be transferred to the Soviet Union, provided the University was suitably reimbursed. Following negotiations between various English and Soviet parties ably guided by Cockcroft, a deal was agreed upon, and most of the equipment was crated and shipped to the newly established Institute of Physical Problems in Moscow during the winter of 1935/36. Kapitza's two former technical assistants in Cambridge were given leave of absence to join Kapitza in Moscow to help him set up the equipment.

Rutherford was still preoccupied with the Kapitza affair when Chadwick announced his own departure for Liverpool. Chadwick, who with Rutherford had first met Lawrence at the Solvay Conference in 1933, had become convinced that the cyclotron would be essential for future progress in nuclear physics. He also knew Rutherford was equally firmly against a cyclotron for the Cavendish; moreover, he was beginning to find academic life in Cambridge somewhat circumscribed [12-6]. Just as he feared a showdown with Rutherford was inevitable, he was offered the chair of physics at the University of Liverpool, and accepted forthwith. The physics department's George Holt Laboratory had hardly changed since it first opened 30 years earlier; the galvanometer mirrors in the teaching laboratories were illuminated by little oil lamps, and the electricity supply was still direct current. However, far from being discouraged by what he saw on his first visit, Chadwick saw his appointment as the opportunity to build a modern laboratory virtually from scratch, including a cyclotron in the commodious laboratory basement. And so ended his long and productive collaboration with Rutherford — one that had been a key factor in elevating the Cavendish to one of the world's finest research laboratories.

No sooner had Chadwick arrived at Liverpool than he received a telegram from Stockholm, informing him that he had been awarded the Nobel Prize in Physics for 1935 for the discovery of the neutron. (The prize had not been awarded the year before, in part because the Nobel Committee was deadlocked on whether the prize should go to Chadwick or the Joliot-Curies. Both Chadwick and the Joliot-Curies were candidates again in 1935, and with Rutherford's influential backing the prize went to Chadwick. However, Rutherford also suggested the Joliot-Curies for the Chemistry Prize that year, and the Nobel Committee agreed.) Among the many letters of congratulations was one from Lawrence, who had recently met in Berkeley with Arthur Fleming, the Research Director of

Metropolitan Vickers. Fleming had confided in Lawrence his firm's willingness to undertake building a cyclotron for Chadwick's refurbished laboratory, and Lawrence offered to help with 'detailed recommendations, drawings and specifications' [12-7]. With that, serious plans for a 36-inch cyclotron for Liverpool soon materialized on the drafting table, with Bernard Kinsey of Berkeley slated to be in overall charge of erecting the facility. The magnet was completed in 1936, though the first beam was not circulated until mid-1939. Part of the cost of the cyclotron was paid for by Chadwick himself from his Nobel Prize money—equaling about £8350 in 1935.

Chadwick had brought with him Norman Feather from Cambridge as his senior lecturer and second-in-command. In the event, Feather returned to Cambridge after one year, where he succeeded Charles Ellis both at Trinity College and on the Cambridge University faculty. Ellis succeeded E V Appleton as Wheatstone Professor of Physics at King's College, London. Gradually, Rutherford's team was dissolving. Ernest Walton, Cockcroft's partner, had left Cambridge in 1934, returning to Trinity College, Dublin for a Fellowship in experimental physics. Allibone had long since returned to Metro-Vick. Patrick Blackett, too, was gone, having accepted a chair at Birkbeck College of London University.

The last to leave, and the one closest to Rutherford personally, was Oliphant. In 1936 he was offered the chair of physics at Birmingham University. Although Rutherford himself had suggested that Birmingham consider Oliphant when the Dean of the Birmingham Faculty of Science came looking for a new professor, Oliphant departed on a highly emotional note. Rutherford had told Oliphant that he intended to retire, four years hence, and Oliphant, uncertain about what to do, had sought Chadwick's advice. Chadwick told Oliphant that it was up to him. If he wanted a university chair and a laboratory of his own, arrangements could and would be made that satisfied his obligations to Cambridge as well as to himself. If he felt lukewarm about the offer, it would be a mistake to leave Cambridge. When Oliphant went to see Rutherford about the matter, Rutherford grew red in the face, thumped the table and shouted for him to 'go and be damned with you! First Chadwick deserts me. Now you, who took his place, also desert me' [12-8]. Within the hour, true to form, Rutherford apologized profusely, and discussed how he might help Oliphant in his new post.

Of Rutherford's key people, only Cockcroft remained at Cambridge, though from 1935 he found little time for experiments, being increasingly preoccupied with administrative affairs at St John's College, where he was a Fellow. When he did appear at the laboratory to start up the apparatus, more often than not he would race back to St John's on his bicycle for some administrative task, and then furiously cycle back.

Though still opposed to a cyclotron, Rutherford was mindful of the need for a new Cavendish wing, partly for housing a new high-voltage laboratory, in light of the evident importance of the ongoing disintegration experiments. In 1934 he appointed Cockcroft and Oliphant members of a building committee, and Cockcroft was also made responsible for the design of the high-voltage laboratory. To raise funds for such an expensive undertaking at a time of lingering global economic depression, Rutherford and his staff came up with a scheme for rallying support from likely donors—industrialists and other friends of science and of Cambridge University. The centerpiece of the campaign would be a pamphlet extolling the distinguished record of the Cavendish over the past 60 years, with stress on the last two years, and making the case for a new building and equipment. Realizing its author had to be a prominent Cambridge scholar well known in public circles, yet not someone with close ties to the Cavendish, Rutherford cunningly turned to Sir Arthur Eddington, Plumian professor of astronomy, fellow of Trinity College and best-selling author. (Rutherford was often at odds about Eddington's science, but appreciated his penmanship as better than his own.) Eddington agreed, and came up with a 17-page pamphlet. Titled simply *The Cavendish Laboratory*, it dwelt in moving prose on the 'Mecca of physics for the Empire' [12-9].

> A period of about twelve months in 1932–1933 was an *annus mirabilis* for experimental physics. For some years previously the centre of advance had been in theoretical physics while experimental physics plodded patiently on. Then in rapid succession came a series of experimental achievements, not only startling in themselves but presenting immense possibilities for further advance. The laboratories of the world are now pressing forward in an orgy of experiment which has left the theoretical physicist gasping—though not entirely mute.

Cockcroft and Oliphant, meanwhile, toured several laboratories on the Continent, and were particularly impressed with the Philips Laboratory at Eindhoven and their cascade generators. By the spring of 1935, however, the two had reached a state of disagreement over the design of the high-voltage laboratory, other than its overall dimensions of 45 feet in height and width to accommodate a two-million-volt generator for producing X-rays and positive ions. Oliphant thought Cockcroft's ideas for the laboratory were far too elaborate, and so wrote Rutherford, then on vacation. Too much attention had been paid to providing a pleasant exterior, when what was needed was something more like an airship hangar, which could be easily altered and adapted to changing programmatic requirements [12-10]. (The Round Hill installation in America was a case in point.) They also disagreed on the type of high-voltage generator to be used. Oliphant preferred a Metro-Vick generator with homemade rectifiers immersed in oil; Cockcroft disliked the latter feature, feeling the cumbersome apparatus would frequently have to be dismantled for maintenance.

In the end two generators were acquired from Philips, since technical difficulties precluded going directly to two million volts. A 1.2-million-volt cascade generator with an accelerating tube made in the Cavendish shops was erected in one corner of the newly completed high-tension hall in late 1936, and somewhat later a generator designed for two million volts was installed in the center. However, trouble with its vacuum tube and flashover to the walls prevented the latter generator from operating reliably at its full design voltage.

Rutherford was still adamantly against a cyclotron for the Cavendish in the spring of 1936. However, Cockcroft nevertheless went ahead with a design for a cyclotron, based on a 36-inch diameter magnet deemed capable of generating 17.5 kG, and mailed detailed blueprints to Lawrence for his comments. Lawrence, in turn, wrote Rutherford that they had recently improved the cyclotron performance, allowing beams of hydrogen and deuterium ions to be extracted out of the vacuum chamber through a thin platinum window.

> I assure you that we have got quite a thrill out of seeing the beam of six million-volt deuterons making a blue streak through the air for a distance of more than twenty-eight centimeters. [12-11]

The Cavendish possessed one advantage over, say, Chadwick at Liverpool: the roubles received from the Soviet Government in exchange for the magnetic and cryogenic equipment from the Mond Laboratory — approximately £30,000. With the automobile magnate Lord Austin also responding to the Rutherford-Eddington campaign by contributing no less than £250,000 to the Cavendish upgrades, Rutherford's penny-pinching came to an abrupt end. However, he was still not convinced about the advisability of a cyclotron, as Chadwick learned when he came down from Liverpool in June, to be the external examiner in the Cambridge Tripos examination. Staying several days with the Rutherfords, he and Rutherford were sitting in the garden at Newnham Cottage when Chadwick brought up the importance of a cyclotron at Cambridge. Rutherford blew up, thundering that 'I won't have a cyclotron in my laboratory!' [12-12]. Nevertheless, by the end of that month, Rutherford acquiesced, and gave Cockcroft permission to proceed with the cyclotron. Rutherford had initially proposed a hybrid magnet 'for general purposes, and also probably for use as a cyclotron.' Lawrence countered that a cyclotron required a dedicated magnet, and also rejected Cockcroft's initial choice of a magnet with pole faces mounted vertically, not horizontally as at Berkeley, for it to be adaptable to cosmic-ray work.

By the summer of 1936 three medium-energy European cyclotrons were in the design stage, at Liverpool, at Cambridge and at Copenhagen. They were essentially copies of Berkeley's 37-inch cyclotron, which had been enlarged from the 27-inch machine. (Joliot's cyclotron in Paris, also on the drawing board, would be somewhat smaller.) The two English machines were virtually identical, with 36-inch diameter magnets weighing 50 tons each, and built by Metro-Vick and Brown-Firth of Sheffield [12-13]. Cockcroft had the opportunity of inspecting Lawrence's large cyclotron first-hand, when he was invited to give a course of lectures on low-temperature physics and transmutation experiments at Harvard in the spring of 1937. While there, he was offered a professorship in nuclear physics by James Conant, the President of Harvard—a tempting offer, to be sure, but one he had to decline on account of his various commitments at Cambridge.

The highlight of Cockcroft's visit to America, as in 1933, was four days in Berkeley, where he stayed with Donald Cooksey while 'getting the hang of the cyclotron.' Returning to Cambridge, he found that magnet coil winding at Metro-Vick was proceeding at an alarmingly slow pace; still, the magnet was installed in the Austin Wing by the end of the year. They obtained the first faint evidence of a deuteron beam in August 1938, after months of struggling, but nothing yet on target [12-14]. Switching to protons, they had 12 μA of 5 MeV protons circulating by Christmas. However, Oliphant, while visiting Berkeley as well that year, expressed his opinion that the Cavendish cyclotron was too small, and found that Lawrence was of the same opinion. The machine was unlikely to compete successfully with the Americans; Berkeley's 60-inch cyclotron had been on the drawing board since the spring of 1936, and Merle Tuve, too, had his eyes on a similar machine.

Oliphant, in fact, expected to have a cyclotron of his own, and one superior to Cockcroft's. While the Cambridge machine was a near copy of Lawrence's 37-inch cyclotron, Birmingham's machine, paid for by Lord Nuffield, the chairman of Morris Motors, would exceed Lawrence's 60-inch ('Crocker') cyclotron. In the event, the onset of World War II stopped construction of the Birmingham cyclotron, and it was obsolete when it started up in 1950 [12-15].

Cockcroft disagreed, at any rate, with Oliphant and Lawrence's assessment of the new high-voltage laboratory at the Cavendish, claiming it produced excellent physics under Philip Dee's leadership [12-16]. As for Cockcroft himself, he wound up having his hands full, as noted, between the cyclotron, as Assistant Director of the Mond, and as Junior Bursar of St John's. In the latter capacity he was responsible for the domestic administration of the College, which included managing the staff and overseeing the maintenance of buildings and grounds. The most exacting work as Junior Bursar was probably his duties concerned with the restoration of the crumbling old buildings and with the construction of a new extension to the College. As Anna Kapitza wrote him in a sympathetic letter in 1939:

> The Cavendish is nothing compared to the College [of St. John's] to rebuild. All the styles to consider and all the Fellows to please. God help you! [12-17]

Rutherford, too, found much less time for active research, in the last years of his life. He was head of the Academic Assistance Council, created as a result of a massive public rally in London's Royal Albert Hall in October 1933 for assisting scientific refugees reaching Britain, primarily but not exclusively from Germany. Another activity in which he was much involved was with the Department of Scientific and Industrial Research, created in 1915 for providing financial support to investigations in both pure and applied science. 'The primary objective is to encourage scientific research which may ultimately benefit the industries, but there is no intention of interpreting this very strictly,' wrote Arthur Schuster, who got Rutherford his Manchester chair in 1907, in urging him, too, to apply for a grant. In 1930 Rutherford was appointed chairman of the Advisory Council of the Department. He also spoke occasionally on scientific matters before the House of Lords, of which his elevation to the peerage in 1931 made him a member.

Then again, despite his natural vigor and energy, Rutherford appeared to tire more easily during his final years. He spoke of retiring at 70, which would be in 1941. On the death in 1934 of his old Manchester friend, the mathematician Horace Lamb, he had written that Lamb was 'one of the few men that grew old gracefully. So many are inclined to hang on and to grasp for power when they ought to be dandling their grand children' [12-18].

Rutherford was not simply growing tired, however. For years he had suffered from a small umbilical hernia. Though it seemed not to worry him, he wore a truss on account of it. On 14 October 1937 he became sick in the night—seriously enough that he was taken to a hospital. He was operated upon for a strangulation of the hernia, and at first appeared to be recovering well. There was even talk of his embarking on a scheduled trip to India to preside over a joint meeting of the British Association and the Indian Association of Science—a trip he had eagerly looked forward to. Alas, he began deteriorating rapidly, and on 19 October he died from intestinal paralysis. Philip Dee cabled the news to Cockcroft and Oliphant, who were attending a congress in Bologna, Italy honoring the two-hundredth anniversary of Luigi Galvani's birth. They passed it on to Niels Bohr, who wept openly as he informed the audience from the congress podium. Rutherford's body was cremated two days after his death, and his remains were interred in Science Corner in Westminster Abby, next to Newton, Herschel, Darwin, and Kelvin.

Rutherford's successor as Cavendish Professor was Sir (William) Lawrence Bragg, the crystallographer, and his election hastened the change in emphasis of research at the Cavendish, from nuclear physics to solid state, the crystal structure of complicated molecules, as well as radio astronomy. Among other things, as Cavendish Professor Bragg participated in the work of M F Perutz and J C Kendrew on the crystal structures of proteins, and also interested himself in the work of F H C Crick and J D Watson on the structure of DNA, thereby assisting in the foundation of molecular biology [12-19]

## 12.2. Another dispute gives way to American success

Whereas the Cavendish, following one brilliant discovery after another under Rutherford, Oliphant and Cockcroft, was turning to entirely new areas of research, the American laboratories were poised to take center stage in nuclear physics. At Berkeley, with Lawrence having learned a lesson from drawing unwarranted conclusions, his laboratory was becoming a beehive of experimental activity with larger cyclotrons manned by a cadre of graduate students and assistants. At the Carnegie Institution, the two-meter generator was running reasonably well during the dry winter months, in support of a more narrowly focused program. Curiously, while Tuve was devoting much of his energy, during the spring of 1934, to refuting Lawrence's claims for an 'unstable deuteron,' he also found himself embroiled in another dispute with Lawrence's rival at Caltech. And this time Tuve, too, found himself proven wrong. To his credit, however, he was able to rise above the fray and turn what had the potential for a serious blunder into the opening salvo for what became a major subfield of nuclear research.

The latest controversy had its origins in January 1934. That month, as may be recalled, the Joliot-Curies announced their discovery of artificial radioactivity, found by bombarding certain nuclei (Al, B, and Mg) with $\alpha$-particles. The unstable nuclei thus created decayed by positron and neutrino emission to stable nuclei [12-20]. In the case of boron, for example, $\alpha$-bombardment went as follows:

$$_5B^{10} + _2He^4 \longrightarrow _7N^{13} + _0n^1,$$

in which $_7N^{13}$ is radioactive, decaying to the stable isotope $_6C^{13}$ by positron emission. In their article in *Comptes Rendus*, the authors ended with the observation that

...long-lived radioactivity, analogous to what we have observed, no doubt can be created by several nuclear reactions. For example, the nucleus $_7N^{13}$ which is radioactive on our hypothesis, could be obtained by bombarding carbon with deuterons, following the emission of a neutron. [12-21]

Early the next month, following up on the French suggestion, Charles Lauritsen at Caltech with Richard Crane, a graduate student of his, reported a very penetrating radiation from carbon bombarded with deuterons, which they concluded were very 'hard' $\gamma$-rays in the reaction [12-22]

$$_6C^{12} + {}_1H^2 \longrightarrow {}_6C^{13} + {}_1H^1 + \gamma. \tag{12.1}$$

Later in the month they had more substantive news to report, namely a 10-minute activity from deuterons on carbon that yielded positrons as well, confirmed by their colleagues Carl Anderson and Seth Neddermyer with the Caltech cloud chamber. The activity appeared associated with the same radioactive end product as Joliot and Curie had obtained by bombarding boron with $\alpha$-particles, obtained as follows:

$$_6C^{12} + {}_1H^2 \longrightarrow {}_7N^{13} + {}_0n^1, \qquad {}_7N^{13} \longrightarrow {}_6C^{13} + e^+. \tag{12.2}$$

Moreover, by comparing the intensity of the ionization as a function of time due to positrons and $\gamma$-rays from carbon, they found identical half-periods (10 minutes) for the two emission processes. Ergo, the same radioactive process was responsible for both the positrons and $\gamma$-rays, and the $\gamma$-rays had their origin in the *annihilation* of the positrons when colliding with nearby electrons [12-23]. Apparently carbon could be transformed by deuteron bombardment via either reaction (12.1) or (12.2), with the first reaction ten times as frequent as the second.

Merle Tuve, in addition to his $(d, d)$-sleuthing on Lawrence's behalf, lost no time in testing the reported discovery of 'induced radioactivity' in Paris, and in checking the more recent results from Pasadena. As of February 1934, the DTM team had little if any evidence for 'such a delayed gamma-ray effect' [12-24]. From the paucity of such effects in their own ionization chamber, Tuve was inclined to ascribe the purported observations by their Danish colleague of induced radioactivity to some form of contamination, 'not necessarily carbon' [12-25]. Contamination, it seemed, was the bane of laboratory experiments on the West Coast. By April he was less dogmatic.

...the reaction...responsible for the induced radioactivity characteristic of most targets after deuton bombardment has not been established, but it appears at least probable that deutons on carbon are responsible as assumed by Lauritsen and others. [12-26].

Cockcroft, for his part, begged to differ with Tuve, having obtained, with Walton and a new collaborator, C W Gilbert, essentially the same results as Lauritsen. 'We differ with you [he wrote to Tuve on 30 April] that the carbon effect is real and connected with the transformation of $C^{12}$ to $C^{13}$. [12-27]

Despite these disagreements, Odd Dahl, Tuve's machine-builder and partner in all of this, was well received by Cockcroft during his stopover in Cambridge in May, en route to Norway to look into physics opportunities in his native land.

> Found Rutherford pleasant, and was permitted to have 3/4 hours talk with him. Did not like Chadwick, and am sure he does not think very much about work done any place else. [12-28]

Mainly, however, Dahl spent his time with Oliphant, soaking up the latter's ion source experience. Rutherford, on the other hand, expressed considerable interest in the Wilson chamber developments at DTM, in particular their use of sylphon bellows, not available in England.

> He said that the sad state of affairs today is that a man spends ninety five percent of his time monkeying with [the Wilson chamber] and five percent working with it.... He said jokingly that he would personally give five dollars for a good one. I told him to watch our smoke. [12-29]

Tuve, replying to Dahl's letter, noted that 'it appears from what you write that our efforts are not so far behind what even the good Lord and his men do' [12-30].

Lauritsen, meanwhile, had continued their work on induced radio-activity, finding that bombarding carbon (and boron oxide) with *protons*, as with deuterons, gave rise to appreciable delayed activity as well [12-31]. The same was found at Cambridge. Tuve was still not convinced, however, blaming the latest evidence at Pasadena and at Cambridge for radioactivity by proton bombardment on 'deuterons contaminating their so-called proton beam' [12-32]. He stuck to his position while attending the meeting of the American Physical Society in Berkeley during 19–23 June 1934, where claims for deuterium contamination had only recently been retracted. Tuve, Lawrence, and Lauritsen all read papers in Le Conte Hall, introduced by Robert Van de Graaff. Tempers flared when Tuve mentioned Lawrence's discredited deuteron dis-integration theory in the discussion period following Lawrence's own report, until Raymond Birge stepped in to restore calm [12-33]. To Tuve's further irritation, no mention was made of contaminants by the scientific secretaries who reported on the symposium in *Science*. Their summary left the impression that he, Tuve, had shown that

...any previous outstanding discrepancies in findings were almost entirely to be ascribed to the differences in energies of the incident particles utilized in the respective laboratories, together with the measuring techniques used in the identification of the disintegration products. It appeared that the results were not contradictory in the least, but rather supplementary. [12-34]

When back in Washington, Tuve dispatched an angry 'correction' to *Science*, in which, among other things he pointed out that

...resolution of the outstanding discrepancies between our findings and those of the Berkeley investigators came about through their abandonment several months ago of a striking hypothesis with quantitative consequences that they had claimed were demonstrated by their observations, but which they were notified could not be substantiated in Pasadena, in Cambridge nor here in our laboratory, and by their admission at that time and recently in a paper on another subject that spurious effects have been present in their observations, due to various contaminations. [12-35]

If Tuve was unhappy in Berkeley, his next stop in California proved no less troublesome and downright awkward for him personally. On his way back from the APS meeting, he stopped over in Pasadena, where he had a long talk with Lauritsen and went over the Caltech data with great care. Tuve had to admit that he was wrong and Lauritsen right, as he wrote Hafstad on the train 'somewhere in Kansas en route to St. Paul.'

Just a note to say we are evidently wrong on the carbon radio-activity question. This...has been worth the two weeks of indigestion, if it serves only to keep us from being too cocky again....
Am nearing the end of a hell of an uncomfortable trip. Nuclear physics isn't physics yet, and symposiums on a subject that isn't born yet are premature....I'll go into the details when I reach [Washington], but this is just a tardy warning that we haven't got the world by the tail yet. [12-36]

Hafstad promptly wrote back, attempting to put a reverse spin on the unhappy business.

Cheer up Merle....We may not be right but somebody will have to do some good experiments to prove us wrong. Remember our position is analogous to the Cavendish position in the Vienna controversy viz the skeptical viz the gullible position. [12-37]

Back in Washington, 'difficulties with high humidity completely incapacitated the high-voltage equipment during most of August,' forcing Tuve to put off pursuing further experiments with protons and deuterons on carbon. It was just as well. He had, instead, the opportunity of consulting with theorists over their program. George Gamow had attended the APS meeting in Berkeley, where he confided that he felt it unwise for him to return to Russia. Lawrence was unsuccessful in persuading the University authorities to hire him, despite Oppenheimer's concurrence that Gamow was a top-notch theorist. In the event, Gamow wound up teaching at the University of Michigan's summer school in Ann Arbor that summer, and later in the year Tuve helped him secure a faculty position at George Washington University, conveniently close to the Carnegie Institution. In September 1934, Tuve arranged for a conference at DTM with Gamow and Gregory Breit, where they went over the DTM nuclear physics program in considerable detail. Both Breit and Gamow concurred that their results from the carbon radioactivity experiments were compatible with theoretically expected values for proton capture with $\gamma$-ray emission [12-38].

No sooner had their discussion with Breit and Gamow ended, than Larry Hafstad sailed for London, to attend the International Conference on Physics, which, as we learned, began at the Royal Institution and continued in Cambridge. In his own copy of Hafstad's report on the Conference, we find Tuve's marginal inscription: 'This is a <u>most excellent</u> report.'

> While I was impressed with the wide knowledge of the Cambridge people and their familiarity with the literature, both experimental and theoretical, and especially with their uncanny skill in drawing correct conclusions from comparatively few observations, I am well satisfied with the soundness of our own position. I was particularly pleased to discover that, in spite of the advantage of the larger Cambridge group in permitting one man to specialize on a given type of observing apparatus, our observing technique is almost on a par with theirs. The close cooperation possible in our small group plus the privilege of devoting our full time to research is apparently nearly sufficient to offset the advantage of their greater specialization. The larger group, however, retains the advantage of having all varieties of apparatus in running order at all times. It is a well-known rule in experimental work that apparatus not in use invariably disintegrates. An appreciable fraction of our time must therefore be charged to the process of 'tuning up' auxiliary observing apparatus.
>
> Another point which I noticed in my comparison of the Cavendish method with our own was that students were at work on several subsidiary experiments which we had discussed in our own laboratory and had dismissed as being too uncertain of

results to warrant interrupting the main program. It is clearly part of the Cavendish method to use time of students in reconnaissance, and then calling on the experts of the regular staff for help when a 'strike' appears to be particularly promising.

It was my impression that the Cavendish is especially well organized to follow up hints in order to make a few outstanding discoveries. This impression was independently verified during my visit with Dr. Blackett ... who described the Cavendish method as 'picking up nuggets, and leaving more time-consuming researches for other workers.' My conclusion from all these things is that we need not be too discouraged by the fact that many of our pet experiments are first performed in Cambridge. The Cavendish Laboratory is an organization which has been gradually evolved for the last thirty years for that express purpose.

In deliberately choosing to work at 1,000 kV and above, since the lower voltages are available in any university laboratory with sufficient initiative, we have sacrificed the recognition our work might have had from more frequent publications in the lowest energy range.... Our contribution to nuclear physics should be properly compared only to those of the Berkeley investigators, who are at present the only other group pioneering in both the method of attack and in the analysis of reactions. [12-39]

Tuve concurred heartily with Hafstad's subtly nuanced observations and recommendations. After much effort devoted to coping with the ever problematic charging belt of the two-meter generator, with Dahl's latest high-intensity, low-voltage ion source running well, and the tube vacuum seemingly under control, the generator was fired up once more. Almost at once, in attacking the carbon radioactivity problem anew, a crucially important discovery was made.

All the observations combine to give strong evidence for a resonance effect at about 500 kilovolts, an effect readily accounted for by theory, and further, such that the apparently contradictory observations of Lauritsen are exactly those which would be expected from his apparatus in which alternating potentials must be used. [12-40]

What they had observed was early experimental evidence for *nuclear resonance* effects: conditions under which protons, and nuclear projectiles in general, are much more effective in causing reactions at specific bombarding energies—an effect Breit had urged them to look for. As Breit had anticipated, it was soon apparent that the Van de Graaff generator, with its highly stable, monoenergetic beam, was much better suited for observing narrow resonances than cascade generators or cyclotrons.

By early 1935, Tuve, Hafstad, and Dahl had embarked on a fresh program of mapping resonances below 1000 kV under proton bombardment of carbon, lithium, and fluorine. In effect, they were inaugurating a period of nuclear physics focusing on *nuclear spectroscopy*, paralleling the development of atomic and molecular spectroscopy by the previous generation of physicists. These studies would be of both fundamental and practical importance. On the one hand, they revealed a wealth of discrete resonance energy levels or excited states associated with *compound nucleus* formation. On the other hand, the well-defined energy values of emitted particles or $\gamma$-rays, such as $\gamma$-radiation resonance peaks from the Li(p, $\gamma$) reaction, with energy spreads of typically a few kilovolts, made them handy for beam-energy calibration — that is, providing fixed calibration points for the voltage scale of the accelerator.

Three major manuscripts were prepared for submission to the *Physical Review* during the spring of 1935, reviewing and summarizing the ongoing experimental and developmental work at DTM. One, authored by Tuve, Hafstad, and Dahl (in that order), covered the one-meter and two-meter generator in considerable detail. The second (Tuve, Dahl, and Hafstad) covered their work on high-intensity positive ion sources and the electrostatic focusing of ion beams in high-voltage tubes. The third (Hafstad and Tuve) covered the latest experiments on carbon radioactivity and resonance reactions by protons [12-41].

19 April began what would be the first of eight Washington Conferences on Theoretical Physics, organized jointly by Tuve and by Gamow in their respective Washington institutions. Behind the scene of the organizers were Gregory Breit and Edward Teller, the latter having joined George Washington University the same year. In a letter to Dr. Fleming, Hans Bethe praised the Washington meeting as 'the most ideal conference [on nuclear theory] which I ever attended' [12-42].

When the heat and humidity again put a temporary halt to much further Carnegie experimentation in the early summer of 1935, Tuve decided it was time to address the question of the DTM program over the next 10 to 20 years. In particular, a more powerful accelerator would clearly be needed to sustain a longer-range program. But what kind of accelerator? Tuve was not yet prepared to jump on the cyclotron bandwagon. Considerably higher voltage than the two-meter, open-air generator could deliver (say up to 10 million volts) was certainly desirable, but beam homogeneity and stable operation remained equally important, in his opinion. One possibility was an alternative to the 'direct attack' being explored by Jesse Beams and Leland B Snoddy at the University of Virginia, based on electrical wave-front propagation along a transmission line, and some cooperative tests with Beams and Snoddy were started. However, there was also the possibility of increasing the electrical breakdown strength of the air or other gases surrounding the

Van de Graaff generator—a promising technique being pursued by Raymond G Herb, a graduate student at the University of Wisconsin.

Initially, Herb had attempted operating a small generator in vacuum, with marginal results. In the spring of 1933, he decided to pressurize the vacuum tank instead. With D B Parkinson and D W Kerst, he pressurized the tank with air. 'Measuring with a meter stick, when the flat ends of the tank bowed out about 3/16 of an inch, Herb decided it was time to stop,' and raised the potential [12-43]. The air pressure was 3.3 atmospheres, and the potential rose to 500 kV without difficulty. Herb and his companions immediately reconfigured the generator to accommodate an acceleration tube, and the little generator ran reliably at 400 kV, earning Herb his PhD with the Li(p, $\alpha$) reaction in 1935.

With his degree in hand, Herb joined Tuve and company in June as a 'summer postdoc,' building a high-resistance voltmeter for the two-meter generator. However, his presence at DTM that summer undoubtedly lent force to Tuve's argument for a pressurized generator for the Department. He and Tuve differed in one important respect, it seems [12-44]. Tuve favored a sizable generator housed in a large tank under relatively low pressure, whereas Herb preferred a compact generator operating under high pressure. Herb's approach proved right. Returning to Wisconsin, he and his students developed a succession of highly successful generators over the years, setting the standard for the worldwide proliferation of university and industrial Van de Graaff accelerators in the post-World War II years. Charles and Tom Lauritsen were early proponents of Herb's design principle, as noted in section 7.1.

A radically different electrostatic generator was also pursued at DTM in the fall of 1935, but dropped as a needless complication [12-45].

Tuve had initially proposed a 60-foot diameter steel tank for the pressure vessel, similar to the large welded tanks used for storing natural gas. However, that raised objections by the local zoning board and surrounding community over the industrial appearance of the facility, and by CIW management over its cost ($150,000). Not until the spring of 1936 did Tuve obtain official permission to proceed, after the tank had been redesigned as a pear-shaped pressure vessel, 30 feet in diameter, with an estimated cost of $23,650. By the fall of 1937, the Chicago Bridge and Iron Works had completed the vessel, and by the year's end it dwarfed the Experiment Building on the DTM campus. Mounted with its lower half enclosed by a pleasant brick structure, it gave the appearance of an astronomical observatory, as we see in Figure 12.1. Indeed, the Atomic Physics Observatory, as it was named, was said to have *improved* the Chevy Chase neighborhood. However, the generator itself was not installed until 1939.

In the spring of 1936, meanwhile, the group began a new series of investigations, this time on the scattering of protons by protons—Tuve

**Figure 12.1.** *The Atomic Physics Observatory, January 1938. The Experiment Building is to the left. DTM Archives B-931.*

and Breit's original objective of the high-voltage work started at the Carnegie a full decade earlier. It had seemed to them to be the most fundamental experiment that could be undertaken. Indeed, this latest investigation was probably the crowning achievement in experimental nuclear physics by Tuve and his little group. In it, Tuve and Hafstad were joined by Norman P Heydenburg, who had arrived at DTM the previous September. Born in 1908 in Big Rapids, Michigan, Heydenburg had earned his PhD in physics from the State University of Iowa in 1933. He then held a National Research Council fellowship at New York University, where he continued his dissertation work on the measurements of nuclear moments. In 1935 he moved to the University of Wisconsin, where he took up nuclear physics with Herb's electrostatic generator; later that year he joined Tuve and company at DTM.

Basically, the new experiments went as follows. The beam of mono-energetic protons from the two-meter generator was shot into a cylindrical chamber filled with hydrogen gas—effectively into a collection of nearly stationary protons, as far as the impinging protons were concerned. The number of incident protons bouncing off the target protons at various angles with respect to the beam direction were measured by rotating a detector (ionization chamber) about the chamber axis. The measurements were repeated at progressively higher bombarding energy, beginning at about 600 kV. The experiment confirmed the existence of an attractive force between the two colliding protons, which overpowered their mutual electric (Coulomb) repulsion, when they were virtually in contact, or ten-trillionth of a centimeter ($10^{-13}$ cm) apart.

This short-range, attractive p–p force was found, later in the year (after correcting for the Coulomb force), to be about the same as that between a proton and a neutron, and represents the specifically 'nuclear' or 'strong' force which bind together the basic constituents (protons and neutrons) of atomic nuclei [12-46]. Submitted at the same time to the *Physical Review*, and appearing immediately after the experimental paper by Tuve, Heydenburg, and Hafstad, was the analysis of the proton–proton scattering data by Breit, Edward U Condon and R D Present [12-47]. These early p–p scattering experiments were subsequently extended, first at DTM and then by Ray Herb, with Donald Kerst, David Parkinson, and Gilbert Plain, to 2.4 MeV at Wisconsin shortly before World War II.

The proton scattering experiments, as well as the earlier Van de Graaff experiments at DTM, had caught the attention, among others, of Ernest Rutherford, who wrote Tuve in November 1936.

> I have read with great interest your papers on the transmutation of the isotopes of lithium and your accurate determination of the scattering of protons, described in the last two numbers of the *Physical Review*. I congratulate you and your collaborators on two excellent pieces of work.
>
> . . .
>
> I am very pleased to see that the rush period of work on transmutation has come to an end, and as your papers show, results of real value can only be obtained by accurate and long continued experiment. [12-48]

The equal strengths of the p–p and n–p nuclear forces led to the conjecture known as the charge-independence hypothesis: that the n–n, p–p, and n–p forces (again after correcting for the Coulomb interaction) are identical [12-49]. As discussed in great detail by L Brown, the mathematical formulation of this notion relied on what are known as Heisenberg's 'isospin operators' in writing an interaction that is invariant by the substitution of a proton for a neutron, or vice versa [12-50].

After the war the p–p scattering experiments were resumed under Herb, still at Wisconsin, using a new Van de Graaff generator and a new crop of graduate students. The revised scattering cross sections confirmed not only marked deviations from the ordinary Coulomb forces, but, in addition to the non-classical 'pure s-wave scattering' seen earlier, revealed a small contribution from higher-order, quantum-mechanical effects. As before the war, the person most responsible for extracting the fundamental significance out of these data was the irrepressible Gregory Breit, the theoretical linchpin at the DTM, and for a period at Wisconsin. Breit's name is also associated with the analysis of the nuclear resonance phenomena discussed earlier, summarized in the famous Breit–Wigner formula for the shape of a nuclear resonance absorption line.

As for Tuve and the DTM in about 1936, we end on a less happy note with Odd Dahl's departure for his homeland that year. It was Harald Sverdrup, as we recall, who had brought Dahl to the Carnegie Institution in 1927. By 1934 Sverdrup was a Member (Fellow, we would say) of the Christian Michelsen's Institute (CMI) for applied physics in Bergen, Norway. During one of his periodic stays at the Carnegie that year, he asked acting director Fleming if he might borrow Dahl for a month or so in Bergen, helping him with some oceanographic equipment — Dahl's original expertise. Fleming concurred, and it was on his way to Bergen that Dahl stopped in Cambridge and had a chat with Rutherford and his people. Once in Bergen, he also had a chat with Professor Bjørn Helland-Hansen, head of the Geophysical Institute in Bergen and chairman of the board of directors for the CMI. Helland-Hansen, and everybody else in Bergen, hinted at their wish to see Dahl join them on a permanent basis at the two Institutes. Dahl was flattered, but needed time to think it over. And so he did. The upshot was that in 1936, after much introspection and soul searching, he accepted a post at the CMI in Bergen, to bring back to Norway experience the self-made physicist had gathered abroad, and help build up modern laboratory practice in Bergen.

Though a blow to Tuve and his colleagues, Dahl said simply, 'I must go' [12-51]. At the DTM, Robert Meyer replaced him as instrument maker and machine physicist on Tuve's team. Like Dahl, Meyer had a strong technical background, having a degree in electrical engineering from the University of Kansas and a master's degree in the same subject from M.I.T. Since 1933, he had been employed by the Raytheon Production Corporation in the design and construction of equipment used in the manufacture of radio tubes.

## 12.3. The Rad Lab in action

Somewhat to his chagrin, Lawrence, too, had read the Joliot-Curies' suggestion for creating radioactive elements artificially by deuteron bombardment, in *Comptes Rendus* and *Nature* in January 1934. After all, bombardment with deuteron projectiles was perceived in Berkeley as their own strong card. Making doubly sure with both 1.5 MeV protons and 3 MeV deuterons, Malcolm Henderson, Livingston, and Lawrence lost no time in bombarding 14 elements, from lithium through calcium.

Their letter to the editor of the *Physical Review* [12-52] was mailed off on 27 February, the same day Lauritsen, Crane, and Harper submitted their own substantive paper on carbon activation to *Science*. Except possibly for carbon, no activity was induced with protons, whereas every target exhibited activation under deuteron bombardment. They gave off both $\gamma$-rays and ionizing radiation, presumably from $(d, e^+\gamma)$ reactions in light of the French experiments. Rough half-life estimates were given in their letter, ranging from 40 seconds for calcium fluoride to 12 minutes for carbon. However, in an unpublished lecture Lawrence conceded that none of the measurements could match the 'very significant experimental findings' of Crane and Lauritsen [12-53]. Nevertheless, Lawrence and his team pursued the new subject with gusto, firing deuterons as fast as targets could be mounted in the cyclotron chamber.

As noted earlier, someone else who had pored over the Joliot-Curie announcement was Enrico Fermi in Rome. Fermi was the same age as Lawrence, and, like his American colleague, had been the only representative of his native land, Italy, at the Solvay Conference in 1933. He was born in Rome in 1901, the son of an administrative employee of the Italian railroads [12-54]. He received the traditional education in the public schools of Rome; however, he owed his scientific development as much to the books he avidly read as a boy, as to any particular personal contact, save for Enrico Persico, a schoolmate and lifelong colleague. The two of them performed several experiments with homebuilt apparatus, such as determining the density of Roman tap water. After high school, Fermi competed successfully for a fellowship at the Scuola Normale Superiore in Pisa. There he was soon considered the resident authority on relativity, statistical mechanics, and quantum theory. He obtained his doctorate from the University of Pisa in 1922, based on an X-ray diffraction experiment. (Though he was basically considered a theorist, he was an equally gifted experimentalist. Moreover, theoretical physics was not yet recognized as a university discipline in Italy in 1922, and would have been out of the question for a dissertation.)

Returning to Rome, Fermi encountered Orso Mario Corbino, Senator of the Kingdom and director of the Physics Institute of the University of Rome. Corbino recognized Fermi's talents, seeing him as one who could help vitalize the rebirth of Italian physics—in the doldrums for nearly a century—and became his devoted patron. Fermi, for his part, felt the need to establish closer contact with fellow physicists abroad, and spent the winter of 1923–24 on a traveling fellowship at Max Born's Institute in Göttingen, and the fall of 1924 with Paul Ehrenfest at Leiden. Curiously, the period in Göttingen was not a very happy one for Fermi; Heisenberg, Pauli, and Jordan were too engrossed in their own problems, and Fermi was too introverted to speak up. He found a much more congenial atmosphere in Leiden, thanks largely to Ehrenfest's personality. Following an interim stint as lecturer at Florence, in 1927 Corbino saw to it that he was appointed to a newly established chair in theoretical physics at the University of Rome. Before long, Fermi, approaching his peak as a theorist [12-55], was joined by Franco Rasetti, Emilio Segrè, and Edorado Amaldi, and therewith had the foundation for a crack team of experimental physicists in the Eternal City.

In the beginning, Fermi and his budding team confined themselves to optical spectroscopy and atomic physics, highly traditional subjects, but they soon perceived, as had Corbino, that the future lay in nuclear physics. This was made all the more evident with the remarkable discoveries of 1932 — deuterium, the neutron, the positron, and artificial acceleration — underscored by the Solvay Conference the next year, followed soon afterwards by artificial radioactivity. The announcement in Paris, in particular, provided the impetus needed for Fermi's fledgling group to get properly started in nuclear research. Why not look into similar effects as those observed in Paris, but with Chadwick's neutron instead of $\alpha$-particles? The Joliot-Curies had observed one disintegration for every million alpha incidents on aluminum. With neutrons, reasoned Fermi, there would be no electric repulsion, and the reaction yield should approach unity. On the other hand, the neutrons had to be produced by $\alpha$-bombardment of certain elements, and since only one in 100,000 alphas found their mark in, say beryllium, neutron sources were correspondingly weak. Only experience would show if neutrons were actually good projectiles.

By happenstance, the Rome Physics Department had access to a plant for producing radon (the gas formed in the first step of radium decay) owned by the city's Public Health Department. When mixed with beryllium powder in a sealed glass capsule, Fermi had a much more powerful neutron source than the polonium–beryllium mixture then in common use. With it, he and his group began bombarding elements in the spring of 1934. Starting with hydrogen, they worked their way up through the periodic table. All elements gave negative results until they reached fluorine, when a few counts in their Geiger–Müller tube announced the presence of artificial radioactivity. Aluminum was next. By late spring they had numerous radioactive elements all the way up to uranium, produced by $(n, \alpha)$, $(n, p)$, or $(n, \gamma)$ reactions and with the respective carriers of the activity identified by radiochemical analysis. Some of the half-lives were very short, and the time to cover the length of the corridor between the irradiation and counter rooms had to be reduced by swift running, with the athletic Fermi usually beating Amaldi.

In the case of uranium, which showed several forms of activity, none of the activities could be ascribed to elements of atomic number greater than that of lead. Hence, they plausibly concluded, the activity was most likely due to a transuranic element produced by an $(n, \gamma)$ reaction followed by $\beta$-emission, perhaps forming element-93 or one heavier yet. In this they were seriously mistaken, as time would tell; indeed, for unknown reasons they failed to heed an article mailed to them by the German chemist Ida Noddack, which offered a quite different suggestion [12-56]. At any rate, their activation results in general, reported almost weekly in short letters to the Italian journal *Ricerca Scientifica* and mailed to prominent colleagues abroad, were of sufficient importance to warrant a full paper to the Royal Society; it was delivered personally to Rutherford by Amaldi and Segrè during their summer visit to Cambridge in 1934 [12-57]. Rutherford, in turn, wrote Fermi, congratulating him on his experiments and on his 'successful escape from the sphere of theoretical physics!' [12-58]. On Segrè's polite inquiry of Rutherford whether the paper might receive speedy publication, Rutherford retorted, 'What do you think I was the president of the Royal Society for?' [12-59].

In Berkeley, meanwhile, Lawrence was aware, not only of the Paris results, but of the ongoing experiments in Rome as well. Though they may not have read Italian all that well, a letter to the editor of *Nature*, dated 10 April, described the activities of some two dozen elements following neutron bombardment. For that matter, Don Cooksey had suggested the effectiveness of neutrons as projectiles a year earlier. 'I suppose that the neutron in the $H^2$ [he had written Lawrence] is the boy that when given an introduction in the company of a proton raises all this merry hell' [12-60]. And at the Solvay Conference later that year, Lawrence had stressed the importance of deuteron bombardment as a source of neutrons, for example with the $Be(d, n)$ reaction; as little as $0.01\,\mu A$ of deuterons on beryllium should produce far more neutrons than Fermi could possibly obtain from natural radioelements [12-61]. Indeed, silver coins, outside the cyclotron vacuum chamber but close to an internal beryllium target under deuteron bombardment, became in a few minutes active enough to set a Geiger–Müller counter crackling at a furious rate [12-62].

By April 1934, Lawrence, Livingston, and Henderson were in full swing verifying and extending the Italian measurements. With $0.7\,\mu A$ of 3-MeV deuterons from a larger ion source in the 27-inch cyclotron, they counted in excess of 500 million neutrons per second from a beryllium target. The cyclotron may not have been the right instrument for doing precision proton–proton scattering *à la* Tuve, Heydenburg, and Hafstad; however, it was the ideal workhorse for grinding out radioactive isotopes. Lawrence's group soon had on hand large amounts of radioactive aluminum, copper, silver, and fluorine.

Aside from its intrinsic as well as utilitarian interest, Lawrence had another, more ambitious reason for pursuing these substances, as he confided in Howard Poillon, president of Frederick Cottrell's Research Corporation and someone with an eye for practical applications:

We are not unmindful of the possibility that we may find a substance in which the radioactivity may last for days instead of minutes or hours, in other words, a substance from which we could manufacture synthetic radium. [12-63]

The closest thing to synthetic radium was discovered in Berkeley in September 1934. Possibly unaware that Fermi had produced the radioactive isotope of sodium, $Na^{24}$, by the $Al^{27}(n, \alpha)Na^{24}$ as well as the $Mg^{24}(n, p)Na^{24}$ reaction, Lawrence did so by the $Na^{23}(d, p)Na^{24}$ route instead. $Na^{24}$, or 'radiosodium,' had several properties useful for biomedical applications. Its half-life was 15 hours, it was nontoxic, and, being an isotope of ordinary sodium, it was not necessary to separate it from the parent elements before application [12-64].

Lawrence demonstrated the activity of radiosodium before audiences in Berkeley and in the East during late spring of 1934, at Virginia, Columbia, Carnegie Tech, Princeton, and many other places. He used a weak salt solution or 'cocktail,' sometimes drinking it himself, sometimes feeding it to willing subjects from the audience (including Oppenheimer in Berkeley). The clicks of a Geiger counter showed how long it took for the activity to reach various parts of the body. Edwin McMillan airmailed fresh samples to Lawrence for each performance in the East, timed to reach him while the activity was still effective [12-65].

The most important property of radiosodium was a highly energetic $\gamma$-ray accompanying its disintegration, emitted from magnesium in the following reaction sequence:

$$Na^{23} + H^2 \longrightarrow Na^{24} + H^1$$

$$Na^{24} \longrightarrow Mg^{24*} + e^-$$

$$Mg^{24*} \longrightarrow Mg^{24} + \gamma.$$

At first they estimated over 5 MeV for the $\gamma$-ray energy, or over three times the energy of the hardest $\gamma$-rays from radium. The $\gamma$-ray energy estimate was soon reduced by 30%, to slightly below 3 MeV. Still, it was quite enough for laboratory applications, including electron–positron pair production and the photodisintegration of the deuteron, as well as for medical uses, as Lawrence wrote in December 1934:

We have succeeded in producing radioactive substances that have properties superior to those of radium for the treatment of cancer, and probably before long we shall make available to our medical colleagues useful quantities of radiosodium. [12-66]

Within two months of his first batch, Lawrence had a full mCi (millicurie) of radiosodium on hand, and two years later could make 200 mCi per day [12-67]. 'Magnets of science produce radiation equal to

$5,000,000 worth of radium,' proclaimed *Science Service* in its summary of a talk by Berkeley biophysicist Paul Aebersold on the production levels at the meeting of the American Physical Society in December 1936. Alas, radiosodium failed to fulfill Lawrence's hope of supplanting radium, when its clinical potency in treating tumorous tissue proved less effective than expected. If anything, other isotopes turned out to be more effective as tracers, in cancer diagnosis, and in treatment, including radio-phosphorus and radioiodine. And the neutrons from the cyclotron would prove considerably more effective than X-rays in destroying malignant tissue.

While the production and study of artificial isotopes by cyclotron bombardment with neutrons, protons, deuterons, and $\alpha$-particles dominated the research program at the Rad Lab until the onset of World War II, it also focused attention on the desirability of a larger 'medical' cyclotron dedicated to biomedical uses. In 1936 work had begun on enlarging the pole pieces of the highly productive 27-inch cyclotron to 37 inches. The same year, however, preliminary design also began on a 60-inch medical cyclotron. The Chemical Foundation, which had supported the 27-inch cyclotron, now pledged $68,000 for the big machine, which would be in charge of John Lawrence, Lawrence's younger brother. John Lawrence, a medical scientist, became interested in the biological effects of neutrons during a stay in Berkeley in the summer of 1935, convalescing from an automobile accident. University Regent W H Crocker provided what was needed to house the machine, and with his munificence the Crocker Radiation Laboratory came into being.

It was John Lawrence who alerted the Berkeley scientists to the importance of enforcing radiation protection measures, especially from neutrons. He put a mouse in a little box built into the cyclotron wall, and bombarded it with neutrons. Seemingly, mice agreed even less with neutrons than with heavy water; after five minutes the mouse was dead. Only later did they discover that the mouse died from lack of air! All the same, the fizzled experiment made a deep impression at the time, and planted the seed for the laboratory's excellent reputation in matters pertaining to radiation safety.

In contrast to the earlier cyclotron magnets, a scale model preceded the 60-inch magnet, crafted by Luis Alvarez with input from Robert Wilson. (Livingston was no longer around, having left for Cornell University in 1934, as the first 'cyclotron missionary' from Berkeley.) It fell on William Brobeck, a mechanical engineer, to scale up the model to the full-size magnet, taking over from Donald Cooksey; in so doing he set the pattern for a collaborative effort between accelerator physicists and engineers that would henceforth govern accelerator projects. The order for steel was placed with Columbia and Carnegie–Illinois Steel in March 1937, after much dickering, and a year later the 150-ton rectangular yoke with its circular pole-pieces was in place in the Crocker Laboratory. Gardner Electric supplied the energizing coils, and the magnet was completed in August 1938, at a cost scarcely more than the Poulsen magnet in 1930, thanks to the laminated (rather than cast) steel construction.

The Radiation Laboratory was reorganized into a number of machine groups headed by Brobeck: vacuum chamber and ion source under Cooksey, low-voltage power and wiring (McMillan), radiation protection and electronics (Alvarez), controls and instruments (Arthur Snell), magnet and mechanics (Brobeck), oscillator (Winfield Salisbury), and John Lawrence looking after medical matters. All systems were completed and integrated into a finished cyclotron by the end of 1938— 'truly a colossal machine,' as exclaimed by Edorado Amaldi who passed through Berkeley in September 1939. And so it was; the magnet weighed 220 tons, and the entire laboratory staff could pose before the camera inside the yoke before the pole-pieces and coils were added—a design, notes Heilbron and Seidel, purportedly wasting efficiency in pole-to-pole flux in favor of openness and accessibility for biomedical research [12-68]. It took all of the spring of 1939 to locate and circulate the first beam, but on 10 June the blackboard in the Rad Lab proclaimed 0.6 μA of 19-MeV deuterons, soon boosted to 15 μA.

The blackboard recorded another laboratory milestone when, on 9 November, someone scrawled in chalk, 'Assoc'd Press—Unconfirmed— E.O.L. has Nobel Prize.' There had been rumors from Stockholm that, because of the war, the Nobel Prize in chemistry and physics would not be awarded in 1939 [12-69]. Lawrence received confirmation of the happy news while on a court of the Berkeley Tennis Club. It would be the first Nobel Prize awarded to a University of California faculty member. It was out of the question for him and Molly to travel to Sweden to accept the Prize from the King. On 3 September, the day the war began, John Lawrence had sailed from Europe on the SS *Athenia*, which was torpedoed that evening off the Hebrides; he was the last surviving passenger to board a lifeboat. Ernest Lawrence accepted the medal and diploma from the Swedish Consul General in a moving ceremony in UC-Berkeley's Wheeler Hall, having risen that afternoon from a sickbed for the occasion.

The first cyclotron to be constructed outside Berkeley was a modest 16-inch machine commissioned by Livingston at Cornell in 1934. The next year, no fewer than eight cyclotrons were completed, ranging in pole tip diameter from 13 inches (Washington University in St Louis) to 42 inches (Michigan), and by 1940 there were 22 cyclotrons completed or under construction in the US [12-70]. The first working cyclotron abroad was, curiously, not in a major European laboratory, but at the Institute for Physical and Chemical Research (Riken) in Tokyo under Yoshio Nishina, who had studied both under Rutherford and Bohr

[12-71]. The Riken 26-inch cyclotron was built with technical guidance from Berkeley. It first circulated beam in April 1937, and was followed by a similar machine in Leningrad that fall. The same year, Nishina's laboratory began planning a 60-inch cyclotron, again with Lawrence's help; it became operational in 1941 or early 1942 [12-72]. By 1940, seven additional foreign cyclotrons were operational. The first was Cockcroft's 36-inch cyclotron at Cambridge, and the largest were two 40-inch machines, one in Osaka under Ken Kikuchi and one in Heidelberg under Bothe and Wolfgang Gentner. The largest U.S. cyclotron, after Lawrence's 60-inch medical Crocker machine (and Japan's forthcoming carbon copy), would be the Carnegie Institution's own 60-inch machine. It was designed specifically for research in the physical and biological sciences, not as a medical treatment facility. Medical research was included on the agenda, but was conferred 'no preference whatsoever.' [12-73]

# Chapter 13

# FISSION: RETURN OF LIGHTFOOT

### 13.1. Slow neutrons in Rome; nuclear fission in Berlin

One morning in October of 1934, Edoardo Amaldi and Bruno Pontecorvo, a newcomer to Fermi's little group, made a singularly important discovery. Bombarding a silver target with neutrons, they noticed that the silver became more active when it was irradiated on a wooden table than when it was irradiated on a marble shelf. Curious, they made a series of measurements, first with the source and silver target placed inside a small lead box, then with the source and target outside the box. Outside the box the activity decreased considerably with increasing distance from the source, while inside the box it did not. The next day Fermi joined them, and they decided to pursue the matter further by interposing a lead wedge between the source and target. Since preparing the wedge would take some time, Fermi suggested they try an attenuating filter of some lighter material first. Scrounging around, they located a piece of paraffin, placed it where the lead should be, and activated a Geiger counter next to the target. Behold! The counter crackled madly, nearly going off scale. It was the morning of 22 October, an important date for mankind, as time would tell.

Returning from lunch that afternoon, Fermi had the explanation for the puzzling behavior of filtered neutrons. Paraffin, like wood, contains a great deal of hydrogen, and hydrogen nuclei, protons, are virtually the same size as neutrons. Hence, like a billiard ball being stopped by colliding with a ball of its own size, protons are particularly effective in slowing down neutrons by elastic collisions [13-1]. Slowing down a neutron by repeated collisions with protons in its path gives it more time to spend in the vicinity of, for example, a silver nucleus, with a greater chance of being captured.

The same evening, at Amaldi's home, they prepared a short letter to *Ricerca Scientifica* [13-2], Fermi dictating, Segrè writing, and Amaldi, Rasetti, and Pontecorvo excitedly milling about. Among other things, they had found that slow neutrons produced one activity by the $(n, \gamma)$ reaction, while fast neutrons produced another activity by $(n, 2n)$. That is, the only elements affected by the hydrogenous materials were those which under bombardment gave rise to activities due to isotopes of the starting element. The discovery of the hydrogen effect had an immediate impact on Fermi's research program; for the next several years they concentrated on the slow neutrons, not on the substances produced by them. Corbino, sharp as always, insisted that they take out a patent, both for the production of radioisotopes by neutron bombardment, and also for the enhancement effect by slow neutrons [13-3]. Nobody could then foresee that slow neutrons held the key to nuclear energy—the doom and gloom of *Wings over Europe*. Nor does it seem to have been immediately evident how important the production of heavy water, under way at Norsk Hydro as Fermi's group made their latest discovery, would be in this connection. Deuterium, it turned out, is even more effective than hydrogen in slowing down neutrons.

Alas, the deteriorating political situation in Italy, exacerbated by the Fascist racial laws of 1938 that threatened Fermi's wife, convinced Fermi that he must leave the country. He passed word to Columbia University, New York, where he had spent the summer of 1936, that he would accept a position there, and they were receptive. Providentially, in December of 1938 he learned (tipped off by Niels Bohr) of his having received the Nobel Prize, both for his discovery of new radioactive substances, and for his discovery of the selective power of slow neutrons. On 6 December, the entire family, including their maid, proceeded to Stockholm for the ceremony, and from there directly to New York. By then the rest of the team in Rome was widely scattered, with Rasetti also in New York, Segrè in Berkeley, and Pontecorvo with Joliot in Paris. Only Amaldi stayed on, inheriting the chair in Rome and doing his best to keep some semblance of Italian physics going in the troubled years ahead.

The developments in Rome in 1934 had not been lost on Irène Curie in Paris. She and Frédéric Joliot, too, had received the Nobel Prize, jointly in the tradition of Pierre and Marie Curie, for chemistry in 1935, 'for the synthesis of new radioactive elements.' Returning from Stockholm, Joliot found himself increasingly preoccupied with administrative academic matters, hastened by his election to a vacant chair in chemistry, renamed 'nuclear chemistry,' at the Collège de France. Irène, for her part, was elected professor at the nearby Sorbonne. However, she continued actively in research, now alone, as research director of the Radium Institute, despite deteriorating health from tuberculosis. In particular, she brought her radiochemical skills, inherited from her mother, to bear on the complex broth of substances observed in Rome from bombarding thorium and uranium with neutrons.

Irène and her students first concentrated on thorium, and on one element of the broth with a pronounced half-life of 3.5 hours. They added lanthanum (element 57, a rare earth) as a 'carrier' (stable isotopic impurity) to the solution, then precipitated it back out. The 3.5-hour activity accompanied the lanthanum, instead of remaining in the solution, and must have come from a substance chemically similar to lanthanum. They did not for a minute believe that the substance *was* lanthanum, whose atoms are only a little over half the weight of thorium atoms. Perhaps actinium, element 89, closer in weight to thorium and very radioactive? [13-4].

In 1937 Curie and the young visiting Yugoslavian physicist Pavel Savitch turned to the mixture of isotopes resulting from irradiating uranium. Filtering the radiation from the solution with metallic foils, they again were left with radiation having a half-life of 3.5 hours, which they this time simply ascribed to an isotope of thorium [13-5]. They had not given up on actinium, however. In 1938 they resorted to an elaborate procedure (fractional crystallization) for separating actinium from lanthanum. The 3.5-hour activity followed the lanthanum, not the actinium. They still didn't know what to believe. 'It seems [they concluded] that this substance cannot be anything except a transuranic element, possessing very different properties from those of other known transuranics, a hypothesis which raises great difficulties for its interpretation' [13-6].

Just as Curie and Savitch were reporting their latest results to the Paris Academy, Joliot ran into Otto Hahn at the Tenth International Congress of Chemistry in Rome. It was the first time they met. Hahn and his partner, Lise Meitner, had also been on the trail of Fermi's heavy elements, as Joliot was only too aware. They began in 1934, irradiating thorium and uranium with neutrons, and were peeved with Mme Irène Curie for not having cited them properly [13-7]. They professed, as well, to a certain skepticism over Irène Curie's latest results, which they dubbed 'Curiosum,' and set out to clarify the situation, by repeating the French experiments.

Otto Hahn and Lise Meitner had reason to feel as they did, having collaborated in radiochemistry off and on for over 30 years. Hahn, Meitner's junior by a few months, was born in 1879, the son of a father who had made a successful transition from farming to the glazing business in Frankfurt [13-8]. He obtained his doctorate in organic chemistry at Marburg University. After two more years with his major professor, Theodor Zincke, he went to England at his own expense in 1904, where Zincke had secured a temporary place for him with Sir William Ramsay at University College, London. Ramsay, famous for his discovery of several 'inert' gases, put him to work as his radiological assistant, despite Hahn's ignorance of the basics of radioactivity. Hahn caught on fast; before long he had isolated radiothorium, a new radioelement.

Impressed, Ramsay urged him to forget about his intention of entering industry, and take a post in Emil Fischer's Chemical Institute at the University of Berlin. Hahn agreed, but first wished to spend an additional year abroad, sharpening his skills in radioactivity at the University of McGill in Montreal under Rutherford, the unquestionable guru of the young field. So he did, and added radioactinium to his dossier during his year in Canada. Within a year of returning to his homeland, he was appointed *Privatdozent* in Fischer's Institute, where he teamed up with Meitner.

The third of eight children of a Viennese lawyer, Lise Meitner was the second woman to receive a doctorate in science at the University of Vienna [13-9]. After graduating, she remained in Vienna for a time, during which she was introduced to the intricacies of the new subject of radioactivity by Stefan Meyer. She was also keen on obtaining a deeper understanding of theoretical physics, and with her mentor Ludwig Boltzmann's suicide in 1906, she left for Berlin to study with Max Planck for a year or two, while casting about for a place to do experimental work. There she ran into Hahn, who was looking for a physicist to help him in his work on the chemistry of radioactivity. They joined forces, with Hahn primarily interested in discovering new elements and determining their properties, and Meitner more concerned with their radiations. When the newly formed Kaiser Wilhelm Gesellschaft opened its Institut für Chemie in Berlin-Dahlem in 1917, Hahn was made head of a small department of radioactivity, and Meitner stuck with him.

At the time, Emil Fischer was notoriously biased against women in his laboratories, but Hahn did his radiochemistry in a basement room with a separate outside entrance, and Meitner was tolerated there. This enforced restriction may have been a blessing in disguise, as it resulted in an unusually close collaboration between the two. Their work was interrupted by World War I, with Hahn serving in the chemical corps under Fritz Haber, and Meitner, like Irène Curie, volunteering as a radiological nurse. In 1918 Meitner was appointed head of a newly established physics department in the Institute, while also maintaining a tenuous academic connection with the University of Berlin. She became *extraordinary* professor in 1926.

After 1918, Hahn and Meitner largely went their separate ways. He worked on various aspects of radiochemistry, while she continued her work on the relationship between $\beta$- and $\gamma$-rays, and on the attenuation of $\gamma$-rays in matter. Only in 1934 did they resume their collaboration, following up on Fermi's neutron-bombardment experiments in Rome [13-10]. In particular, they sought to isolate and identify what seemed to be transuranic decay products for uranium, due to the accompanying $\beta$-activity. (Nuclei of lower atomic number might result from the ejection of protons or $\alpha$-particles.) Using precipitation procedures, they separated out elements 90, 91, and 92; apparently elements beyond 92 remained in the irradiated solution. In the spring of 1935, Hahn, as director of the KWI for Physical Chemistry, found it necessary to ask Fritz Wilhelm Strassmann, one of his assistants, to join him and Meitner in their investigation. Soon they had 10 half-lives on their hands, some apparently due to uranium itself, and some ascribed to transuranics: EkaRe (element 93), EkaOs (94), and EkaIr (95). (The prefix 'Eka' indicated that they were higher homologues of rhenium, osmium, and iridium.) Some were initiated by fast neutrons, some by slow, and complicating matters further (simplifying, actually), some were *isomeric* states (single isotopes having more than one half-life).

Curie and Savitch's latest paper on the mysterious 3.5-hour substance reached Hahn on about 20 October 1938. Not having seen that particular activity himself, his first reaction was that the Paris team was 'muddled up' [13-11]. Unfortunately, Meitner was no longer with them. Three months earlier, being of Jewish extraction, she had fled Berlin for Holland in the wake of the annexation of Austria, and thence to Copenhagen. There Niels and Margrethe Bohr received her warmly, and Bohr made final arrangements for her prearranged exile in Sweden with Manne Siegbahn in his new Nobel Institute for Experimental Physics nearing completion outside Stockholm. Though a 35-inch cyclotron was also under construction there, working conditions in Siegbahn's Institute would be far from those she had known at the KWI for Chemistry in Berlin.

On second thought, Hahn decided he couldn't simply ignore the French paper, and passed it on to Strassmann, a skilled physical chemist and by 1938 acknowledged as an equal partner in his collaboration with Hahn and Meitner. Strassmann went over the paper with care, suspecting that the purported 3.5-hour substance was real. He soon convinced Hahn that, due to the French choice of a lanthanum carrier, their precipitate contained a mixture of radium and its decay products. Here then, if really true, was a case of a remarkable jump from uranium, element 92, to radium (88), involving an unprecedented double $\alpha$-particle emission. Hahn and Strassmann proceeded on several radiochemical paths, only one which need concern us here: using a barium carrier to precipitate alkaline-earth elements such as radium. On 8 November they reported the gist of their preliminary findings to *Naturwissenschaften* [13-12]. Seemingly, a multi-step process was at work. First, the uranium, absorbing a slow neutron, decayed by $\alpha$-emission to a short-lived thorium isotope. The thorium isotope, in turn, decayed by $\alpha$-emission to an isotope of radium. The radium isotope, finally, decayed by $\beta$-emission to an actinium isotope which, it being a $\beta$-emitter as well, led back to thorium, completing the isomer loop. If true, it was also the first observation in Berlin of $(n, \alpha)$ reactions with slow neutrons.

Now came the hardest work yet, a grueling round of tests for the alleged radium, in an effort to separate it from the barium carrier by fractional crystallization—a procedure devised by Marie Curie. The problem was the feebleness of the radiations involved, representing no more than a few thousand atoms. What few emitters there were could easily be carried away by the overwhelming number of inactive barium atoms without detection. Accordingly, Hahn diluted some natural (off-the-shelf) radium down to the intensity of the preparations separated by Strassmann using barium chloride as a carrier, and passed it through the whole radiochemical process. It separated without difficulty. Slowly it dawned on Strassmann that the alleged radium not only behaved like barium, but *was* barium—element 56, slightly more than half the atomic weight of uranium. One more test, completed on 19 December, clinched the matter. If the supposed radium isotope was truly radium (element 88), it should decay by $\beta$-emission to actinium (89). If it was actually barium, it should decay by $\beta$-emission to lanthanum, element 57, which could be separated from actinium by fractionation. The test showed without doubt that the alleged radium was indeed barium. Still cautious, Hahn wrote Meitner the same day. 'We more and more come to the terrible conclusion: Our radium isotopes do not behave like radium, but like barium....' [13-13].

The day after the final actinium–lanthanum experiment was completed, the annual Christmas party was held at the KWI, with the Institute scheduled to close the day after that. Hahn hurriedly revised his and Strassmann's paper, largely written by then. Originally titled 'Concerning the Existence of Radium Isotopes Resulting from Neutron Irradiation of Uranium,' they substituted 'Alkaline Earth Metals' for 'Radium Isotopes,' thus avoiding committing themselves fully to barium. The short manuscript was rushed to the publishing house Springer-Verlag on Linkstrasse on Thursday 22 December. The essence of the communication read as follows:

> We come to the conclusion that our 'radium isotopes' have the property of barium. As chemists we should actually state that the new products are not radium, but rather barium itself. Other elements besides radium or barium are out of the question. [13-14]

Meitner, having received Hahn's letter of 19 December in Stockholm on the 21st, replied immediately.

> Your radium results are very startling. A reaction with slow neutrons that supposedly leads to barium! By the way, are you quite sure that the radium isotopes come before actinium?... And what about the resulting thorium isotopes? From lanthanum

one must get cerium. At the moment the assumption of such a thoroughgoing breakup seems very difficult to me, but in nuclear physics we have experienced so many surprises that one cannot unconditionally say: it is impossible. [13-15]

While deeply preoccupied with these matters, Meitner was obliged to leave town. It was just as well, as it turned out. She had made arrangements to spend the Christmas holidays in Kungälv (King's River) outside Göteborg on Sweden's southwest coast, with a friend, the physicist Eva von Bahr-Bergius who had helped with getting her out of Germany. Joining them would be Otto Robert Frisch, Meitner's nephew, then in Copenhagen. We met Frisch briefly in Vienna, in connection with the troublemakers Pettersson and Kirsch. After receiving his doctorate under Karl Przibram in 1926 [13-16], Frisch gravitated to the Physikalisch-Technische Reichsanstalt in Berlin, and then to the University of Hamburg, spending three years there with Otto Stern. In 1933 the racial laws of the National Socialist government forced him, like many other Jewish physicists including Stern, to leave Germany. Supported by Rutherford's Academic Assistance Council, Frisch worked for a short period with Patrick Blackett at Birkbeck College, London, before settling in for the duration at Niels Bohr's Institute in Copenhagen. There he took up experimental nuclear physics, mainly studying slow neutrons in the company of Hans von Halban. He was still there when Tante Lise passed through in July of 1938. He had shown her the new cyclotron in care of L Jackson Laslett, on loan from Lawrence in Berkeley, and they relaxed with the Bohrs at their Tisvilde summer villa by the sea.

Now, five months later, Frisch joined Meitner for breakfast at an inn on Västra Gata in Kungälv, where she was poring over Hahn's letter. She handed it to Frisch.

I had to read that letter. Its content was indeed so startling that I was at first inclined to be skeptical. Hahn and Strassmann had found that those three substances were not radium…[but] barium. [13-17]

They took a long walk in the fresh snow, across the Kungälv and into the woods beyond—he on skis and she keeping up on foot. Finally they sat down on a log, and pondered the results [13-18]. Gradually they came up with an explanation.

Instead of chipping a few protons or $\alpha$-particles off the uranium nucleus, the explanation seemed to lie in the liquid-drop model of the atomic nucleus, proposed by Gamow in 1929. It was utilized by Bohr, in his 1936 paper on the compound nucleus in nuclear reactions. The model treats nuclear matter as consisting of protons and neutrons interacting via strong, short-range forces in a collective manner [13-19]. As a drop of liquid, the model is characterized by a certain temperature, surface tension, and modes of vibration. The total energy, or mass, of the droplet is determined by the mean binding energy per nucleon, modified by a surface energy and an electrical (Coulomb) energy between the protons alone. The surface tension holding the drop together is opposed by the electrical repulsion of the protons; the heavier the element, the more intense is the repulsion. The uranium nucleus is so heavy as to be on the verge of instability, easily upset by the addition of a single neutron. Upon capture of the incident neutron, the imparted energy is shared among the constituents of the nucleus. The nucleus goes into an unstable, oscillatory mode, one of which is an elongated ellipsoid. When the dumbbell-shaped deformation reaches a critical point, the system breaks into two fragments of roughly equal mass, which fly apart with great energy. The energy driving the fragments apart is given by Einstein's $E = mc^2$, alias the 'packing fraction' for the system: the two smaller nuclei weigh less combined than the parent nucleus plus neutron. To be sure, the 'mass defect' is small — only about 0.1%; however, multiplying by the factor $c^2$ gives a sizable energy output. It should amount, Frisch estimated, to about one-fifth of a proton mass, or about 200 MeV.

A copy of the Hahn–Strassmann manuscript had been mailed to Stockholm, where it sat, Hahn not realizing that Meitner was away on vacation. Upon receiving Meitner's letter of 21 December, however, and perhaps reading between the lines, he, too, came to the same conclusion, that the uranium nucleus had split in two. At the eleventh hour, he managed to alter the page proofs from Springer-Verlag [13-20]. Nevertheless, their article as published is hesitant and tentative, ending as it does with the comment that 'there could perhaps be a series of unusual coincidences which has given us false indications.'

On 1 January 1939 Meitner returned to Stockholm and Frisch to Copenhagen. The same day Meitner received a copy of Hahn's revised proofs, and acknowledged them on an optimistic, yet melancholy note.

> Dear Otto! I am now almost <u>certain</u> that the two of you really do have a splitting to Ba and I find that to be a truly beautiful result, for which I most heartily congratulate you and Strassmann.... Both of you now have a beautiful, wide field of work ahead of you. And believe me, even though I stand here with very empty hands, I am nevertheless happy for these wondrous findings. [13-21]

Frisch, meanwhile, got in touch with Bohr, explaining his and Meitner's momentous conclusion. We shall hear of Bohr's enthusiastic reaction, and its aftermath, in the next section.

In his careful analysis of 'internal and external conditions' in the background of the discovery of fission by the Berlin team, Fritz Krafft is adamant in awarding credit for the discovery equally among the team partners, Otto Hahn, the organic and nuclear chemist, Fritz Strassmann, the analytical and physical chemist, and Lise Meitner, the theoretical and experimental nuclear physicist [13-22]. Meitner's claim to collaborator rests, in Krafft's opinion, not only on her correct interpretation, with Frisch, of the Hahn–Strassmann results, but on her original instigation in 1934 of the joint study of the 'transuranics' that led to the final discovery. Strassmann, moreover, with his excellent knowledge of physical chemistry, provided the missing link between the chemistry and the physics. And here we seemingly have the reason why the discovery of fission was made in Berlin, not in Rome or Paris, namely the interdisciplinary forces brought to bear on the problem of the transuranics by the well-rounded team in Berlin [13-23]. Fermi's team in Rome, and Irène Curie's in Paris, were by and large physicists led by physicists.

## 13.2. News of fission reaches America

On 30 November 1938, Merle Tuve, George Gamow, and Edward Teller sat down and drew up a program for the Fifth Washington Conference on Theoretical Physics, sponsored as usual by the Carnegie Institution of Washington and George Washington University [13-24]. To be held sometime during January 1939, the subject this time would be 'Magnetic, Electric, and Mechanical Properties at Very Low Temperatures.' In the end, about 40 participants were invited, of whom only some would have their expenses paid, and 20 more local physicists were expected as well. Formal letters of invitation went out on 22 December – the very day Hahn and Strassmann rushed their paper to *Die Naturwissenschaften*. Not all were experts in low-temperature physics to be sure, and at least one leading US low-temperature physicist (William F Giauque of Berkeley) could not attend. At any rate, the conference is remembered mainly for its historic opening session on 26 January 1939, appropriated, in effect, by Niels Bohr and Enrico Fermi for discussing Hahn and Strassmann's remarkable finding, as we shall get to shortly.

If not for Hitler, the latest international exchange on nuclear developments would have been in Brussels in late October 1938. The Eighth Solvay Conference had been scheduled to convene that month, on 'Elementary Particles and the Mutual Interactions.' Alas, dark political clouds were drawing over Europe. Hitler had annexed Austria the previous March, and in late August, Winston Churchill warned his countrymen that 'the whole state of Europe and of the world is moving steadily towards a climax which cannot long be delayed' [13-25]. The climax came one month later when, on 29 September, Hitler and Neville

Chamberlain dismembered Czechoslovakia. The next day, while Chamberlain flew home with 'peace in our time,' Hahn's laboratory in Berlin-Dahlem was half deserted because most had gone to a reception for Hitler [13-26]. The Solvay Conference was canceled, and the small conference that Bohr had planned in Copenhagen prior to it was canceled as well.

While Hahn and Strassmann's radiochemistry ran its low-keyed course at the KWI for Chemistry in Berlin-Dahlem, impressive high-voltage installations were springing up in many places, including a cascade generator in the stark, silo-like 'Tower of Flashes' of the KWI for Physics across town from Hahn's Institute. Many of these installations contributed little of note to exploratory nuclear physics, for all their seeming might. One exception was the Carnegie Institution's Atomic Physics Observatory, housed in its own intricately brick-patterned silo. Roughly the same size as the KWI Tower but far more handsome, the APO would house a more powerful, pressurized Van de Graaff generator, being installed under the supervision of Robert Meyer (Odd Dahl's successor) during the spring and summer of 1938. The 'mountain of steel and porcelain,' as Tuve called it, was ready for initial test runs at atmospheric pressure during that fateful month of September 1938 [13-27]. In October the generator exceeded 5 MeV under pressure, but without the accelerating tube yet in place [13-28]. The tube itself, plus ion source, was installed by early December, allowing several micro-amperes on the target in time for the Trustees' visit on 9 December [13-29].

The day after the APO demonstration was a bigger day for physics in Stockholm, as the day Enrico Fermi received his Nobel Prize from King Gustavus V in the city's Concert Hall. Following the festivities, the Fermis — Enrico, his wife Laura, the two children Nella and Giulio, and a maid — traveled to Copenhagen for the usual warm welcome by Niels and Margrethe Bohr. Leaving Copenhagen for Southampton, they boarded the Cunard liner *Franconia* on 24 December. They arrived in New York on 2 January 1939, greeted on the Cunard pier by George B Pegram, chairman of Columbia's physics department, and G M Giannini, a prominent Italian-American business man. Giannini was the agent for Italian patent holders, including the Rome group's patent on slow neutron bombardment. That patent was, of course, the basis of the moderator in nuclear reactors, and years later the US Government paid Laura Fermi, by then Enrico's widow, an ex-gratia sum for the plutonium production reactors at Hanford, Washington.

The day after the Fermis landed in New York, Otto Frisch, back in Copenhagen from his time with his aunt in Kungälv, had a hasty conversation with Bohr, who was himself preparing to sail for the United States as well. Frisch hurriedly briefed him on Hahn and Strassmann's results, and his and Meitner's interpretation of them. Bohr listened attentively, agreeing fully that this splitting of a heavy nucleus into two big pieces is practically a classical phenomenon, which does not occur below a certain energy [13-30]. Only later did Frisch dramatize his historic conversation with Bohr. Frisch had hardly begun speaking when Bohr struck his forehead and burst out:

> Oh, what idiots we all have been! Oh, but this is wonderful! This is just as it must be! Have you and Lise Meitner written a paper about it? [13-31]

Frisch said no, but that they would do so right away. On 5 January he and Meitner outlined a note to *Nature* over the telephone. The next day Frisch scribbled a first draft, which on Bohr's request he brought by trolley that Friday evening to Bohr's *Aeresbolig* (Residence of Honor) on the Carlsberg brewery grounds. The two of them discussed the problem in greater depth, and Bohr, who was leaving town the next day, made several minor clarifying corrections to their note. The next morning (Saturday 7 January), there was only time for Frisch to type part of the manuscript draft. He delivered two pages to Bohr at the central Copenhagen train station, from which Bohr and his 19-year-old son Erik were departing for Göteborg on the boat train at 10:29 a.m. Bohr had no time to read it, but pocketed it, promising not to say anything about the note to his American colleagues until he received word from Frisch that it was actually in print.

Frisch phoned Meitner on Sunday 8 January, recounting his discussion with Bohr. He told her that he had shared the eagerly awaited paper by Hahn and Strassmann, that had appeared in print only the day before, with George Placzek, a Bohemian theorist whose opinion counted. Placzek, a great wit and skeptic of most things, was also enjoying the Bohr Institute's hospitality just then. (That same day Placzek left for Paris, and within a month he would catch up with Bohr in Princeton, and quiz him as well about uranium fission, with important consequences.) Placzek urged Frisch to look himself for the telltale, fast-moving fragments signaling uranium breakup. 'Oddly enough, that thought hadn't occurred to me,' muses Frisch [13-32]. All it required was a piece of uranium-coated foil, exposed to one of the Institute's neutron sources, and an ionization chamber and proportional amplifier biased to exclude the natural $\alpha$-particles from uranium in favor of the ionization bursts from the massive fission fragments. On 13 January, 'pulses at about the predicted amplitude and frequency' were observed flickering on a cathode-ray oscilloscope without difficulty [13-33].

Frisch spent another day or so refining his data and writing up his results. While at it, he asked an American biochemist working with Georg von Hevesy on a Rockefeller grant at the Institute, William A Arnold, what biochemists call the process whereby cells divide. 'Binary fission' was the answer. By that time Frisch's joint paper with Meitner was largely written, and Frisch proposed that they avail themselves of the new term, shortened to 'fission' [13-34]. On 16 January Frisch mailed the two manuscript notes, his and Meitner's joint, theoretical note and his own experimental note, to *Nature* [13-35]. Ideally the theoretical contribution should have been submitted first. In the event, the theoretical paper, purely by happenstance, preceded the experimental note into print by a week.

16 January, it turned out, proved even more consequential for the Frisch–Meitner–Bohr collusion. At 1 p.m. on that day, Bohr arrived in New York, greeted on the Swedish-American Line's pier by Enrico and Laura Fermi, in the company of John Archibald Wheeler—a colleague who was looking forward to interacting with Bohr at Princeton's Institute for Advanced Study. Bohr and young Erik had sailed from Göteborg on the liner *Drottningholm* on 7 January, accompanied by Léon Rosenfeld, another close colleague with whom Bohr was to collaborate in Princeton [13-36]. En route to New York, Bohr and Rosenfeld worked steadfastly on the uranium fission mechanism, despite heavy seas all the way and Bohr feeling miserable. Bohr failed to warn Rosenfeld of his promise of confidentiality to Frisch about his and Meitner's paper, which Rosenfeld assumed had already been submitted for publication. On the very day they disembarked in New York, 16 January, one of the regular Monday-evening meetings of Princeton's Physics Journal Club was scheduled, with John Wheeler in charge. Bohr and Erik went off with the Fermis, but Wheeler brought Rosenfeld along to Princeton. During the evening session, Wheeler asked Rosenfeld if he had anything to report from Europe, and Rosenfeld spilled the beans, telling the gathering about the splitting of uranium—the term 'fission' not yet having reached American shores.

> The effect of my talk on the American physicists [recalled Rosenfeld] was more spectacular than the fission phenomenon itself. They rushed about spreading the news in all directions. [13-37]

Bohr, to his distress, learned of Rosenfeld's unfortunate revelation when he arrived in Princeton the next day to take up residence as visiting professor with the Institute for Advanced Study. Nor was there any news from Frisch awaiting him, as he had expected. He immediately set about drafting a letter to *Nature* outlining his and Rosenfeld's deeper understanding of the fission process amassed during their Atlantic passage, and also serving as an explicit defense of Frisch and Meitner's priority in the fission interpretation among the flood of papers from physicists that was sure to come. On 20 January, still not having heard from Frisch, he mailed his own note to Frisch, along with a covering letter asking him to have his secretary, *Frøken* Betty Schultz, forward it to

*Nature*, 'if, as I hope, Hahn's article has already been approved, and the note by you and your aunt has already been sent to *Nature*' [13-38]. P.S., he added, 'I have just seen Hahn and Strassmann's article in *Naturwiss.*' Four days later, with still no word from Frisch, he wrote him again; when might he see their final note to *Nature*, and what was their reaction to his own note? His letters must have crossed Frisch's long overdue letter to himself, written and mailed on 22 January. Frisch excused his tardiness on being exhausted after burning midnight oil on his experiment, which he didn't think was all that important, simply furnishing evidence for a discovery already made. Moreover, he felt, the very idea of cabling Bohr was too presumptuous.

Isidor Rabi of Columbia University was among those who heard Rosenfeld's report before the Princeton Journal Club on 16 January, and carried the news of Hahn and Strassmann's discovery back to Columbia. Fermi himself seems to have learned the news from Willis Lamb, another Columbia man who was spending some time at Princeton that week [13-39]. On Wednesday 25 January, Fermi suggested to John Dunning, the resident expert on neutron physics at Columbia, that they look for fission fragments with the 35-inch cyclotron in the basement of Pupin Hall, built primarily by Dunning and Herbert Anderson, his graduate student. Anderson happened to have on hand an ionization chamber with a linear amplifier; all that was needed was to prepare a layer of uranium oxide on an electrode and insert it in the chamber. With deuterons on beryllium, they had a far more potent source of neutrons than Fermi's radon–beryllium sources in Rome. (They were, of course, unaware of Frisch's similar experiment in Copenhagen.) As they were assembling the equipment, Bohr came calling, looking for Fermi. Knowing the cat was out of the bag, he had to see Fermi about Hahn and Strassmann; *something* would have to be said about fission at the Washington Conference on Theoretical Physics, set to open the next day, 26 January. Not finding Fermi, he had a brief chat with Anderson in the cyclotron vault, telling him what he already knew, and then caught the train for Washington.

Returning from somewhere, Fermi helped Anderson and Dunning get started on the experiment before he, too, left for Washington. Carrying on, Anderson and Dunning had trouble with the cyclotron, and switched to a Rn–Be source instead [13-40]. However, by that time it was getting late and they gave up for the day. Later in the evening, Anderson had second thoughts and returned to the basement of Pupin Hall. Around 9 p.m. he had the neutron source in place next to the ionization chamber, and soon was able to observe a number of large spikes on the face of the oscilloscope, obviously due to energetic fission fragments, amidst the numerous smaller pulses from the spontaneous $\alpha$-radiation from the uranium decay. When Dunning, too, showed up later that evening, he promised to cable Fermi right away. If he did, Fermi missed it, and did not learn of the evening's success at Columbia until the close of the Washington meeting.

And so it came about that atomic energy was liberated on a large scale in New York's Morningside Heights, exactly ten years after it was daringly suggested in *Wings over Europe* by the New York Theater Guild in downtown Manhattan, a few miles away.

Bohr looked up Gamow as soon as he arrived in Washington. Gamow, in turn, called Teller. 'Bohr has just come in. He has gone crazy. He says a neutron can split uranium.' Teller, for his part, 'remembered that when Enrico Fermi...had bombarded uranium with neutrons, a great variety of radioactive substances were produced.... I suddenly understood the obvious.' Fermi, arriving in Washington, found Bohr and learned to his disappointment that Frisch had presumably anticipated Columbia in an experiment similar to the one in progress in Pupin Hall [13-41].

The Fifth Washington Conference on Theoretical Physics opened at 2 p.m. on 26 January, in Room 105, Building C of George Washington University [13-42]. The prestige of the event is obvious from a group photograph of 51 of the 56 participants, including Fermi, Bohr, Gamow, Teller, Otto Stern, Urey, Breit, Rabi, Rosenfeld, Fritz London, Hans Bethe, and Vannevar Bush (CIW's new president). Conspicuously missing were Lawrence, Lauritsen, and other physicists from the University of California and Caltech. Gamow opened the meeting, by introducing Bohr, who was followed on the dais by Fermi. Bohr and Fermi's bombshell, pre-empting opening talks on low-temperature physics, riveted the audience with news, in the words of Merle Tuve,

> ... of the remarkable chemical identification by Hahn and Strassmann in Berlin of radioactive barium in uranium which had been bombarded by neutrons. Professors Bohr and Rosenfeld had brought from Copenhagen the interpretation by Frisch and Meitner that the nuclear 'surface-tension' fails to hold together the 'droplet' of mass 239, with a resulting division of the nucleus into two roughly equal parts. Frisch and Meitner had also suggested the experimental test of this hypothesis by a search for the expected recoil-particles of energies well above 100,000,000 electron-volts which should result from such a process. The whole matter was quite unexpected news to all present. [13-43]

The first reference to low-temperature physics, and to an interesting connection between nuclear physics and low temperatures, as pointed out by Teller in his own presentation, next on the agenda, was the fact that the balance between 'zero point energy' [energy retained by atoms and molecules at the temperature of absolute zero — a quantum-mechanical

199

concept] and potential energy is very similar in the liquid-droplet model of the atomic nucleus and in liquid helium-II.

At the end of the Bohr–Fermi discussion, several participants looked at each other and stole out of the auditorium, heading for the nearest phone to alert their respective laboratories to take Frisch and Meitner up on their suggestion to look for the expected recoil particles. One of them, from Johns Hopkins University, got in touch with his colleagues R D Fowler and R W Dodson, who confirmed the discovery of fission on Saturday morning, 28 January, with 2.4 MeV d–d neutrons from their high-voltage set. They, too, bombarded a layer of uranium nitrate deposited on a paper disk in contact with the mesh-covered opening into an ionization chamber, and observed massive kicks on the oscilloscope from heavily ionizing particles. Similar results were observed under neutron bombardment of thorium oxide [13-44].

Fowler and Dodson barely beat Richard Roberts, Robert Meyer, and Larry Hafstad to the punch. As Bohr addressed the Conference, Tuve had whispered something to Hafstad, who nodded to Roberts. Hafstad and Roberts, too, stole out and dashed across town from Foggy Bottom to the DTM campus in Chevy Chase. There they joined Meyer, intent on getting an experiment going. They hoped to use the trusty two-meter generator, but its ion-source filament was burned out. The new 5-MeV generator, in the Atomic Physics Observatory next door to the Experiment Building, was operational at long last; however, it had a leak in the vacuum tube. Finding the leak promised to be quicker than replacing the filament, and Roberts and Meyer went to work in the APO. It took longer than expected, however, and only on Saturday did they have the apparatus up and running. Hafstad, when things were finally under control, went off skiing for the weekend, 'confident that Roberts could run the show' [13-45]. As had their colleagues elsewhere, they mounted an ionization chamber below the neutron source, with interchangeable copper disks on the collector, which was connected to a linear pulse-amplifier. The upper faces of these disks were coated with the materials to be tested—not only uranium and thorium, but all the heavy elements they could lay their hands on.

Richard Brooke Roberts' ancestors fought in the Indian wars of colonial times, and his grandfather and father made fortunes in Pennsylvania oil and banking, resulting in Robert's financial independence. He received the best of schooling and earned his PhD in experimental nuclear physics on the d–d reaction at Princeton under Rudolf W Ladenburg in 1936, and, like Tuve, had aspirations for a postdoctoral year under Rutherford at Cambridge. With passport in hand, he stopped by at DTM to say goodbye to Tuve, who had provided much essential advice to the Princeton Van de Graaff group on ion sources and acceleration tubes. The ensuing conversation echoed Tuve's own chat with Fleming in 1926. 'I told Tuve of my plans and the experiments that I had in mind for Cambridge when he said "Why not do them here? We have better equipment"' [13-46]. With that, Roberts reported for work at DTM in January 1937. His first project, with Tuve and Meyer, was an experiment on the magnetic deflection of the neutron, as a measure of its magnetic moment. In the event, the experiment was left uncompleted, due to the leakage of slow neutrons out of the scattering equipment. Instead, Roberts turned to the transmutation of lithium isotopes with the two-meter generator and Lynn Rumbaugh of the Bartol Research Foundation, while Tuve remained preoccupied with plans for the new APO facility. During the balance of 1937 and into 1938, Roberts busied himself with the scattering of protons, neutrons, and deuterons, and on the production of radioisotopes, *à la* Berkeley. However, he, too, became more and more sidetracked into assisting in the completion and voltage calibration of the APO.

Thus it was that Roberts was a key member of the APO team when he and Meyer found themselves, seemingly, to be the first attempting to observe the actual splitting of the uranium nucleus. The setup was ready Saturday afternoon, and around 4:30 Roberts and Meyer saw the first pulses on their scope due to 100-MeV fission fragments flying apart from Li$(d, n)$ neutrons striking the uranium-oxide coated disk; the huge vertical spikes on the face of the tube stood out sharply next to the pulses from neutron recoils and background fuzz from $\alpha$-particles. They promptly ran through a full sequence of tests, including thermalizing the neutrons with blocks of paraffin on top of the ionization chamber, and cadmium for filtering out the thermal neutrons. Thorium also 'split,' but no effect was observed from bismuth, lead, thallium, mercury, gold, platinum, tungsten, tin, or silver. Uranium was split, by apparently different processes, by fast *and* slow neutrons, but only fast neutrons were effective in thorium [13-47]. After supper Roberts informed Tuve of their results, and he immediately called Bohr and Fermi.

The Conference had ended, but Bohr, with his son Erik and Fermi, came that evening; soon they were joined by others in the APO control room. All eyes were glued on the circular face of the oscilloscope, where brilliant green lines shot up, as Tuve, Roberts and Meyer 'were privileged to demonstrate [the fission fragments] to Professors Bohr and Fermi' [13-48]. DTM director Fleming had had the sense of bringing along a photographer for the momentous event. His historic photograph (Figure 13.1) shows the small party crowded into the small, 10-foot diameter circular APO target room: Meyer, Tuve, Fermi, Roberts, Rosenfeld, Erik Bohr, Breit, and Fleming, caught in no particular order around the box for the ionization chamber and pre-amplifier. (Missing from the photograph is Teller who, with his wife Mici, were exhausted from the strain of acting as host and hostess for the meeting [13-49].)

**Figure 13.1.** *Verification of uranium fission at the Carnegie Institution on 28 January 1939 was witnessed by (left to right) R C Meyer, M A Tuve, E Fermi, R B Roberts, L Rosenfeld, E Bohr, N Bohr, G Breit, and J A Fleming. DTM Archives 16219.*

Only Fermi and Breit are grinning broadly, perhaps more aware than most of the importance of the occasion; Bohr looks serious, and we know why. 'I had to stand and look at the first [sic] experiment,' he wrote Margrethe, 'without knowing certainly if Frisch had done the same experiment and sent a note to *Nature*' [13-50].

Thus it happened, as Tuve capped his report on 'droplet fission,' that 'the measurements on this extremely interesting new process in uranium and thorium were the first experiments carried out with our new 5,000,000-volt equipment for nuclear physics' [13-51].

'Atom Explosion Frees 200,000,000 Volts; New Physics Phenomenon Credited to Hahn,' ran the headline in the *New York Times* the next morning, Sunday January 29, apparently quoting Fermi in Washington, and without mentioning Strassmann or Frisch's presumed confirmation experiment. Science Service, quoting Tuve, put it more dramatically in a fuller synopsis the next day. 'IS WORLD ON BRINK OF RELEASING ATOMIC POWER? THIS QUESTION ASKED AS BOMBARDMENT OF URANIUM RELEASES MILLIONS OF VOLTS OF ENERGY; EXPERIMENT MAY BE AS IMPORTANT AS DISCOVERY OF RADIO-ACTIVITY.' Referring to Frisch's, as well as to Columbia's experiment, the article continued, picking up the thread at DTM.

> The Carnegie Institution of Washington's Department of Terrestrial Magnetism... got into action with their new atom smasher as soon as they heard of the Berlin experiment and in a historic midnight experimental conference on Saturday (January 28) demonstrated to Professor Bohr and Professor Enrico Fermi... the reality of the energy's release.
>
> Convinced of the reality of the energy release from uranium, there will be a great rush to complete science's current mystery problem. Neutrons will be turned—are being turned this moment—upon other heavy elements. Perhaps some even cheaper element will yield such energy. But uranium costs only a few dollars a pound. [13-52]

Bohr returned to Princeton on Sunday. There he learned to his relief of Frisch's successful experiment from a casual remark in a letter waiting from his son Hans. At long last, on 2 February, he received Frisch's apologetic letter of 22 January, with Frisch's and Frisch and Meitner's notes enclosed. Jumping with joy, he replied to Frisch in a long letter the next day, and at the same time revised his own note to *Nature* accordingly. He enclosed the revised note in his letter to Frisch, who passed it on to *Nature*; it appeared in print on 25 February [13-53].

On 7 February Bohr again sketched the history of the discovery and interpretation of fission in a letter to the *Physical Review* [13-54]. The main purpose of the new letter, however, was to discuss an important new finding with regard to the fission process, brought about by a timely visit to Princeton by George Placzek a few days earlier. As a result of Placzek's penetrating questions over breakfast at the Princeton Club, Bohr began struggling with the problem of explaining the peculiar dependence of the fission cross section on neutron energy [13-55]. Trudging through the fresh snow to his office in Fine Hall, he concluded that it was not the heavy uranium isotope $U^{238}$ that was responsible for slow-neutron fission, but the rare isotope $U^{235}$, which makes up only 0.7% of natural uranium. That, of course, remained to be proved, and few physicists accepted Bohr's conclusion at first. A test would necessitate enrichment of $U^{235}$ in a natural uranium sample, which then ought to exhibit an enhanced rate of fission by slow neutrons [13-56]. That experiment was only accomplished one year later [13-57].

Within days of their demonstration of 'droplet fission,' Roberts and colleagues at DTM obtained a very important experimental result of their own. It had long been clear that neutrons were emitted during the fission process, but not clear how many; nor was it clear whether they came off simultaneously with the fission event, or were somewhat delayed, as were, for example, $\alpha$-particles following the beta decay of $Li^8$. Indeed, they soon verified the emission of neutrons from uranium, following the fission process, with a period of 15 seconds [13-58]. Delayed neutrons, it would turn out, made control of a nuclear chain reaction possible; only a bomb is possible with prompt neutrons.

### 13.3.  Fission in Berkeley, and Cockcroft again in action

News of fission reached Lawrence's laboratory through the *San Francisco Chronicle* on 31 January 1939. '200 Million Volts of Energy Created by Atom Explosions,' announced Tuesday's *Chronicle* in a small inside article easily missed. Luis Alvarez didn't miss it, for one. He was idly skimming through the paper that morning, while having a haircut in Stevens Union on the Berkeley campus, across from Le Conte Hall. We met Alvarez fleetingly in connection with the design of the 60-inch cyclotron. The son of Dr. Walter Clement Alvarez of the Mayo Clinic, Luis Alvarez was Ernest Lawrence's brother-in-law, in a manner of speaking, since his sister happened to be Lawrence's secretary. Young Alvarez had arrived in Berkeley in 1936 with a fresh PhD on cosmic rays from the University of Chicago under Arthur Compton. Other than that, he had 'essentially no knowledge at all of nuclear physics' [13-59]. He learned fast, however. While rising from a paid assistantship to an unpaid instructorship in the Berkeley physics department, he ended up doing a definitive experiment on hydrogen-3 and helium-3 [13-60]. At any rate, on that Tuesday morning, coming across the brief Associated Press bulletin of what physicists in Washington had heard of certain important developments under Dr. G Hahn [sic] in Berlin, Alvarez reacted at once.

> I stopped the barber in mid-snip and ran all the way to the Radiation Laboratory to spread the word. The first person I saw was my graduate student Phil Abelson. I knew the news would shock him. 'I have something terribly important to tell you,' I said. 'I think you should lie down on the table.' Phil sensed my seriousness and complied. I told him what I had read. He was stunned; he realized immediately, as I had before, that he was within days of making the same discovery himself. [13-61]

Philip Abelson was one of Alvarez's best graduate students. He had arrived in Berkeley in 1935 with a Master's degree from Washington State University. In addition to receiving a teaching assistantship, he was soon virtually a full-time member of the cyclotron crew (while taking courses and preparing for his own PhD orals and thesis research).

Abelson, too, remembered well the day news of fission reached the Rad Lab. As usual, that morning he was at the control console, operating the 37-inch cyclotron.

> About 9:30 a.m. I heard the sound of running footsteps outside, and immediately afterwards Alvarez burst into the laboratory.... When Alvarez told me the news, I almost went numb as I realized that I had come close but had missed a great discovery. [13-62]

Abelson's painful realization came about as follows. Having a strong background in chemistry from his undergraduate training at Washington State, Lawrence had suggested, when Abelson had settled in at Berkeley, that he look into the radiochemistry of the transuranics produced under neutron bombardment of uranium — on a hunch that there was something wrong with the decay pattern. In light of the complexities of the decay curves, a better method of identifying the irradiation products than the standard procedure involving inactive carriers and precipitates, was clearly desirable. And such a method was seemingly in hand. In 1937, Alvarez had demonstrated that many radioisotopes emitted characteristic X-rays [13-63], and Abelson thought he might be able to determine the atomic numbers of the transuranium radioisotopes from the X-ray wavelengths emitted.

For a starter, Abelson chose a radioisotope with a 72-hour half-life, separated it chemically from the rest of the brew, and set about recording the X-ray spectral lines on a photographic plate. Assuming he was looking for a transuranic, he expected to see X-rays of energy even higher than those from uranium. In fact, all he saw were X-rays of much *lower* energy. *Voilà!* Once he knew of fission, it made sense. What he had observed were K X-rays from tellurium (atomic number 52) and its daughter element iodine ($Z = 53$). He had stumbled across another mode of uranium fission, since tellurium-52 + zirconium-40 = uranium-92.

With Abelson looking over his shoulder, Alvarez wrote a very brief letter to the editor of the *Physical Review*, John Tate of the University of Minnesota, which Abelson signed and dropped in the mail on 3 February [13-64]. Alvarez himself, meanwhile, had not been idle after bolting from the barbershop and haranguing Abelson. He wired George Gamow for more details, and learned that Tuve's group had detected fission fragment with their generator. Teaming up with G Kenneth Green, a National Research Fellow in Lawrence's group, he set about repeating the experiments of Roberts *et al.*, with a thin-walled ionization chamber, one of whose plates was dusted with uranium oxide ($U_3O_8$), connected to a linear amplifier and oscilloscope. With a certain amplifier gain, they not only confirmed the presence of heavily ionizing 'Hahn–Tuve particles' in a background of natural alphas from uranium, but also, with a pulsed neutron beam, determined that the time delay between irradiation and fission was less than 0.003 seconds [13-65].

Alvarez and Green's letter to the *Physical Review*, mailed on 31 January, appeared in the same issue as Abelson's letter, and several others verifying fission at Berkeley. One, by Dale Corson and Robert Thornton, revealed the fission fragments as oppositely recoiling fragments in a cloud chamber photograph [13-66]. In another approach by Ed McMillan, the fission fragments were caught some time *after* the neutron bombardment had ceased. A thin layer of uranium, deposited on a paper backing, was placed next to a stack of aluminum foils and irradiated by neutrons from the cyclotron. One of the recoiling fission fragments would pass through several aluminum foils before being stopped, and the relative activities of the individual foils were subsequently measured, two hours later, with a Geiger counter [13-67].

As soon as Alvarez had heard from Gamow, he had tracked down Robert Oppenheimer in his office in Le Conte Hall, who 'instantly pronounced the [fission] reaction impossible and proceeded to prove mathematically to everyone in the room that someone must have made a mistake' [13-68]. The next day, Alvarez invited Oppenheimer over to the Rad Lab to view the tall spikes from fission fragments on their oscilloscope, and Oppenheimer needed no further convincing.

> In less than fifteen minutes he not only agreed that the reaction was authentic but also speculated that in the process extra neutrons would boil off that could be used to split more uranium atoms and thereby generate power or make bombs. [13-69]

Added Oppenheimer in a letter to George Uhlenbeck at Columbia on 5 February,

...I think it [is] really not too improbable that a ten cm cube of uranium deuteride (one should have something to slow the neutrons without capturing them) might very well blow itself to hell. [13-70]

Abelson, we might add, received his PhD in May 1939, and in June accepted an offer from the Carnegie Institution to join Tuve, Hafstad, Roberts, and Heydenburg at DTM that fall, along with G K Green (who had received his PhD earlier under P G Kruger at Illinois). Roberts, Green, and Abelson were assigned the task of overseeing the construction of the 60-inch cyclotron at DTM, authorized by Vannevar Bush, CIW's new president, with Roberts in charge. The justification for the cyclotron, basically a copy of Lawrence's Crocker installation, was the influx of biologists to DTM during 1937–39, wishing to irradiate everything from fruit flies to rats, and eager for tracer isotopes. Tuve's guiding principle, not wishing to be drawn into a cyclotron race, was that he wanted a cyclotron as large as, but no larger than, the biggest that already worked [13-71]. Roberts designed the r.f. system, and much else. Before starting, he had a 'final fling' with fission during a short period at Berkeley during the summer of 1939, working with Abelson and the 37-inch cyclotron. They were attempting to find which constituents were responsible for the delayed neutrons from uranium.

We bombarded uranium in the 37″ cyclotron, then tossed the uranium across the room for chemical precipitations, and finally tossed the precipitates to the neutron detector. In the few days we worked we did find which major groups were involved but the main result was getting acquainted with Abelson. I liked him and recruited him for our cyclotron project at CIW. [13-72]

Abelson deserted his newfound colleagues in the spring of 1940, when he returned to uranium, the transuranics, and the problem of separating uranium isotopes. As the clouds of war grew over Europe, the possibilities of uranium fission as a source of power or as a weapon grew as well. Although Tuve, Hafstad, Roberts, and Heydenburg continued making fission measurements needed by others, Tuve remained cautious in assessing the practicability of an atomic bomb project in the escalating crisis; he felt the need for his group to make a more immediate contribution to the nation's military preparedness. His connection with the Navy, established during the ionospheric work, became the basis for a far-reaching activity in the form of Section T (for 'Tuve') of Vannevar Bush's new National Defense Research Committee, concerned with developing the radio proximity fuse for antiaircraft artillery [13-73]. It began with Roberts firing a bullet at a cheap vacuum tube attached to a lead brick suspended from the ceiling (to determine if a vacuum tube could withstand the shock of being fired from a cannon). By the spring of 1942, Section T had outgrown the DTM facilities on the Broad Branch campus in Chevy Chase, and the project was reorganized as the Applied Physics Laboratory, with Tuve its first director, under contract with Johns Hopkins University in Silver Spring, Maryland. The APL became the nation's third largest military-technical project, after the MIT Radiation Laboratory and the Manhattan Project. The proximity fuse first saw action in January 1943, when the cruiser USS *Helena* downed a Japanese bomber with an industrially produced shell; the most spectacular use of the highly successful fuse was against German V-1 flying bombs aimed at London, and in stemming German advances in the Battle of the Bulge.

However, we have strayed from the subject at hand, nuclear fission *circa* 1939. By February of that year it was abundantly clear that uranium undergoes fission during neutron bombardment, splitting into two fragments of roughly equal mass. The combined mass of the two fragments is slightly lower than that of the original uranium nucleus plus neutron. The difference, $\Delta m$, is not much, but multiplied by the prodigious factor of $c^2$ it yields an energy of approximately 200 MeV, shared between the two recoiling fragments. It still remained for this theoretical estimate to be verified by experiment, and this was done at Columbia University during March and April of 1939. Eugene T Booth, John R Dunning, and F G Slack showed that only about 175 MeV of kinetic energy was imparted to the fission fragments, with the missing 25 MeV carried off by other reaction products, perhaps including extra neutrons [13-74].

If, indeed, extra neutrons were released during fission, the possibility of a *self-sustaining* chain reaction could not be discounted. The promise of extra neutrons hinged on the fact that while in light nuclei, the number of protons and neutrons are nearly equal, in heavy nuclei, say $_{92}U^{235}$, there are 92 protons and 143 neutrons, or a considerable 'neutron excess.' When a heavy nucleus splits into two lighter nuclei, the fragments will have too many neutrons to be stable. One way to get rid of the excess neutrons is by liberation of one or more free neutrons. The actual number of neutrons liberated was a constant of great interest; more than one was necessary for the chain reaction to be self-sustaining. Four experimental teams attacked the problem nearly simultaneously: one in Paris (Joliot, Hans von Halban, and Lew Kowarski in Joliot's Laboratory of Nuclear Chemistry), two at Columbia in New York (Anderson, Fermi, and Hanstein, as well as Leo Szilard and Walter Zinn), and one at Imperial College, London (G P Thomson, son of J J, with J L Michiels and G Parry). Let it suffice to note the first results, from Paris as it happened. Joliot himself, in his Nobel Prize address of 1935, had predicted that

> Scientists, disintegrating or constructing atoms at will, will succeed in obtaining explosive nuclear chain reactions. If such

transmutations could propagate in matter, one can conceive of the enormous useful energy that will be liberated. [13-75]

Indeed, a fairly simple calculation, assuming the theoretical prediction of 200 MeV per fission, shows that the energy released by uranium fission in a pound of uranium fuel would be a million times greater than burning a pound of conventional fuel like oil or coal [13-76].

The problem, as Joliot saw it, was how to detect an occasional liberated neutron amidst the torrent of bombarding neutrons needed to induce fission in the first place. In early 1939, he undertook the tricky experiment, assisted by von Halban (his resident expert on slow neutron physics, who had learned radiochemistry under Irène Curie and nuclear physics with Frisch in Copenhagen) and Kowarski (a huge Russian physicist who began with Joliot as his part-time assistant, who had learned radioactivity under Irène Curie and nuclear physics with Frisch in Copenhagen). They immersed a neutron source in a tank of water with and without dissolved uranium nitrate, and plotted the neutron intensity as a function of distance from the source. Their initial result, $3.5 \pm 0.7$ neutrons per fission, was somewhat high, but indicative of the feasibility of a chain reaction [13-77]. (The true value is about 2.5.)

By mid-March, the other teams, under Fermi, Szilard, and G P Thomson, had come to about the same conclusion, that two or three neutrons are released per fission. Just then, German troops invaded what remained of Czechoslovakia after the Munich accord and, largely on the initiative of Szilard, the Americans attempted to have the three journals, the *Physical Review, Nature,* and the *Proceedings* of the Royal Society, delay publishing fission results in the interest of international security. However, Joliot, whose report had been rushed to *Nature* via Le Bouget airport, refused to go along with the idea of voluntary self-censorship. He objected in part because the initial results by Roberts *et al.* at DTM (reporting, to be sure, on *delayed*, not prompt neutron emission) had already been published. Consequently, the early publication of the French results caused quite a stir, precipitating officially supported nuclear energy research in, among other countries, Germany [13-78]. Nevertheless, while Fermi and Szilard relented and the Columbia results also appeared in print, by the end of 1939 a *de facto* blanket of secrecy had settled over most fission research. Thus, further work on the vital constant $\nu$, the number of neutrons produced per fission, was only declassified 10 years after the close of World War II, at the First International Conference on the Peaceful Uses of Atomic Energy in Geneva [13-79].

The literature abounds with accounts of wartime fission research, and of the making of the atomic bomb [13-80]. For our purposes, it suffices to return to the ill-reputed transuranics. After a number of scientists had demonstrated that the supposed transuranics were really fission fragments near the middle of the periodic table, Emilio Segrè, once in Berkeley, resumed the search for transuranics — should they in fact exist. He wound up focusing on a 2.3-day activity, and concluded it was *not* associated with a transuranic element [13-81]. Ironically, in this case he would be proven wrong; the 2.3-day activity was indeed due to a transuranic element, as was demonstrated in 1940 by Ed McMillan, joined briefly by Abelson during the latter's subsequent visit to Berkeley. During 1940-41 the formation of element-93 (neptunium) and 94 (plutonium) was correctly worked out by Berkeley scientists, including Segrè, Lawrence, McMillan, Abelson, and Glenn Seaborg. They are obtainable in the following reaction and decay sequence, starting with fast-neutron fission of $U^{238}$:

$$U^{238} + n^1 \longrightarrow U^{239}$$

$$U^{239} \ (23.5\,\text{min}) \longrightarrow Np^{239} + \beta \qquad \text{(first beta decay)}$$

$$Np^{239} \ (2.35\,\text{day}) \longrightarrow Pu^{239} + \beta \qquad \text{(second beta decay)}$$

Plutonium, it turned out, is, like $U^{235}$, fissionable with slow neutrons, and thus two potential bomb or power-producing materials were suddenly in sight, $U^{235}$ and $Pu^{239}$. The first sample of plutonium, all of $0.5\,\mu g$, was isolated from 3 pounds of irradiated uranium during March 1941, and a larger sample was prepared in Berkeley's 60-inch cyclotron that summer. Uranium-235, on the other hand, had to be isolated from natural uranium by isotope separation in a mass spectrograph, the device originated by J J Thomson and perfected by Aston in 1919; in it, a stream of ions is passed through suitably arranged electric and magnetic fields in such a manner that all ions with the same mass but with different speeds and directions from the ion source are brought to a common focus. By early 1940 the mass spectrograph of Alfred O C Nier at the University of Minnesota had collected enough $U^{235}$ for the Columbia group to confirm Bohr's prediction of its fissioning with thermal neutrons.

Whereas it took about a microgram of $U^{235}$ to verify Bohr's prediction, Fermi's celebrated pile in the University of Chicago Squash Court, which went critical on 2 December 1942, utilized about 6 tons of uranium metal and uranium oxide, embedded in a graphite moderator. (Fermi viewed scarce heavy water as 'too expensive' for CP-1, the first Chicago pile.) As everybody knows, the chain of events in the US that led to Fermi's pile and culminated in the bomb, began with the famous letter from Albert Einstein to President Roosevelt, dated 2 August 1939. Not so well known is a letter from Paul Harteck, the German heavy-water chemist, dated 24 April 1939, to German Army Ordnance. Harteck's letter reads eerily like Einstein's:

We take the liberty of calling to your attention the newest developments in nuclear physics, which, in our opinion, will probably make it possible to produce an explosive many orders of magnitude more powerful than the conventional ones.... That country which first makes use of it has an unsurpassable advantage over the others. [13-82]

The upshot of Harteck's warning was that by the time war was declared in September 1939, Germany, alone among the feuding world powers, had a military office exclusively devoted to wartime applications of nuclear energy. Nor is it well known that while the Manhattan Project was effectively launched at the start of 1942, in June of the same year, with the Soviet counteroffensive gaining momentum on the Eastern Front, Germany officially gave up on the atomic bomb; Albert Speer, among others, had become convinced it could not be produced in time to affect the course of the war. Germany's 'Uranium Club' was not dissolved, but from that time on was only concerned with developing a 'uranium machine' for some vague purpose, perhaps submarine propulsion [13-83].

What, finally, about our British friends in all of this? British military technology had the advantage of a head start in the mid-thirties with 'radio direction finding' spearheaded by the researches of Robert Watson-Watt and his colleagues at Bawdsey Manor on the Suffolk coast. In due course Metro-Vick became heavily involved in the production of CW-power tubes for a chain of radar stations, called Chain Home or CH stations, that sprouted prominently along the southeast coast; they were linked to filter centers that evaluated the picture and relayed it quickly to the fighter squadrons. John Cockcroft's initial involvement with preparations for war had been in preparing St John's College at Cambridge against air raids. However, after Munich he, too, became caught up in radar and related areas of radio technology — mainly spotting and encouraging younger academic scientists to get involved in this work. At the Cavendish, meanwhile, his main preoccupation remained the cyclotron. He was well aware of Hahn and Strassmann's work in Berlin, and of Frisch and Meitner's interpretation of it; Cockcroft had tried unsuccessfully to find a haven for Meitner at Cambridge [13-84]. At any rate, when the war began in earnest, Cockcroft was appointed Assistant Director of Research in the Ministry of Supply, in which capacity he shuttled back and forth between scattered sites comprising the Air Defense Experimental Research Establishment. Before long he was personally engaged in the construction and installation, not simply the administration, of Chain Home sets in the Shetland–Orkney area.

After Dunkirk, Mark Oliphant, still in charge of the Physics Department at Birmingham University, had become involved both in radar (with John Randall and Harry Boot in his department inventing the resonant-cavity magnetron) and in the development of a British version of the proximity fuse. What was clearly needed at this point was an exchange of technical information with the Americans. If their colleagues in the US could obtain plans for secret British equipment, they could also manufacture it on a large scale, and ship it across the Atlantic in what became known as a Lend-Lease arrangement. Consequently, in late summer of 1940, Cockcroft, among other senior uniformed and civilian officials, accompanied Henry Tizard, chairman of the Committee for the Scientific Survey of Air Defense, on the so-called Tizard Mission to Washington, carrying with them the celebrated magnetron in a wooden box secured with thumb screws. Meeting with US Signal and Air Corps representatives, as well as scientists from the Naval Research Laboratory, one evening Cockcroft and R H Fowler also met with Merle Tuve and Larry Hafstad at Tuve's home in Chevy Chase. There the four discussed proximity fuses well into the night. The upshot of that particular meeting was the eventual adaptation of the American fuse for British use in countering the flying bomb threat.

British physicists viewed the possibility of a uranium bomb with cautious alarm. On 10 April 1940, the day after the invasion of Norway and Denmark ended the 'phony war,' a subcommittee of Tizard's Committee held its first meeting at the Royal Society headquarters in London, to weigh prospects for a bomb. The subcommittee's charter members were G P Thomson, its chairman, Cockcroft, Chadwick, Oliphant, and Philip Moon. Known initially as the Thomson Committee, it became better known as the MAUD Committee after a curious telegram from Lise Meitner in Stockholm. Meitner was at Bohr's Institute when the Germans struck Denmark, there to use the Copenhagen cyclotron, since Siegbahn's was not yet operational. Bohr managed to slip her back into Sweden, asking her, once safely back, to cable Owen Richardson at King's College, London, that all was well with the Bohrs. The cryptic telegram caused a stir among Thomson's committee members.

MET NIELS AND MARGRETHE RECENTLY BUT UNHAPPY ABOUT EVENTS PLEASE INFORM COCKCROFT AND MAUD RAY KENT; MEITNER. [13-85]

The reference to Cockcroft was clear enough, in view of his recent concern over Meitner. However, 'Maud Ray Kent' was obviously a code, perhaps having to do with secret German ray research? Only later was it learned that Miss Maud Ray of Kent had once been a governess of the Bohr children.

The ostensible reason for the MAUD Committee was a communication of a more serious nature, namely a memorandum from Otto Frisch and Rudolf Peierls to Oliphant on the practicability of a bomb based on $U^{235}$ [13-86]. Frisch had left Copenhagen for Birmingham University in the summer of 1939. There he struck up with Rudolf Peierls, a German refugee physicist who had been on the Birmingham faculty since 1937. By early 1940 the two had crafted a memorandum on the bomb, including the separation of $U^{235}$, its critical mass [13-87], and the effects of radiation from the blast. Oliphant passed it on to Tizard who, in turn, gave it to Thomson, who took it up with his committee. (Frisch and Peierls could not participate in the Committee's deliberations, being foreign nationals barred from classified Government meetings.)

The same day the Thomson Committee first met, a Frenchman, Lieutenant Jacques Allier, called on Cockcroft in his office in the Ministry of Supply on the Strand. Allier was a senior official in the Banque de Paris et des Pays-Bas, the bank that coordinated the French holdings in Norsk Hydro. He was also an agent in the Deuxième Bureau, France's secret service, and had been entrusted in a mission of the highest importance on behalf of Frédéric Joliot and Raoul Dautry, the French Minister for Armaments. Allier told Cockcroft of his daring mission to Oslo, shortly before the invasion, in which the existing Norwegian stocks of heavy water had been smuggled out of the country by a clever ruse before hoodwinked German agents [13-88]. Flown to Scotland, the metal canisters filled with 185 kg of the precious liquid were brought to France, for use by Joliot and his assistants. Alas, no sooner had the material arrived in Paris than France, too, was on the point of being overrun by General von Rundstedt's Army Group A. The heavy water, accompanied by Halban and Kowarski, was evacuated again, via a succession of French villages to Bordeaux. From there it went by a Scottish coal-carrying steamer to London, and thence to Cambridge. At the Cavendish, Halban and Kowarski continued subcritical chain reaction studies, begun in Paris, with the heavy water as moderator [13-89].

In July 1940 the MAUD Committee concluded that both a uranium 'boiler' and a bomb were indeed practicable. However, in view of the load on the British war effort, it was eventually decided that Britain's uranium research—re-dubbed Tube Alloys—should be transferred to Canada. There a team under Halban set out, in late 1942, to design a pilot plant for the production of plutonium in a reactor, at the University of Montreal. However, the prickly Halban, who had had an earlier falling out with Kowarski in Cambridge, turned out to be the wrong choice for laboratory director, being 'impetuous and vacillating in decisions and unreasonable in his demands of the administrative staff in Ottawa and unfair in criticizing them' [13-90]. Cockcroft, with his high reputation and ability to work with the Americans—an essential point as the Manhattan Project was gaining momentum across the border—was the obvious choice for Halban's replacement, and he flew to Montreal in April 1944. Reorganizing the unhappy laboratory, he relocated it to Chalk River, in a wooded locale 130 miles northwest of Ottawa, where one or more reactors could be safely built. He put Kowarski in charge of building a simple, stopgap heavy-water reactor designated ZEEP (Zero Energy Experimental Pile). The first nuclear reactor to go into operation outside the United States, ZEEP went critical on 5 September 1945 [13-91].

# Chapter 14

# EPILOGUE

## 14.1. Later years

It is not the purpose here to dwell on subsequent developments, in the 1940s and beyond, in the laboratories singled out, or on the post-World War II careers of their scientists in any depth. However, a few additional remarks about the team leaders, and their less well-known lieutenants, are probably appropriate. Since rather less has been written about Merle Tuve and his DTM colleagues, than about their Cambridge or Berkeley counterparts, we may as well start with them.

At the conclusion of World War II, Merle Tuve returned to the DTM, and in 1946 Vannevar Bush, president of the Carnegie Institution of Washington, appointed him to succeed John A Fleming as DTM director [14-1]. In agreement with Bush, Tuve did not wish the department to become involved in 'big science,' including the development of the next generation of large accelerators for high-energy physics. Instead, some members of the rather small department used the cyclotron for molecular biology and radioisotope production, while a few others continued to use the two Van de Graaffs for nuclear physics. Tuve's first personal research in the postwar years involved geophysics: mapping the earth's crustal structure with explosion-generated seismic waves (as opposed to depending on earthquake-generated seismic waves) [14-2]. Field trips took Tuve and his DTM co-workers to the Mesabi Range, Puget Sound, the Wasatch and Unita Mountains, the Colorado Plateau, and to Alaska and the Yukon Territory. As part of the CIW's participation in the International Geophysical Year (IGY) of 1957 and 1958, he also led an expedition to the Andes and altiplano of South America.

Starting in the early 1950s, Tuve's second major line of research was radio astronomy; in particular, observing radio emissions from interstellar hydrogen clouds [14-3]. A 23-foot German radar dish was installed on the DTM grounds, and later a 60-foot parabolic antenna at DTM's Derwood field station in Maryland. Tuve and colleagues devoted much of their time to developing photoelectric image detectors and multi-channel recorders for optical telescopes as well. Tuve also contributed toward the establishment and development of the National Radio Astronomy Observatory, and his group was a frequent user of its 300-foot transit telescope in Green Bank, West Virginia. They also contributed to a 100-foot antenna near La Plata, Argentina, for studying hydrogen clouds in the southern sky.

Tuve served on the executive committee of the United States national committee for the IGY, chaired the physics section of the National Research Council's Committee on Growth, was a member of the US National Commission for UNESCO, and in late 1965 he succeeded Hugh L Dryden as home secretary of the National Academy of Sciences. He retired as DTM director in 1966. His awards included the Presidential Medal of Merit, Honorary Commander of the Order of the British Empire, the National Academy of Science's Comstock Prize, the American Geophysical Union's Bowie Medal, and the Cosmos Club award. He received seven honorary degrees.

Tuve was married to Dr. Winifred Gray Whitman, a practicing psychoanalyst who worked for a period at DTM in the early 1930s, studying the biological hazards of ionizing radiation (the effect of $\gamma$-rays on rats). He died on 20 May 1982.

Gregory Breit, Tuve's first collaborator and the scarcely appreciated guru of physics theory and experiment alike, held a succession of academic positions after he left DTM in 1929, at New York University, the University of Wisconsin, and Yale University, where he was professor for 21 years. The last ten years in New Haven, he held the Donner chair in physics. He completed his career as distinguished professor of physics at the State University of New York in Buffalo, from which he retired in 1978. He died in Salem, Oregon, on 13 September 1981.

Breit's role and attitude at the start of World War II is telling in many ways, and bears recalling. In April 1940 he was the second person to be appointed to the Advisory Committee on Uranium, chaired by Lyman Briggs of the Bureau of Standards. Breit's responsibility was the coordination of Rapid Rupture: fast-neutron bomb studies and overall evaluation of potential weapons problems. However, 'his principal concern,' in the opinion of one, 'seemed to be that they should be kept quiet. Talk of bombs aroused in him the same uneasiness that Briggs felt about spending money' [14-4]. When Arthur Compton organized the Metallurgical Laboratory in Chicago in January 1942, Breit came along—among the few survivors of the nearly extinct Uranium Committee. Again he supervised bomb studies, and lectured the Met Lab staff on bomb theory that might guide plans for a plutonium reactor. Soon Oppenheimer arrived as well, and Compton made him a consultant, formally under Breit. The two did not get along, with Breit's concern for bomb theory leaking out, and Oppenheimer's concern that the theory was not disseminated fast enough. Breit resigned, replaced by Oppenheimer, and left the fledgling Manhattan Project for Aberdeen Proving Grounds and the APL, where he, too, joined the proximity fuse effort [14-5].

Larry Hafstad's post-war career steered him into ever-higher administrative positions in nuclear reactor circles. After five years with the APL, including several years as its director, he headed the Institute for Cooperative Research during 1947–49. Between 1949 and 1955 he served as America's reactor boss as the director of the Nuclear Reactor Development Division of the Atomic Energy Commission. During 1955 he directed the Atomic Energy Division of Chase Manhattan Bank, and from that year to 1969 he held the post of Vice-president in charge of research with the General Motors Corporation. He served as Chairman of the General Advisory Committee to the AEC, and as Chairman for the Committee on Underseas Warfare of the National Research Council.

What about Odd Dahl, who returned to Norway in 1936? He remained a Fellow with the Christian Michelsen's Institute in Bergen for the rest of his career, though his professional undertakings over the years were far from routine. Starting with oceanographic and meteorological instrumentation for geophysical colleagues, his first major project, early in World War II, was the design and construction of a 1.5-million volt Van de Graaff installation for cancer therapy in a municipal hospital in Bergen – the 'world's largest medical apparatus' when completed. That it was completed at all during those difficult wartime years is itself quite remarkable. Late in the war, the Norwegian underground delivered a set of plans for the construction of a betatron, courtesy of Tuve and Hafstad, and it gave Dahl something else to think about while riding out the war. In the immediate post-war years, a 50-MeV betatron, plus a second Van de Graaff, became the basis for a nuclear physics laboratory in the brand new University of Bergen. (The University, incidentally, presented Dahl with its first honorary doctorate.)

Next came JEEP (Joint Establishment Experimental Pile) at Kjeller near Oslo. Actually a joint Norwegian–Dutch undertaking, it was the first reactor constructed outside the 'major' nuclear nations. It was designed by Dahl with physics input by Gunnar Randers, the ex-astrophysicist. Fueled with Dutch uranium moderated by Norwegian heavy water, JEEP went critical in 1951. A second reactor followed, the world's first 'boiling heavy water reactor,' ostensibly built for powering a saw-mill in Halden on the Swedish border east of Oslo; in actuality it became an important test-bed for reactor technologies. As a diversion from atoms, Dahl (and his CMI colleagues, we should always add) also designed a state-of-the-art solar tower observatory, erected in the hill country north of Oslo. The optical components ('celeostat') were supplied by Zeiss in West Germany.

The largest project in Dahl's career was the 25-GeV Proton Synchrotron at CERN, the European Center for Nuclear Research outside Geneva. Dahl's adoption of the 'alternating gradient' or 'strong focusing' principle, newly discovered at Brookhaven National Laboratory [14-6], for the CERN machine, made it the forerunner for a new generation of proton synchrotrons that would dominate physics in years to come. Typically, having started the CERN PS, Dahl left it to others (mainly John Adams of England, in this case) to finish the job in 1959, while he turned to new projects at home. With Haakon Mosby, later rector of the University of Bergen, he initiated a project involving instrumented buoys for oceanographic research. Some were released into the Gulf Stream in the North Atlantic, while others were set adrift in the Weddell Sea of Antarctica by the American icebreaker USS *Glacier*.

Dahl's last project was helping to establish Andøya Rocket Range — which became an international facility — in the Vesterålen Islands north of the Arctic Circle, from which instrumented sounding rockets are launched into the auroral belts. The first rocket, a Nike/Cajun with a Ferdinand I payload, was launched in 1962. Dahl retired in 1968, and passed away in 1994.

Richard Roberts, too, spent part of the war on the proximity fuse project. Tuve put him in charge of the radio fuse, while others got variants such as photoelectric and acoustic fuses. Later Roberts went on to guided missiles and ramjets. After the war he stayed on for a while with a much reduced Applied Physics Laboratory, before returning to DTM in early 1947 [14-7]. However, he was quickly sidetracked into several military and defense-related consulting jobs, dealing variously with anti-submarine warfare, Polaris missiles, and civil defense. He also became involved in establishing a subcommittee to the Democratic Advisory Council on Science and Technology, with Ralph Lapp, Ernest Pollard, Harold Urey, and others. The subcommittee initiated something called the National Peace Agency, which after much wrangling evolved into the Arms Control Agency.

Back at DTM, meanwhile, a biophysics group had grown out of the wartime cyclotron-oriented biomedical work, with Tuve convinced it was good science and skeptical approval by Bush; Philip Abelson and Dean Cowie took the early lead in this. Tuve, as the new DTM director, was at first 'very busy getting rid of people interested in magnetism and electricity' [14-8]. Although he felt nuclear physics had changed from a sport into a business [14-9], he did allow a small continuing effort in nuclear physics in the department. At first it was mainly by Norman Heydenburg and Roberts, who had 'the two Van de Graaffs and the cyclotron to play with,' with Steve Buynitsky helping run the cyclotron. Roberts soon bowed out, and joined Abelson and Cowie in the new biophysics work, which profited from the cyclotron even after it was decommissioned and removed at great expense. 'In retrospect the cyclotron was a fine machine that came at the wrong time' [14-10]. The permanent benefit was the cyclotron building which housed the biophysics group for nearly 30 years [14-11].

When Roberts left Heydenburg, he was replaced by George M Temmer, and the two set about exploring a new form of nuclear interaction known as Coulomb excitation. However, they soon found that it demanded an electrostatic accelerator of higher beam energy than available at the DTM. Such a generator soon became available commercially from High Voltage Engineering Corporation, the company started by Robert Van de Graaff, in the form of the tandem Van de Graaff [14-12]. The third such accelerator had been purchased by Florida State University, and in late 1959 Heydenburg and Temmer went there, initially for three years. Eventually Heydenburg ended up professor and department head at FSU, while Temmer became head of a new tandem laboratory at Rutgers University.

So much for the Carnegie Institution. While C C Lauritsen did not figure prominently in our account, he deserves a final word as well. Unfortunately, in concentrating on X-rays in Pasadena, we have said less about Charlie Lauritsen's subsequent program in nuclear physics and astrophysics in the Kellogg Laboratory, before and after World War II. Following Cockcroft and Walton's as well as Chadwick's success in Cambridge, Lauritsen and H R Crane, a graduate student, converted one of the old X-ray tubes in the High Voltage Laboratory into a tube for positive ion acceleration, using helium gas and a primitive ion source [14-13]. Thus equipped, they began a program of neutron physics, with neutrons generated by $\alpha$-bombardment of beryllium. Soon, however, after Gilbert Lewis supplied them with a sample of heavy water, they found that deuteron bombardment not only produced neutrons more copiously than $\alpha$-particles, but radioactive nuclei as well. Pretty soon they, too, extended their program to proton-induced reactions.

Some of this we have touched on before. Here we merely note one of Lauritsen's most important discoveries, the capture of protons by carbon, with the emission of $\gamma$-rays—a process known as radiative capture. The full significance of this discovery became clear in 1939, when Hans Bethe at Cornell and Carl F Von Weizsäcker in Germany suggested that hydrogen is converted to helium in stellar interiors by a catalytic process involving isotopes of carbon and nitrogen, later known as the CN cycle [14-14]. Several reactions in the cycle involved radiative capture of protons by C and N isotopes. Another cycle, suggested by Bethe and Critchfield, involved hydrogen conversion to helium by a proton–proton cycle. In time it became clear from measurements in Lauritsen's laboratory that the p–p cycle dominates in 'main sequence' stars like the sun, and the CN cycle becomes operative in stars much hotter than the sun.

In the summer of 1940, Lauritsen went to Washington, joining the National Defense Research Committee, newly formed by Vannevar Bush, as vice chairman on Armor and Ordnance. During 1940–41 he participated actively in the proximity fuse work under Merle Tuve. The summer of 1941 found him back at Caltech where, with his son Tom and other members of the Kellogg group, he organized their own project for developing artillery rocket technology, under the aegis of the Office of Scientific Research and Development and the US Navy. The project developed a number of different service weapons, many of which were adopted and used by the armed services. Lauritsen spent much of the last two years of the war at Los Alamos. When the war ended, he, along with Tom and William A Fowler, resumed their study of hydrogen-and-helium burning nuclear reactions believed operative in stellar interiors. C C Lauritsen remained active as well on a number of advisory panels with the Department of Defense, and was influential in the establishment of the Office of Naval Research—the organization that played such an important role in the recovery of scientific research after the war, and set the pattern for its federal support.

In Lawrence's laboratory, the historic contribution of the 60-inch cyclotron to the burgeoning fission project early in World War II, was its production of plutonium in 1941. Two years earlier, not long after the 60-inch circulated its first beam in mid-1939, the Rockefeller Foundation gave the University of California 1.25 million dollars for a 184-inch cyclotron, to be erected on the hill overlooking the Berkeley campus [14-15]. However, before work could be started on the giant cyclotron, Lawrence journeyed to the East Coast for a meeting with his longtime friend Alfred Lee Loomis, a former New York banker and amateur scientist who had played a key role in the discussions with the Rockefeller Foundation officials. Loomis introduced Lawrence to members of the Tizard Mission, in the US just then. From his old friend John Cockcroft, Lawrence learned to his amazement of their outstanding scientific contributions to the British war effort. At a meeting in Loomis's private laboratory in Tuxedo Park, New York, Lawrence agreed to assume the responsibility for recruiting young nuclear physicists to assist in the British radar effort. Returning to Berkeley, he persuaded Lee DuBridge to leave his cyclotron at Rochester University in the hands of colleagues and take charge of the fledgling Radiation Laboratory at MIT—a radar laboratory established by a committee chaired by Loomis. From his own laboratory, he also recruited Ed McMillan, Winfield Salisbury, and Luis Alvarez for the MIT project.

Lawrence himself joined the Microwave Committee under the National Defense Research Committee, also chaired by Loomis, and became involved in anti-submarine warfare as well. At Berkeley, meanwhile, he converted the 37-inch cyclotron into a mass spectrometer for separating $U^{235}$. It proved so successful that he also committed the 184-inch cyclotron to the separation task. With a D-shaped mass-spectrometer tank replacing the usual dees, it became the first 'calutron.' However, to separate 30 kg of $U^{235}$ for an efficient bomb would require 2000 calutrons running 300 days. Even so, electromagnetic separation appeared more promising than gaseous or thermal diffusion. And so it proved, with the gigantic calutron plant at Oak Ridge producing the bulk of the $U^{235}$ for the Hiroshima bomb [14-16].

Once the war ended in 1945, Lawrence persuaded General Groves to authorize Manhattan District funds for the completion of the 184-inch cyclotron. It had been the intention to build the machine as a 'conventional cyclotron,' but a problem stood in the way. As may be recalled, the 'cyclotron condition,' or equation (5.2), assumes the constancy of $m$, the particle mass. In fact, any accelerating particle experiences an increase in mass, in accordance with the principles of relativity. As a consequence, the orbiting particles in the cyclotron will gradually lag behind the constant radiofrequency. In the event, the effect is small, providing the mass increase does not exceed 1%, as was the case in the early cyclotrons. However, the effect did not bode well for the 184-inch cyclotron. Fortuitously, a remedy was at hand. A new principle, known as 'phase stability,' developed independently by McMillan while at Los Alamos late in the war, and by Vladimir Veksler in the Soviet Union, appeared capable of overcoming the relativistic limitation of the conventional cyclotron. The idea was to *reduce* the r.f. voltage across the dees as the particles gained energy, and hence mass, so as to continuously match the angular velocity of the spiraling particles [14-17]. To be on the safe side, Lawrence and McMillan modified the 37-inch cyclotron to operate as a model 'synchrocyclotron,' as the frequency-modulated cyclotron would be called. The 37-inch workhorse worked fine in this guise as well, and by November 1946 the 184-inch synchrocyclotron was accelerating deuterons to 195 MeV. In due course it routinely delivered 350-MeV protons.

With the successful operation of the 184-inch machine, Lawrence became an active partner in a new round of experiments, whenever he could find the time. Among other things, he personally discovered the delayed neutron activity that he and his colleagues soon showed was due to $N^{17}$. The question also arose, was the big machine, with roughly twice the beam energy visualized in 1940, capable of producing the newly discovered $\pi$-mesons? The rare, unstable particles had been detected in cosmic ray tracks in nuclear emulsions by Cecil Frank Powell and his group in Bristol [14-18]. Encouraged by Edward Teller, Eugene Gardner placed a carbon target and a stack of photographic emulsions in the chamber of the synchrocyclotron, and exposed it to a beam of $\alpha$-particles. Alas, no tracks were found. However, shortly afterwards, in February 1948, Cesare M G Lattes arrived in Berkeley on a Rockefeller Fellowship. He had been a member of the Bristol team, knew what to look for in the emulsions, and in no time they spotted pions in copious numbers. With that, the creation and study of mesons, rather than the manufacture of exotic isotopes, became the principal achievement and justification of the new machine [14-19]. It also marked the birth of a new era in science—that of high energy physics [14-20].

There is another way around the relativistic limitation of the conventional cyclotron. Both McMillan and Veksler had shown that, besides modulating the radiofrequency, the guiding magnetic field may also be varied, by increasing it in synchronism with the increasing particle energy. In fact, they also pointed out, as had Mark Oliphant in England, that by simultaneously decreasing the frequency and increasing the field, the orbiting particles can be held on a circular orbit of constant radius, rather than spiraling outward, as in the conventional or FM cyclotron. You now have a *synchrotron*, in which the magnet only has to enclose a doughnut-shaped vacuum chamber, and an accelerator capable of much higher beam energy than possible with either form of cyclotron. As the new experiments were barely under way on the Berkeley hill in 1948, William Brobeck convinced the laboratory staff that a proton synchrotron could be built, capable of reaching the multibillion-volt energy range. Lawrence authorized a quarter-scale model, and they had it working within nine months. Lawrence then gave the 'go-ahead' for the full-size Bevatron, as it would be called [14-21]. Unfortunately, just then the USSR detonated its first nuclear device, and Lawrence was forced to turn his attention to more pressing matters.

Lawrence had become concerned that all the US nuclear weapons development work was confined to a single government laboratory, Los Alamos, and offered Livermore as a suitable site for a second weapons laboratory. (A former Naval Air Station, Lawrence's laboratory had acquired Livermore as a site for the Materials Testing Accelerator, a huge high-current linac for producing neutrons by deuteron impact on heavy targets. In the event, only a test section of the MTA was built.) Livermore Laboratory was established in 1952, with Herbert York as its first director and Edward Teller a senior staff member. Lawrence spent most of his remaining time and energy on the new laboratory, until his untimely death in 1958, at 57, from complications after surgery for chronic colitis.

After four years in Berkeley, with his PhD and three cyclotrons partly to his credit, Livingston decided it was time for him to move on. He did not leave in the happiest of moods, feeling Lawrence had not given him adequate credit for his contributions to the development of the cyclotron [14-22]. He moved to Cornell, where he built a small but efficient 16-inch cyclotron, the first outside Berkeley. It first circulated a beam in July 1935.

Together with Robert Bacher and Hans Bethe, he also helped inaugurate an excellent nuclear physics program at Cornell. In 1938 MIT persuaded him to relocate there, to supervise the building of a larger, 42-inch cyclotron and assume a professorship in physics. When many of his colleagues took up radar work or joined the Manhattan Project, Livingston stayed on with the MIT cyclotron, contributing to the war effort by producing radioisotopes for medical purposes. He spent the last two years of the war with OSRD and the Navy Department, working on anti-submarine radar and countermeasures.

Not long after Livingston returned to MIT after the war, scientists from the major East Coast universities, led by I I Rabi and Norman Ramsey, set about establishing a government-funded regional nuclear science laboratory to serve the university community. Its centerpiece would be a large research reactor and a particle accelerator — user-facilities on a scale too large for a single university. The result was Brookhaven National Laboratory on Long Island, New York, established in 1946 with Philip Morse of MIT its first director [14-23]. Morse persuaded Livingston, his MIT colleague, to take charge of the accelerator project. But what sort of accelerator should it be? At first Livingston felt he could outdo Berkeley with a synchrocyclotron in the 600–700 MeV range, but on a trip to Berkeley Rabi learned of plans being drawn up there for a 10-GeV proton synchrotron; Rabi argued that this was the way to go at Brookhaven as well. Unfortunately, the AEC was unwilling to underwrite *two* 10-GeV accelerators. In the end, discussions between Leland Haworth, the Brookhaven associate director for projects, Ernest Lawrence, and the AEC authorities led to the decision that both laboratories, instead of competing for a 10-GeV machine, would each build a smaller proton synchrotron, one around 3 GeV and one for 6 GeV (the threshold for anti-proton production). Haworth chose the smaller Cosmotron (which could be completed first), and Lawrence's Radiation Laboratory got the 6-GeV Bevatron [14-24]. The rest is history, as they say, and as we have touched on here and there.

### 14.2. Mostly Cockcroft

So much for the principals in the post-war times of the American laboratories. What about their British colleagues? It mainly boils down to John Cockcroft. We met him last in Montreal and Chalk River, Canada, late in the war, up to his neck in activity, as usual. In 1945 there was great pressure building in the United Kingdom for bringing Cockcroft home, to launch a post-war effort in Britain on atomic energy [14-25]. On his and Mark Oliphant's suggestion, the British Atomic Energy Research Establishment was founded in 1946 at Harwell, a former RAF station, with Cockcroft the director [14-26]. His appointment forced him to resign the Jacksonian Professorship of Natural Philosophy at Cambridge, which he had held since 1939. Among his division leaders at Harwell was Otto Frisch, who returned from Los Alamos to take charge of Nuclear Physics, and Klaus Fuchs, also from Los Alamos, who formed a Theoretical Physics Division. (In 1950 Frisch left Harwell for Cambridge to become the Jacksonian Professor in place of Cockcroft. It came as a great shock, the same year, to Cockcroft in learning of Fuch's wartime role as a Russian spy.)

At Harwell, as elsewhere, Cockcroft is fondly remembered 'for the little black book in which he noted everything requiring attention or that he needed to remember, his microscopic handwriting, and the crisp, terse notes' [14-27]. Stories about Cockcroft's imperturbability are legion as well.

> Edwin McMillan recollected going to Harwell on a visit from Berkeley and finding senior staff concerned, but relieved, after their Director had walked through a glass door without hurting himself. 'I have often walked *into* a door,' mused McMillan on recounting the incident, 'but never through one!' [14-28]

As for the Harwell program, by 1948 two experimental reactors, Gleep and Bepo, were operating on site, while at Windscale, on the Cumbrian coast, a plutonium production reactor was under construction. The next year, the first post-war proton accelerator in Europe, a 180-MeV synchrocyclotron, was completed at Harwell under Cockcroft.

In November 1951 it was announced that Cockcroft and Walton were to be jointly awarded the Nobel Prize in Physics for their disintegration experiments in April 1932. Why it took the Swedish Academy nearly 20 years in the case of Cockcroft and Walton, but only three years (1935) for Chadwick and nine years (1939) for Lawrence, is anybody's guess. On the way to Stockholm, the Cockcrofts stopped in Oslo for another ceremony, the formal opening of the Kjeller reactor by King Haakon on 28 November, in the presence of other leading physicists, including Bohr, Lawrence, Hafstad, Amaldi, Kowarski, and Francis Perrin. Lawrence, too, was on his way to Stockholm to see his colleagues Edwin McMillan and Glenn Seaborg receive the Nobel Prize in Chemistry for their isolation of neptunium and plutonium. (He was also, finally, to present his own long-delayed Nobel lecture that World War II had prevented him from giving in 1939.) From Oslo, Lawrence and the Cockcrofts proceeded to Uppsala for the opening of a cyclotron at the new Swedborg Institute.

> The Swedes had decided to enliven the ceremony by advancing the feast of St. Lucia from 13 December. The party 'sat at tables in the deep cyclotron pit with candles and cakes' while a procession

of girls dressed in white paced round them singing carols. Finally, Lawrence pressed a button in the control room upstairs to start the cyclotron. [14-29]

On the afternoon of 10 December the Cockcroft and Walton families assembled with the other Prizewinners in the concert hall of the Palace to receive their awards from King Gustav Adolph VI. The citation for Cockcroft and Walton, read by a Swedish professor, concluded that

> By its stimulation of new theoretical and experimental advances, the work of Cockcroft and Walton displayed its fundamental importance. Indeed, this work may be said to have produced a totally new epoch in nuclear research. [14-30]

Besides Harwell, Cockcroft took on an impressive number of subsidiary tasks as time went by. He represented the United Kingdom at the United Nations Conference on the Peaceful Uses of Atomic Energy at Geneva in 1955, and again in 1958. He and William Penney (the physicist who directed the British postwar atomic bomb program) alternated as leaders of the British delegation to an East–West conference on nuclear test control. Cockcroft was active in the establishment of CERN, and served on the Nominations Committee as well as on the Scientific Policy Committee. He also allocated three of his Harwell staff members to work on the projected proton synchrotron at CERN, including Frank Goward who became Dahl's deputy, and John Adams, who eventually replaced Dahl. Cockcroft became heavily involved in policy-making in Whitehall, as titular chairman of the Defense Research Policy Committee when Tizard retired, and as Chadwick's replacement on the Advisory Council on Scientific Policy—'part of the time representing the Atomic Energy Authority and part of the time in his own right as a distinguished scientist' [14-31]. Largely on Cockcroft's initiative, the Rutherford High Energy Laboratory was created adjacent to the Harwell site—somewhat in the Brookhaven style. The centerpiece of the Rutherford Laboratory would be Nimrod, a 7-GeV proton synchrotron.

In October 1959, Cockcroft became the first Master of Churchill College, Cambridge. The college was intended originally as a technological institution, somewhat on the lines of MIT, on the urging of Churchill himself. Churchill requested that the Mastership should be a Crown appointment, but it was agreed by the Trustees that the first appointment should be made by Churchill personally [14-32]. Receiving the word from Lord Tedder, the wartime air commander and now Chancellor of Cambridge University, the Cockcrofts left Harwell and returned to Cambridge right away. (By happenstance, Cockcroft, then 61, had been preparing to retire from the Atomic Energy Authority.) Their first official function was to welcome Sir Winston to the future College site on Madingley Road on

17 October, his one and only visit [14-33]. The College was opened, half completed, by the Duke of Edinburgh in June 1964. In May 1966 Peter and Anna Kapitza made their long anticipated visit to the College, staying with the Cockcrofts in the Master's Lodge. It was Kapitza's first visit to England since 1934. A special meeting of the Kapitza Club was held in his honor. In March 1967, by then also Chancellor of the Australian National University, Cockcroft wrote to Sir Ernest Marsden, long retired from government service in New Zealand, of the excellent progress being made at Churchill College. Six months later, on 18 September 1967, Elizabeth Cockcroft awoke to find her husband dead from a heart attack.

As for Ernest Walton, there is not much to be said about him, aside from his sharing the Nobel Prize for Physics in 1951. He is said to have been 'quiet, undemonstrative, and little given to talk' [14-34]. In contrast to Cockcroft's brilliant postwar career, Walton had been content to forsake original research and public service for a quiet academic life as Professor of Experimental Physics at Trinity College in Dublin, Ireland. In 1952 he became chairman of the School of Cosmic Physics and the Dublin Institute for Advanced Studies, and in 1960 he was named Senior Fellow of Trinity College.

Mark Oliphant's wartime contributions spanned both radar and atomic energy, in roughly equal measure [14-35]. The cavity magnetron was the all-important product of his Birmingham Physics Department; producing microsecond pulses at a wavelength of 10 cm, it could light a cigarette in no time flat. In addition, Oliphant spent much time on radar business in the US, and a whole year introducing it in Australia. As a member of the Tizard Mission to the US early in the war, he cajoled Briggs, Conant, Bush, and Fermi in the strongest terms to get seriously going on the uranium bomb; never mind power plants. Said the irrepressible Leo Szilard after the war,

> If Congress knew the true story of the atomic energy project, I have no doubt but that it would create a special medal to be given to meddling foreigners for distinguished services, and Dr. Oliphant would be the first to receive one. [14-36]

In Berkeley, however, Oliphant had no trouble impressing Lawrence with the seriousness of the situation. Oliphant, too, had thought of converting his Birmingham cyclotron, still under construction, to a mass spectrometer for separating $U^{235}$, and here was obviously something he had in common with Lawrence. On the hill in Berkeley, next to the 184-inch magnet and out of hearing range of others, he talked with Lawrence not only about $U^{235}$, but also about element 94 and the MAUD report, repeating what he had emphasized before the lukewarm Uranium Committee under Lyman Briggs.

By 1943 Oliphant ranked second only to Chadwick in the British contingency with the Manhattan Project, and Groves and Oppenheimer wanted him at Los Alamos. Oliphant felt his time would be better spent with Lawrence in Berkeley, mainly on the isotope-separation problem, and so it was to be. In November 1943 he joined Lawrence for the remainder of the war, shuttling back and forth between the Rad Lab and Y-12, the electromagnetic separation plant at Oak Ridge.

Alas, Oliphant's postwar career as a builder of accelerators would prove less successful than his early work with Rutherford. In Birmingham his cyclotron was followed by a 1-GeV proton synchrotron that began operations in 1953 — too late and at too low an energy to compete with Brookhaven's Cosmotron. The Birmingham machine is mainly remembered, due to its awkward magnetic design, for a stray magnetic field that grabbed iron tools and implements from anybody walking past. Before it was finished, Oliphant had returned to his native Australia, where he launched a more ambitious synchrotron project at the new Australian National University in Canberra; it was not completed. Knighted in 1959, Sir Oliphant himself is better remembered for the apex of his career, as a distinguished professor at Canberra from 1950 to 1963, and as Governor of South Australia during 1971–76.

At this point, a closing word or two about our principal theme seems appropriate. To the extent that it can be viewed as a competition, we may recapitulate as to why the British were first in smashing the atom by artificial means. The Cavendish, tucked away on Free School Lane, was at its peak when Cockcroft obtained his PhD and Walton arrived, *circa* 1928, and despite a woefully inadequate research budget. The staff was what counted. There was Chadwick, Ellis, Blackett, Aston, Wynn-Williams, and the incomparable Kapitza, all under Rutherford who, too, was then at the height of his power. The Cavendish was a place of pilgrimage of distinguished visitors from abroad, who often came to give a lecture and chat with the research students, among them Bohr, Frank, Langevin, and Born. Technically, Cockcroft had the advantage of a strong engineering background and benefited as well from the continuing, close collaboration between the Cavendish and Metro-Vick, with people such as Brian Goodlet, C R Burch, Miles Walker, and T E Allibone coming and going. And while Cockcroft may not have been a top-notch experimental physicist, Walton's patience and diligence made up for Cockcroft's lack of finesse. Finally, there was Gamow's timely visit in 1929, with his law for $\alpha$-emission and the ensuing discussion of the reverse possibility of getting particles *into* the nucleus at remarkably low voltages.

In short, there seems to have been two factors at work in the background of Cockcroft and Walton's experiment: a laboratory research atmosphere second to none, and Cockcroft's happenstance interaction with Gamow, illustrating Louis Pasteur's famous dictum that 'chance favors the prepared mind' [14-37].

Lawrence had 1-million volt protons by early 1932, which, as Gamow had shown, he really didn't need for transmutation purposes. Still, Lawrence was well placed to undertake experiments similar to those of Cockcroft and Walton, with his cyclotron and a cadre of excellent graduate students. One reason he lagged behind his British colleagues was that he was late in getting into accelerators, being preoccupied since arriving in Berkeley with building apparatus for continuing his study of the photo-effect. When he did get going on his 'proton spinning merry-go-round,' there were practical difficulties to overcome—particularly restraining the protons to the median plane of the dees. At the same time, Lawrence's overwhelming concern was always on raising the beam energy with a next, larger magnet, less on what to *do* with the cyclotron. As a result, he lacked detectors of adequate sensitivity for spotting reaction products; in the late summer of 1932, when Cooksey and Kurie finally had a good cloud chamber functioning, they, too, saw the alphas from disintegrating lithium without difficulty.

Lauritsen, like Lawrence, was preoccupied with improving apparatus in 1932—in his case, applying higher potentials across their X-ray tubes at Caltech. Of all our team leaders, only Tuve had nuclear experimentation firmly in mind from the very start. Though, like Fleming, he accepted the legitimacy of doing nuclear physics as contributing to a deeper understanding of terrestrial magnetism, he also saw nuclear physics and the study of nuclear forces as an end in itself. However, he, Hafstad, and Dahl spent five years on their Tesla coil, a device wholly unsuited for such experiments. In point of fact, it was not a wasted effort; in coping with the coils' frustrating behavior they developed state-of-the-art cascaded accelerator tubes able to withstand punishing voltages, as well as the necessary auxiliary apparatus: ion sources, the Wilson chamber, ionization chambers, and pulsed electronics. When they finally adopted Van de Graaff's generator, a frustrating period of desultory outdoor operation was necessary before DTM management gave approval for ripping out the Tesla coil from the Experiment Building, and installing the one-meter generator. No sooner had the church bells of Chevy Chase rung in the New Year of 1933, than they, too, repeated Cockcroft and Walton's historic disintegration of the lithium nucleus.

# ABBREVIATIONS

The following abbreviations are used extensively in the notes and bibliography.

## Archives and Collections

BSC    Bohr Scientific Correspondence, Niels Bohr Archive, Copenhagen

CHAD    Chadwick Papers, Churchill College Archive, Cambridge

CKFT    Cockcroft Papers, Churchill College Archive, Cambridge

CUL    Cambridge University Library

EOL    E O Lawrence Papers, Bancroft Library, University of California, Berkeley

GLP    Gilbert Lewis Papers, University of California Archives, College of Chemistry, Berkeley

MAT    M A Tuve Papers, Library of Congress

MPG    Archiv zur Geschichte der Max-Planck-Gesellschaft, Berlin

MTNR    Lise Meitner Papers, Churchill College Archive, Cambridge

NHAN    Norsk Hydro Archives, Notodden

NHAM    Norsk Industriarbeidermuseum Archives, Vemork

NHAO    Norsk Hydro Archives, Oslo

PHP    Paul Harteck Papers, RPI, Troy, NY

RTB    R T Birge Papers, Bancroft Library, University of California, Berkeley

WLTN    Walton Papers, Churchill College Archive, Cambridge

## Journals and Periodicals

*AJP*    *American Journal of Physics*

*CR*    *Comptes Rendus Hebdomadaires des Séances de l'Académie des Sciences*

DSB     *Dictionary of Scientific Biography*
HSPS   *Historical Studies in the Physical Sciences*
JPR     *Le Journal de Physique et le Radium*
NA      *Nature*
NW     *Die Naturwissenschaften*
PM     *The London, Edinburgh, and Dublin Philosophical Magazine*
PR      *The Physical Review*
PRS    *Proceedings of the Royal Society of London*
PT      *Physics Today*
RSI     *Review of Scientific Instruments*

## Organizations and Institutions

The following abbreviations are used extensively in the text.

AAAS   American Association for the Advancement of Science
AAUP   American Association of University Professors
AEC     Atomic Energy Commission
APL     Applied Physics Laboratory, Silver Spring, MD
APS     American Physical Society
BIR      Board of Inventions and Research
CERN   European Center for Nuclear Research, Geneva
CIW     Carnegie Institution of Washington
CMI     Christian Michelsen's Institute, Bergen, Norway
DSIR    Department of Scientific and Industrial Research
DTM    Department of Terrestrial Magnetism, CIW
GE      General Electric Company
KWI     Kaiser Wilhelm Institute, Berlin
LHS     Lawrence Hall of Science, University of California, Berkeley
MAUD  Acronym for sub-committee of the Committee for the Scientific Survey of Air Warfare ('Thomson Committee')
NAS     National Academy of Sciences
NRC     National Research Council
NTH     Norges Tekniske Høyskole, Trondheim
OSRD   Office of Scientific Research and Development
PTR     Physikalisch-Technische Reichsanstalt, Berlin

## Technical Abbreviations

The following abbreviations are also used extensively in the text.

AGS     Alternating Gradient Synchrotron (Brookhaven National Laboratory)

230

| | |
|---|---|
| APO | Atomic Physics Observatory (Carnegie Institution of Washington) |
| CH | Chain Home (radar) stations |
| D | Deuterium |
| Gleep | Graphite Low-Energy Experimental Pile (Harwell) |
| IGY | International Geophysical Year |
| JEEP | Joint Establishment Experimental Pile (Norway) |
| MTA | Materials Testing Accelerator (Livermore) |
| PS | Proton Synchrotron (CERN) |
| ZEEP | Zero Energy Experimental Pile (Canada) |

# NOTES

## Notes chapter 1

**1-1**  MacCarthy, Desmond 1932 'Science and politics' *The New Statesman and Nation* 7 May 1932 pp 584–585. *Wings over Europe* was first performed by the New York Theater Guild in December of 1928 about when atom smashing was first seriously considered in America as well as in Europe

**1-2**  Langer, William L editor 1968 *An Encyclopedia of World History: Ancient Medieval and Modern Chronologically Arranged* fourth ed (Boston: Houghton Mifflin Company). *The World Almanac and Book of Facts* 1933 (New York: World Telegram)

**1-3**  Weiner, Charles 1972 '1932 — Moving into the new physics' *PT* **25** 332-339

**1-4**  Cockcroft J D and Walton E 30 April 1932 letter to *Nature* **129** 649

**1-5**  Kaempffert, Waldemar 1932 'The atom is giving up its mighty secret' in the *New York Times* Sunday 8 May Sect 9 p xx cited also in 'Atomic energy — is it nearer?' The editor 1933 *Annualog* ed Louis S Treadwell (New York: Scientific American Publishing Co) pp 47–51

**1-6**  Rutherford E to N Bohr 21 April 1932 BSC 1930–1945 Reel 25 Rosseland–Teller

**1-7**  Kaempffert (note 1-5) p 51. Brown, Andrew 1997 *The Neutron and the Bomb: A Biography of Sir James Chadwick* (Oxford: Oxford University Press) p 22

## Notes chapter 2

**2-1**  Dahl, Per F 1997 'Dawning of the atomic age' ch 17 in *Flash of the Cathode Rays: A History of J J Thomson's Electron* (Bristol and Philadelphia: Institute of Physics Publishing) pp 321–354. Marsden, Ernest 1962 'Rutherford at Manchester' *Rutherford at Manchster* ed J Birks (London: Heywood & Company Ltd) pp 1–16

**2-2**  Marsden (note 2-1) p 11

**2-3**  Marsden E and Lantsberry W C 1915 'The passage of $\alpha$-particles through hydrogen II' *PM* **30** 240–243 on p 243

**2-4**  The characteristics of the first two rays, 'one that is very readily absorbed and which will be termed for convenience the $\alpha$-radiation and the other a more penetrating character which will be termed the $\beta$-radiation,' were outlined in Rutherford's monumental paper on radioactivity: Rutherford E 1899 'Uranium radiation and the electrical conduction produced by it' *PM* **47** 109-163. Though Paul Villard of Paris is generally given credit also by Rutherford for discovering $\gamma$-rays (penetrating rays with wavelengths comparable with those of X-rays), evidence points to Rutherford as a co-discoverer. Wilson, David 1983 *Rutherford: Simple Genius* (Cambridge, MA: The MIT Press) pp 127-128

2-5  On his return from a British Association meeting in Australia, Rutherford wrote to Marsden, himself then in New Zealand, to take up a professorship at Victoria University College, asking if he 'minded' if Rutherford went on with the experiment as there were no suitable facilities at Wellington. In the event Marsden was soon seconded as Lieutenant in the Royal Engineers, ending up with a Sound-Ranging Section in France. Marsden to Rutherford 28 November 1915 CUL Add 7653 M65 Marsden (note 2-1) p 12

2-6  In Admiralty circles the unhappy BIR initiative went under the name 'Board of Intrigue and Revenge.' Wilson *Rutherford* (note 2-4) p 346

2-7  The relevant notebooks, five in number, are the following: NB 21 (March–September 1917) 'Production of H-atoms & other high speed atoms'; NB 22 (September 1917–February 1918) 'Range of high-speed atoms in air and other gases'; NB 23 (March–December 1918) dealing variously with the theory of scattering of recoil atoms in hydrogen and with the magnetic deflection of hydrogen atoms; NB 24 ( June 1918) 'Calculations on the passage of $\alpha$-particles through hydrogen'; NB 25 ( January–March 1919) 'Experiments on H-particles' CUL Add 7653

2-8  The term was coined by Rutherford, on the suggestion of Aston, Darwin and Fowler, in memory of William Prout, who proposed in 1815 that all elements are made up of the atoms of hydrogen as a primordial substance. It was first suggested by Rutherford in a talk before the British Association at Cardiff in 1920, and first used by him in print the same year: *NA* **106** (1920) 357. 'Notes on the use of the words 'proton' and 'nucleus' PA 322 (Photographs and Miscellanea) CUL Add 7653

2-9  'Experiments on H-particles' NB 25 (note 2-7)

2-10  Rutherford E 1919 'Collision of $\alpha$ particles with light atoms IV. An anomalous effect in nitrogen' *PM* **37** 581-587 on p 587

2-11  Rutherford E Draft manuscripts of published papers PA 25 CUL Add 7653

2-12  Boorse H A and Motz L eds 1966 *The World of the Atom* (New York and London: Basic Books Inc) p 808

2-13  Rutherford E, Chadwick J and Ellis C D 1930 *Radiations from Radioactive Substances* (Cambridge: Cambridge University Press) pp 215, 246, 249; Blackett P M S 1925 'The ejection of protons from nitrogen nuclei photographed by the Wilson method' *PRS* **A107** 349–360; Badash L 1983 'Nuclear physics in Rutherford's laboratory before the discovery of the neutron' *AJP* **51** 884–889 on p 887

2-14  Rutherford 'An anomalous effect...' (note 2-10) p 587; Rutherford draft manuscript PA 25 (note 2-11). The handwritten addition includes acknowledgment of Mr. William Kay 'for his invaluable assistance in counting scintillations'

2-15  Badash 'Nuclear physics...' (note 2-13) p 885

2-16    Wilson, *Rutherford* . . . (note 2-4) pp 3, 5

2-17    Brown, Andrew 1997 *The Neutron and the Bomb: A Biography of Sir James Chadwick* (Oxford: Oxford University Press) p 22

2-18    Along with the Cavendish, the Physics Department of the University of Leiden (nowadays known as the Kamerlingh Onnes Laboratory) and the Physikalisch-Technische Reichsanstalt in Berlin ranked among the world's leading physics research laboratories

2-19    Wilson *Rutherford* (note 2-4) pp 420–421

2-20    Ibid 410. Sir Joseph Larmor was an elector for the Cavendish Chair in 1919 and had championed Rutherford's candidacy before the Vice-Chancellor

2-21    Rutherford and Chadwick 1921 'The artificial disintegration of light elements' *PM* Ser 6 **42** 809-825. The laboratory notebooks for this period are as follows: NB 27–36 (Nov 1919–Aug 1921) 'Artificial disintegration of light elements'; NB 37–38 (Dec 1921–June 1922) 'Disintegration of elements by alpha particles.' The disintegration chamber is preserved in Section 7 ($\alpha$-Particles and Early Nuclear Physics) of the Cavendish Museum in the Cavendish Laboratory Cambridge; Falconer I J 1980 *Apparatus from the Cavendish Museum* (Cambridge: pamphlet, University of Cambridge)

2-22    Eve A S 1939 *Rutherford: Being the Life and Letters of the Rt Hon Lord Rutherford OM* (New York: Macmillan) p 291

2-23    The concept of *isotopes*, elements of different atomic weight but chemically indistinguishable (i.e. the same atomic number), was initially floated by Frederick Soddy in 1910, and taken up in earnest by him in 1913 when he also introduced the term [More on Soddy's personality—see '*Flash*' (note 2-1) pp 289-90]

2-24    Rutherford 1920 'Nuclear constitution of atoms' Bakerian Lecture *PRS* **A97** 374-400; reprinted in *The Collected Papers of Lord Rutherford of Nelson* ed by Sir James Chadwick vol 3 pp 424–428. The Bakerian Lectures date from 1775 and are named after Henry Baker. This was Rutherford's second of these prestigious lectures; he gave the first on radioactivity in May 1904

2-25    Rutherford: see note 2-24 p 396

2-26    Oliphant M 1972 *Rutherford: Recollections of the Cambridge Days* (Amsterdam: Elsevier)

2-27    Ibid p 73

2-28    Wilson *Rutherford* (note 2-4) p 450

2-29    Rutherford E 1924 'The natural and artificial disintegration of the elements' Franklin Institute *Journal* **198** 725-744; Draft manuscripts of published papers CUL Add 7653 PA26

2-30    Stuewer, Roger H 1986 'Rutherford's satellite model of the nucleus' *HSPS* **16**(2) 321-352 on p 331

2-31    Kirsch G and Pettersson H 1923 'Long-range particles from radium-active deposit' *NA* **112** 394-395

2-32    Cited in Brown *The Neutron* . . . (note 2-17) p 77. The Pettersson–Rutherford correspondence during 1924–27 is found in CUL, Add 7653, P12–P16

2-33    Pettersson 1924 'On the structure of the atomic nucleus and the mechanism of its disintegration' Physical Society *Proceedings* **36** 194-202

2-34    Rutherford and Chadwick 1924 'The bombardment of elements by $\alpha$-particles' *NA* **113** 457

**2-35**  Kirsch and Pettersson 1924 'The artificial disintegration of atoms' *NA* **113** 603

**2-36**  Cited in Stuewer 'Rutherford's satellite model...' (note 2-30) p 333

**2-37**  Chadwick to Rutherford 12 December 1927. CUL, Add 7653, C31. Cited in Wilson *Rutherford* (note 2-4) p 461

**2-38**  Frisch, Otto R 1976 'The discovery of fission: how it all began' *PT* **20** 272–277 on p 273

**2-39**  Ibid

**2-40**  Rutherford, Chadwick and Ellis *Radiation...* (note 2-13) p 296

**2-41**  Financed by the income from surplus funds generated by the Great Exhibition held in London in 1851, the prestigious scholarship amounted to £250 per year in 1927, or enough for room and board and a bit left over. Rutherford himself had been an 1851 Exhibitioner in 1895, as had Chadwick in 1913

**2-42**  Walton, Ernest T S 1984 'Personal recollections of the discovery of fast particles' pp 49–55 in *Cambridge Physics in the Thirties* edited and introduced by John Hendry (Bristol: Adam Hilger Ltd). McMillan E M 1979 'Early history of particle accelerators' in *Nuclear Physics in Retrospect* ed R H Stuewer (Minneapolis: University of Minnesota). Appendix, Walton to McMillan 11 April 1977 pp 141–142. The original letter is found in a file containing notes and correspondence between Walton, McMillan, T E Allibone and A D Wilson: WLTN 1/10, *History of Acceleration and of Lithium Disintegration*

**2-43**  Walton to McMillan *ibid* p 141

**2-44**  For years J J Thomson had sought to eliminate the complications encountered in discharge phenomena due to the presence of the electrodes, by producing an electrodeless discharge in an endless ring. He finally succeeded around 1890 with a scheme in which the path of the discharge was the secondary circuit of an alternating current transformer. Thomson 1891 'On the discharge of electricity through exhausted tubes without electrodes' *PM* **32** 321–336, 445–464

**2-45**  Wideröe, Rolf 1928 'Über ein neues Prinzip zur Herstellung hoher Spannungen' *Archiv für Elektrotechnik* **21** 387–406; translated by G E Fischer, F W Basse, H Kumpfert, and H Hartmann 1966 as 'A new principle for the generation of high voltages' in *The Development of High-Energy Accelerators* ed M Stanley Livingston (New York: Dover) pp 92–115. Walton's proposal to Rutherford for a linear accelerator is found in WLTN 1/2

**2-46**  Walton to McMillan 1 September 1977 in McMillan 'Early history...' (note 2-42) p 147

**2-47**  Allibone T E 1984 'Metropolitan-Vickers Electrical Company and the Cavendish Laboratory' pp 150–173 in *Cambridge Physics...* (note 2-42). Hartcup, Guy and Allibone T E 1984 *Cockcroft and the Atom* (Bristol: Adam Hilger) p 37

**2-48**  Allibone 'Metropolitan-Vickers...' (note 2-47)

**2-49**  Allibone 'Metropolitan-Vickers...' (note 2-47) p 156

**2-50**  Rutherford, Sir Ernest 1927 'Scientific aspects of intense magnetic fields and high voltages' *NA* **120**(3) 809–811

**2-51**  Badash 'Nuclear physics in Rutherford's laboratory' (note 2-13) on p 887. Allibone T E 1964 'The industrial development of nuclear power' Rutherford Memorial Lecture 1963 *PRS* Ser A **282** 447–463

**2-52**   Hartcup and Allibone *Cockcroft...* (note 2-47) p 38

**2-53**   Cockcroft 'Rutherford life and work after the year 1919 with personal reminiscences of the Cambridge period' Lecture to the *Physical Society* 1946 cited in Hartcup and Allibone *Cockcroft...* (note 2-47) p 24

**2-54**   Rutherford to Secretary, Research Board, DSIR 26 October 1927 cited in Hartcup and Allibone *Cockcroft...* (note 2-47) p 28; Wilson *Rutherford* (note 2-4) p 496

**2-55**   Rutherford 'Scientific Aspects...' (note 2-50)

**2-56**   Hartcup and Allibone *Cockcroft...* (note 2-47) p 31

**2-57**   The typewritten manuscript of Rutherford's address, on 28 Feb 1930, with editing in pencil, is preserved in CUL, Add 7653, PA 138. It is cited in Eve A S 1939 *Rutherford...* (note 2-22) p 338

## Notes chapter 3

**3-1**   Cornell, Thomas D 1986 *Merle A Tuve and His Program of Nuclear Studies at the Department of Terrestrial Magnetism: The Early Career of a Modern American Physicist* PhD dissertation Johns Hopkins University; Cornell 1970 'Tuve, Merle Antony' *DSB* **18** Supplement II 936–941

**3-2**   Childs, Herbert 1968 *An American Genius: The Life of Ernest Orlando Lawrence* (New York: EP Dutton & Co Inc) p 25

**3-3**   Childs *ibid* p 61

**3-4**   E O Lawrence to M A Tuve 25 July 1922 MAT Box 2 In his letter of recommendation, on Lawrence's behalf, to the Minnesota Physics Department, South Dakota's Dean of the College of Arts and Sciences credited Lawrence with 'an excellent mind a scientific interest and a genuine enthusiasm for advanced work'. The President of USD added that 'Mr Lawrence is an unusually good student.... He is an unusually industrious and reliable young man and I am positive will make good in any work that he attempts' EOL Reel 15 Items 033191 and 033192

**3-5**   Kevles, Daniel J 1995 *The Physicists: The History of a Scientific Community in Modern America* (Cambridge, MA, and London: Harvard University Press) p 188

**3-6**   Cochran, William 1995 'Solid-state structure analysis' ch 6 in *Twentieth Century Physics* Volume I Ed by Laurie M Brown, Abraham Pais and Sir Brian Pippard (Bristol and Philadelphia: Institute of Physics Publishing; New York: American Institute of Physics Press) p 463. See also Gehrenbeck, Richard K 1978 'Electron diffraction: fifty years ago' *PT* **31** 34–41

**3-7**   Wheeler, John A 1979 'Some men and moments in the history of nuclear physics: the interplay of colleagues and motivations' in R Stuewer *Nuclear Physics...* (note 2-42) pp 217–321 on p 219. See also N Ernest Dorsey 1944 'Joseph Sweetman Ames: the man' *AJP* **12** 135–148

**3-8**   Wheeler *ibid* p 230

**3-9**   Wheeler *ibid* p 234

**3-10**   Tuve to McMillan 21 April 1977 in McMillan 'Early history...' (note 2-42) p 133

**3-11**   Breit to McMillan 5 June 1977 in McMillan 'Early history...' (note 2-42) p 139

**3-12** Tuve to McMillan 21 April 1977 in McMillan 'Early history...' (note 2-42) p 134

**3-13** Ibid

**3-14** Appleton's method was to compare the strength of radio waves received at about 100 km from a transmitter, as the transmitted frequency was slowly varied. His theoretical and experimental work in ionospheric physics earned him the Nobel Prize in Physics in 1947. Brush S G and Gillmor C S 1995 'Geophysics' ch 26 in *Twentieth Century Physics* Volume III ed by Laurie M Brown, Abraham Pais and Sir Brian Pippard (Bristol and Philadelphia: Institute of Physics Publishing; New York: American Institute of Physics Press) pp 1984–2001

**3-15** Tuve, notebook entry 28 July 1922 MAT Box 388 [Cited in Cornell 'Tuve, Merle Anthony' (note 3-1) p 59]

**3-16** Cornell *Merle A Tuve and His Program...* (note 3-1) p 129. Rutherford 'The natural and artificial disintegration of the elements' *Ernest Rutherford: A Bibliography of His Non-Technical Writings* Berkeley Papers in History of Science IV (Berkeley: Office for History of Science and Technology University of California) p 22

**3-17** Tuve to McMillan 21 April 1977 in McMillan 'Early history...' (note 2-42) p 134. Fleming's suggestion that Tuve abandon the ionospheric work, which matched DTM's research goals exceptionally well, and take up transmutation studies at DTM — a subject quite at variance with ongoing research programs at the Carnegie — was a courageous act on Fleming's behalf, who was still only Assistant Director; Louis A Bauer was still the nominal DTM Director, although incapacitated by mental illness. Brown, Louis 'Reports from a golden age of physics' unpublished draft manuscript p 11. Brown's work contains Merle Tuve's monthly reports to Fleming, presented in chronological order with notes and tutorial commentary where appropriate

**3-18** It was not the famous ether drift experiment that convinced the Swedish Academy. Michelson's citation lauds 'the methods which you have discovered for exactness of measurements' and 'the investigations in spectroscopy which you have carried out in connection therewith'

**3-19** The first tentative confirmation of Einstein's equation for the photoelectric effect was achieved by Owen Richardson and Karl T Compton at Princeton in 1912. That same year, Millikan at Chicago began a new round of experiments that conclusively proved the validity of the equation

**3-20** Lawrence, Work Book March–August 1924. EOL Reel 58 Item 12949

**3-21** Childs *Genius...* (note 3-2) pp 72–76

**3-22** Childs *Genius...* (note 3-2) p 75 Perhaps not surprisingly there is no record of the disaster in Lawrence's laboratory notebook

**3-23** The decrease in frequency and increase in wavelength of X-rays or $\gamma$-rays when scattered by free electrons

**3-24** Compton, Arthur H 1956 *Atomic Quest: A Personal Narrative* (New York: Oxford University Press) p 5

**3-25** Lawrence E 1925 'The photo-electric effect in potassium vapor as a function of the frequency of the light' *PM* **1** 345-359

**3-26** Leonard Loeb to Lawrence 23 November 1926, EOL Reel 17 Item 038118; Lawrence to Loeb 10 December 1926 Item 038122; Loeb to Lawrence 5 April 1927 Item 038124

**3-27** Lawrence to Loeb 13 April 1927, EOL Reel 17 Item 038125

3-28    Loeb to Lawrence 20 May 1927, EOL Reel 17 Item 038129; Loeb to Lawrence 21 February 1928 Item 038132
3-29    Loeb to Lawrence 16 March 1928, EOL Reel 17 Item 038142


**Notes chapter 4**

4-1     Gamow, George 1970 *My World Line: An Informal Autobiography* (New York: The Viking Press) p 9
4-2     German custom decreed that every full (ordinary) professor must have an Institute, and since there were already three professors at Göttingen when Born was invited to succeed Peter Debye, there ended up three Institutes under one roof, chaired respectively by Born, James Franck and Robert Pohl
4-3     Bohr N 1913 'On the constitution of atoms and molecules, Part I' *PM* **26** 1–25; 'On the constitution..., Part II, Systems containing only a single nucleus' *PM* **26** 476–502; 'On the constitution..., Part III, Systems containing several nuclei' *PM* **26** 857–875
4-4     Rechenberg, Helmut 1995 'Quanta and quantum mechanics' ch 3 pp 143–248 on p 160 in *Twentieth Century Physics Volume I* edited by L M Brown, A Pais and Sir Brian Pippard (Bristol and Philadelphia: Institute of Physics Publishing)
4-5     Bohr to Sommerfeld 19 March 1916 cited in Bohr N 1972 *Collected Works* vol 2 ed by L Rosenfeld, J Rud Nielsen and E Rüdinger (Amsterdam: North-Holland) p 603
4-6     Sommerfeld, Arnold 1919 *Atombau und Spektrallinien* (Braunschweig: F Vieweg)
4-7     Jungnickel, Christa and McCormmach, Russell 1986 *Intellectual Mastery of Nature: Theoretical Physics from Ohm to Einstein* vol 2 (Chicago and London: The University of Chicago Press) p 355
4-8     Ibid p 358
4-9     Rechenberg 'Quanta...' (note 4-4) on p 186. Rechenberg Helmut 2001 'Werner Heisenberg: The Columbus of quantum mechanics' *CERN Courier* **41** 18–20. Eckert, Michael 2001 'Werner Heisenberg: controversial scientist' *Physics World* **14** 35–40
4-10    Jungnickel and McCormmach *Intellectual Mastery* (note 4-7) p 363
4-11    Pais, Abraham 1986 *Inward Bound: Of Matter and Forces in the Physical World* (Oxford: Clarendon Press; New York: Oxford University Press) p 255. For a recent historical review of the subject, see Daniel Kleppner and Roman Jackiw 2000 'Pathways of discovery: one hundred years of quantum physics' *Science* **289** 893–898
4-12    Rutherford E 1927 'Structure of the radioactive atom and origins of the $\alpha$-rays' *PM* Ser 7 **4** 580–605. Handwritten manuscript: CUL Add 7653 PA 28
4-13    Gamow *My World Line...* (note 4-1) p 60
4-14    Gamow G 1928 'Quantum theory of the atomic nucleus' *ZP* **51** 204-212; Gurney R W and Condon E U 1928 'Wave mechanics and radioactive disintegration' *NA* **122** 439. A typescript of Gamow's paper, with equations and symbols in pencil, is found among the Gamow–Cockcroft correspondence at Churchill College: CKFT 20/10A
4-15    Gamow *My World Line* (note 4-1) p 64. Here I am reminded by Roger Stuewer that Gamow's recollections are badly flawed. He obtained

a stipend from the Rask-Ørsted Foundation, not a Carlsberg Fellowship. Moreover, Gamow had written Bohr as early as July 1928, soliciting his assistance in securing a Danish visit, and enclosing a letter of reference from Abram F Ioffe. Stuewer to the author, 8 November 2001

4-16    Ibid pp 66–67
4-17    Bohr to R H Fowler 14 December 1928, and Rutherford to Bohr 19 December 1928 (BSC)
4-18    The undated memorandum, 'The probability of artificial disintegration by protons,' was unearthed by Mark Oliphant from Rutherford's papers after his death. It then disappeared in the Cockcroft files, until it was resurrected again by Allibone. CKFT 20/28; Hartcup and Allibone *Cockcroft...* (note 2-47) p 41
4-19    Rutherford 1929 'Discussion on the structure of atomic nuclei' Royal Society of London *Proceedings* **123 A** 373–382; abstracted in *NA* **123** 246–247. At the meeting Gamow and Rutherford applied the concept of quantum mechanical barrier 'tunnelling' to α-particle scattering in general. Brown L M and Rechenberg H 1996 *The Origin of the Concept of Nuclear Forces* (Bristol and Philadelphia: Institute of Physics Publishing) p 16
4-20    Gamow *My World Line...* (note 4-1) p 60
4-21    Ibid 83. R H Fowler to Bohr 3 February 1930 cited in Bohr N 1987 *Collected Works* Vol 8 ed Jens Thorsen (Amsterdam: North-Holland) p 678
4-22    Breit to McMillan 5 June 1977 in McMillan 'Early history...' (note 2-42) p 138
4-23    Tuve 1968 'Odd Dahl at the Carnegie Institution, 1926–1936' *Festskrift til Odd Dahl i Anledning av Hans Fylte 70 År, 3 November 1968* (Bergen: A S John Griegs Boktrykkeri) pp 40–46 on p 43
4-24    Cornell *Merle A Tuve and His Program...* (note 3-1) p 171
4-25    Cornell *ibid* pp 206–207
4-26    Breit 'Report for September 1928 to January 1929' 16 January 1929 MAT Box 15 cited in Cornell *Merle A Tuve and His Program...* (note 3-1) pp 207–208
4-27    Tuve 'Radio ranging and nuclear physics at the Carnegie Institution' p 177 cited in Cornell *Merle A Tuve and His Program...* (note 3-1) p 170
4-28    Tuve lecture draft 1928 MAT Box 5
4-29    Breit, Tuve and Dahl 1929 'A laboratory method of producing high potentials' *PR* **35** 51–65
4-30    Tuve, Breit and Hafstad 1929 'Application of high potentials to vacuum tubes' *PR* **35** 66–71
4-31    Tuve Report for November 1928 (December 4 1928) MAT Box 15 File 6
4-32    Breit to Tuve 7 October 1928 MAT Box 4 File 3
4-33    Childs *Genius...* (note 3-2) p 109
4-34    Lawrence to Leonard Loeb 27 September 1927 EOL Reel 17 Item 038131
4-35    Childs *Genius...* (note 3-2) p 111
4-36    Ibid p 113. The Silliman Lectures were established in 1901 at Yale College, New Haven, by a bequest from August Ely Silliman, as an annual lectureship designed to illustrate the presence and providence, the wisdom and goodness of God, as manifest in the natural and moral world. J J Thomson gave the first, in 1903
4-37    Tuve to McMillan 21 April 1977 McMillan 'Early history...' (note 2-42) p 135

4-38    Heilbron J L and Seidel, Robert W 1989 *Lawrence and His Laboratory: A History of the Lawrence Berkeley Laboratory* Volume 1 (Berkeley: University of California Press) p 24

4-39    Childs *Genius*... (note 3-2) p 135

4-40    Lawrence is not very helpful on the date; in his Nobel Lecture he cites 'one evening early in 1929,' Lawrence 'The evolution of the cyclotron' *Les Prix Nobel en 1951*, Imprimerie Royale (Stockholm: P A Norstedt & Söner 1952) pp 127-140; reprinted in *The Development of High-Energy Accelerators* ed M Stanley Livingston (New York: Dover 1966) pp 92–115. The date April 1 appears in a formal statement by Thomas Johnson in 1931, written for possible use in patent arguments; Johnson 'To whom it may concern' 15 September 1931 EOL Reel 14 Item 030433. In a letter to McMillan, Johnson is less helpful, placing the evening 'probably in March or April 1929'; Johnson to McMillan 9 April 1977 in McMillan 'Early history... (note 2-42) p 130

4-41    *The Infancy of Particle Accelerators: Life and Work of Rolf Wideröe* compiled and edited by Pedro Waloschek (Braunschweig/Wiesbaden: Friedr Vieweg & Sohn Verlagsgesellschaft 1994)

4-42    Ibid pp 11–12

4-43    Ibid 13. Rolf Wideröe 1984 'Some memories and dreams from the childhood of particle accelerators' *Europhysics News* **15**(2) 9–11. Aaserud, Finn and Vaagen, Jan 1983 'Et møte med Rolf Wideröe den første akselatordesigner' *Naturen* **5-6** 191–196

4-44    *The Infancy*... (note 4-41) p 15

4-45    Wideröe 'Some memories...' (note 4-43) p 9

4-46    Breit, Tuve and Dahl 1927 'Magnetism and atomic physics' *Carnegie Institution of Washington Yearbook* **27** 208–209

4-47    Ising G 'Prinzip einer Methode zur Herstellung von Kanalstrahlen hoher Voltzahl' *Arkiv för Mathematik Astronomi och Fysik* **18** 1–4; translated by F W Brasse as 'The principle of a method for the production of canal rays of high voltage' in *The Development of High-Energy Accelerators* (note 4-40) on pp 88–90

4-48    Wideröe 1928 'Über ein neues Prinzip zur Herstellung hoher Spannungen' *Archiv für Elektrotechnik* **21** 387–406; translated by G E Fischer, F W Brasse, H Kumfert and H Hartmann as 'A new principle for the generation of high voltages' in *The Development* (note 4-40) on pp 92–115. Wideröe's paper covers both the beam transformer and the resonance accelerator

4-49    Wideröe 'A new principle...' (note 4-48) pp 104–105. Wideröe's r.f. source was limited to a frequency of about $10^7$ Hz, or about ten million oscillations per second, and was consequently limited to the acceleration of heavy ions For cesium ions accelerated to 1 MeV, the apparatus would be about one meter long, whereas acceleration of protons required an evacuated tube many meters in length

4-50    *The Infancy*... (note 4-41) p 37. By an odd coincidence, on the very same day that Wideröe's thesis appeared in print in the *Archiv*, 17 December 1928, Leo Szilard applied for a patent on a linac design at the Reichspatentamt in Berlin. However, the patent was denied. Heilbron J L and Seidel, Robert W 1989 *Lawrence and His Laboratory*... (note 4-38) 78–79; Weart, Spencer R and Weiss Szilard, Gertrude editors 1978 *Leo Szilard: His Version of the Facts: Selected Recollections and Correspondence* (Cambridge, Mass: The MIT Press) p 11

240

**4-51** Burrill, E Alfred 1976 'Van de Graaff Robert Jemison' *DSB* **XIII** 569–571. Burrill 1976 'Van de Graaff, the man and his accelerators' *PT* **20** 49–52. Rose, Peter H 1968 'In Memoriam: Robert Jemison Van de Graaff' *Nuclear Instruments and Methods* **60** 1–3

**4-52** Burrill 'Van de Graaff, the man...' (note 4-51) p 50

**4-53** Van de Graaff R J 1931 'A 1,500,000 volt electrostatic generator' *PR* **38** 1919–1920

**4-54** *The New York Times* Wednesday 11 November 1931, Section 1 pp 1, 17

**4-55** Ibid p 17

## Notes chapter 5

**5-1** Royal Commission for 1851 Exhibition file on E Walton, cited in Hartcup and Allibone *Cockcroft...* (note 2-47) p 43

**5-2** Hartcup and Allibone *Cockcroft...* (note 2-47) ibid

**5-3** Hartcup and Allibone *Cockcroft...* (note 2-47) p 40

**5-4** Cited in Wilson *Rutherford...* (note 2-4) p 560

**5-5** Cockcroft 'The development of high-voltage experiments in Cambridge' pp 2–3 in unpublished ms CKFT 20/80 cited in Heilbron and Seidel *Lawrence...* (note 4-38) p 68

**5-6** Cockcroft and Walton 1930 'Experiments with high velocity positive ions' *PRS* A**129** 477–489. The frustrating experiments are covered in Walton's notebook for Oct 1929–Oct 1930, WLTN 2/2

**5-7** Heilbron and Seidel *Lawrence...* (note 4-38) p 68

**5-8** Cockcroft to Karl Darrow, 1 February 1938; CKFT 20/28

**5-9** Dahl P F 1999 'The neutron' ch 3 in *Heavy Water and the Wartime Race for Nuclear Energy* (Bristol and Philadelphia: Institute of Physics Publishing) pp 11–21

**5-10** Lawrence 'The evolution of the cyclotron' in *The Development...* (note 4-40) pp 136–137

**5-11** Wideröe 'Some memories...' (note 4-43) p 9

**5-12** *Leo Szilard: His Version...* (note 4-50) p 11. (Tape-recorded interviews supplemented by correspondence and other documents.) The application was actually filed on 5 January 1929. Szilard, Leo 1972 *The Collected Works I Scientific Papers* ed by Bernard T Field and Gertrude Weiss Szilard (Cambridge, Mass: The MIT Press) pp 554–563

**5-13** Steenbeck Max 1977 *Impulse und Wirkungen: Schritte auf meinen Lebensweg* (Berlin: Verlag der Nation) p 556

**5-14** Szilard to OS (Otto Stern?) n.d. in Szilard *Collected Works* (note 5-12) p 728, cited in Heilbron and Seidel *Lawrence* (note 4-38) p 82

**5-15** Johnson to McMillan 9 April 1977. McMillan 'Early history...' (note 2-42) pp 130–131

**5-16** J J Brady to McMillan 21 April 1977. McMillan 'Early history...' (note 2-42) pp 131–132

**5-17** Childs *Genius...* (note 3-2) p 147

**5-18** Lawrence and Edlefsen 'On the production of high speed protons' *Science* **72** (10 October 1930) 376–377. If nothing else, this brief article was the first published description of the cyclotron resonance principle. Edlefsen's flattened flask, and his subsequent unit made from plate glass and brass, liberally daubed with red sealing wax, are both on display in the Memorial Hall of the Lawrence Hall of Science in Berkeley. They had been found by Luis Alvarez in 1936 in the back of a drawer in Room 229 of Le Conte Hall. 'It is probably fortunate for the history of science [notes Alvarez] that these treasures were found by someone with archeological instincts, rather than by someone who could easily have been given the job of "cleaning the junk out of Room 229"'. Alvarez, Luis W 1975 'Berkeley in the 1930s' pp 10–21 on p 20 in *All in Our Time: The Reminiscences of Twelve Nuclear Pioneers* ed by Jane Wilson (Chicago: The Bulletin of the Atomic Scientists)

5-19    Livingston 'Work Book' September 1930–January 1931 EOL Reel 59
5-20    Livingston 'Work Book' Reel 59 Item 130764
5-21    Livingston 'Work Book' Reel 59 Item 130776 dated 12/1/30
5-22    Livingston M Stanley 1980 'Early history of particle accelerators' pp 1–88 on pp 26–27 in *Advances in Electronics and Electron Physics* Vol **50** ed by L Marton and C Marton (New York: Academic Press)
5-23    Lawrence to Tuve 4 January 1931 MAT Box 8
5-24    Livingston *The Production of High Velocity Hydrogen Ions Without the Use of High Voltages*, PhD thesis, University of California Berkeley 14 April 1931. Lawrence and Livingston 1931 'A method for producing high speed hydrogen ions without the use of high voltage' *PR* **37** 1707
5-25    Childs *Genius...* (note 3-2) p 151
5-26    Lawrence to Tuve 4 January 1931 MAT Box 8
5-27    Sloan D H and Lawrence E O 1931 'Production of heavy high speed ions without the use of high voltages' *PR* **38** pp 2021–2032; Lawrence E O and Sloan D H 1931 'The production of high speed cathode rays without the use of high voltages' NAS *Proceedings* **17** 64–70
5-28    Breit G 1929 'On the possibility of nuclear disintegration by artificial sources' *PR* **34** pp 817–818
5-29    Breit to Tuve 1 October 1929 MAT Box 4
5-30    Tuve to Breit 5 October 1929 MAT Box 4
5-31    Tuve, Breit and Hafstad 'Application of high potentials...' (note 4-30) p 66
5-32    Tuve to K F Herzfeld 3 February 1930 MAT Box 4
5-33    Tuve Report for May 1930 5 June 1930 MAT Box 15
5-34    Breit to Tuve 4 August 1930 MAT Box 4
5-35    Tuve to Breit 8 September 1930 MAT Box 4
5-36    Tuve to Breit 12 September 1930 MAT Box 4. Tuve is referring to Brasch A and Lange F 1930 'Künstliche $\gamma$-Strahlung ein Vakuum-Entladungsrohr für 2,4 Million Volt' *Die Naturwissenschaften* **18** pp 765–766 submitted in August 1930
5-37    Breit to Tuve 15 September 1930 MAT Box 4
5-38    Breit to Tuve 16 September 1930 MAT Box 4
5-39    Frisch 'The discovery of fission...' (note 2-38) p 274
5-40    Hafstad to Fleming 1 October 1930 MAT Box 15
5-41    Tuve, Hafstad and Dahl 1931 'Experiments with high-voltage tubes' *PR* **37** p 469
5-42    Laurence ,William L 1931 'Giant X-ray tube wins science award' *The New York Times* 4 January 1931 Section 1 p 24
5-43    Time '$1000 Prize' in 'Science AAAS' 12 January 1931 p 44
5-44    'Man's rays stronger than from radium win $1000 prize' *Science News Letter* **19** (10 Jan 1931) Cornell *Merle A Tuve and His Program* (note 3-1) p 249

**5-45**    Lawrence to Tuve 4 January 1931 MAT Box 8

**5-46**    Tuve to Hafstad 4 January 1931 MAT Box 8

**5-47**    Hafstad, Dahl and Tuve to Breit 29 January 1931 MAT Box 4

**5-48**    Brown 'Golden age...' (note 3-17) p 41


# Notes chapter 6

**6-1**    Heilbron and Seidel *Lawrence*... (note 4-38) pp 98–99. The final machine parameters differed somewhat from those originally envisioned; e.g. the magnet pole faces were initially 9 inches in diameter, and grew to 11 inches in the final version

**6-2**    Lawrence and Livingston 'A method...' (note 5-24)

**6-3**    Lawrence to Cottrell 3 July 1931 EOL Reel 7 Item 013952

**6-4**    Childs *Genius* (note 3-2); Heilbron and Seidel *Lawrence*... (note 4-38) pp 103, 107

**6-5**    Tuve M A 1931 'Vacuum tube experiments using Tesla voltages' *PR* **38** 580; Tuve, Hafstad and Dahl 1932 'High-speed protons' *PR* **39** 384–385. Positive rays simultaneously deflected at right angles by magnetic and electric fields revealed themselves by tracing parabolic patterns on a fluorescent screen, and were extensively studied by J J Thomson at the Cavendish. Dahl, Per F 1997 in 'Positive rays' ch 15 in *Flash*... (note 2-1) pp 265–292

**6-6**    Lawrence to Donald Cooksey 17 July 1931 EOL Reel 6 Item 011934

**6-7**    Lawrence to Cottrell 17 July 1931 EOL Reel 7 Item 013953

**6-8**    Lawrence and Livingston 1931 'The production of high speed protons without the use of high voltages' *PR* **38** 834

**6-9**    Lawrence to Cottrell 20 July 1931 EOL Reel 7 Item 013955

**6-10**    Lawrence to Tuve 27 August 1931 MAT Box 8

**6-11**    Rebekah Young to Lawrence c/o Dr George Blumer 3 August 1931. Heilbron and Seidel *Lawrence*... (note 4-38) p 100

**6-12**    Heilbron and Seidel *Lawrence*... (note 4-38) p 114

**6-13**    Lawrence to Tuve 10 Nov 1931 MAT Box 8

**6-14**    Lawrence to Tuve 13 January 1932 MAT Box 4. Lawrence and Livingston 1932 'The production of high speed light ions without the use of high voltages' *PR* **40** pp 19-35

**6-15**    Wilson C T R 1965 'On the cloud method of making visible ions and the tracks of ionizing particles' (Nobel Lecture 12 December 1927) *Nobel Lectures Including Presentation Speeches and Laureate's Biographies; Physics 1922-1941* (Amsterdam: Elsevier) p 194

**6-16**    Blackett 'The ejection of protons...' (note 2-13) p 350; Blackett 1922 'On the analysis of $\alpha$-ray photographs' *PRS* **102** 294–317. The $\alpha$-tracks were often spoken of as 'Shimizu tracks' at Cambridge, in reference to the Japanese researcher at the Cavendish who actually began the study of $\alpha$-particle tracks by this method in the early 1920s

**6-17**    Blackett P M S 1964 'Cloud chamber researches in nuclear physics and cosmic radiation' *Nobel Lectures Including Presentation Speeches and Laureate's Biographies; Physics 1942-1962* (Amsterdam: Elsevier) pp 97–119

**6-18**    Dahl, Hafstad and Tuve 1933 'On the technique and design of Wilson cloud-chambers' *The Review of Scientific Instruments* **4** 373–378

**6-19**    Tuve, Report for June 1931, 2 July 1931, MAT Box 15. Tuve, Hafstad and Dahl 'High-speed protons' (note 6-5) p 384

**6-20**    Cornell *Merle A Tuve and His Program*... (note 3-1) pp 291–293

**6-21**    Harland W Fisk, Acting Director, to John C Merriam, President of CIW, 8 December 1931; Tuve, report for December 1931, 6 January 1932, MAT Box 15

**6-22**    Tuve, Hafstad and Dahl 'High-speed protons' (note 6-5) p 385

**6-23**    Lawrence to Tuve 26 January 1933 MAT Box 12

**6-24**    Tuve to Lawrence 30 January 1933. Lawrence to Tuve 18 February 1933 MAT Box 4

**6-25**    Tuve, Report for September 1931, 3 October 1931, MAT Box 15; Tuve to Lauritsen, 11 December 1931, MAT Box 4. The failure of the Tesla coil as a high-voltage source for particle acceleration, notes Brown, was really an advantage in disguise: its capability of destroying all but the best designed vacuum tubes forced the DTM group to design a robust tube for voltage-drop applications. Moreover, the discovery that vacuum tubes with the electrostatic generator were capable of holding large, steady potentials came as a surprise; hitherto general electrical engineering experience had found alternating voltages easier to handle than direct voltages. Brown 'Golden Age...' (note 3-17) pp 65–66

**6-26**    Tuve, Report for September 1931, 3 October 1931, MAT Box 15

**6-27**    Van de Graaff to Tuve 30 September 1931; Cornell *Merle A Tuve and His Program*... (note 3-1) p 283

**6-28**    Van de Graaff to Tuve October 1931; *ibid* pp 283–284

**6-29**    Cornell *Merle A Tuve and His Program*... (note 3-1) p 284

**6-30**    Tuve 'Special report on the experimental work' 9 October 1931 MAT Box 5

**6-31**    Tuve to Lauritsen 11 December 1931 (note 6-25)

**6-32**    This estimate of 50% of the theoretical maximum was subsequently found by Tuve *et al.* to be overly optimistic; because the presence of corona made voltmeter and spark-gap measurements both indicate higher than true voltage, the highest actual voltage attainable in practice was only about 35% of the theoretical maximum

## Notes chapter 7

**7-1**    Holbrow, Charles H 1981 'The giant cancer tube and the Kellogg Radiation Laboratory' *PT* **34** 42–49. Fowler, William A 1975 'Charles Christian Lauritsen' National Academy of Sciences *Biographical Memoirs* **46** 221–239

**7-2**    Heilbron and Seidel *Lawrence*... (note 4-38) p 56

**7-3**    Lauritsen C C and Bennett R D 1928 'A new high potential X-ray tube' *PR* **32** 850–857

**7-4**    Holbrow 'The giant cancer tube...' (note 7-1) p 43

**7-5**    Du Bridge L A and Epstein, Paul S 1959 'Robert Andrews Millikan' National Academy of Sciences *Biographical Memoirs* **33** 241–282

**7-6**    Lauritsen C C and Cassen B 1930 'High potential X-ray tube' *PR* **36** 988–992

**7-7**    Kellogg to Millikan 16 March 1931 cited in Holbrow 'The giant cancer tube...' (note 7-1) p 45

**7-8**    Holbrow 'The giant cancer tube...' (note 7-1) p 45

**7-9**    'Giant X-ray unit installed' *Los Angeles Times* 4 August 1931

7-10   Lauritsen T, Lauritsen C C and Fowler W A 1941 'Application of a pressure electrostatic generator to the transmutation of light elements by protons' *PR* **59** 241–252

7-11   Fowler, William A and Ajzenberg-Selove, Fay 1985 'Thomas Lauritsen' National Academy of Sciences *Biographical Memoirs* **55** pp 385-394

7-12   Van de Graaff to Karl Compton 20 March 1931 cited in Rose 'In Memoriam...' (note 4-51) p 2

7-13   Cottrell to Lawrence 11 November 1931 EOL Reel 7 Item 013968

7-14   Compton, Karl T 1933 'High voltage II' *Science* **78** 48–52; Van de Graaff R J, Compton K T and Van Atta L C 1933 'The electrostatic production of high voltage for nuclear investigations' *PR* **43** 149–157

7-15   Pang, Alex Soojung-Kim 1990 'Edward Bowles and radio engineering at MIT 1920–1940' *HSPS* **20** 313–337 on pp 317–321

7-16   Kaempffert 'The atom...' (note 1-5)

7-17   'Atomic energy...' (note 1-5) p 50

7-18   Van Atta L C, Northrup D L, Van Atta C M and Van de Graaff R J 1936 'The design operation and performance of the Round Hill electrostatic generator' *PR* **49** 761–776

7-19   Pang 'Edward Bowles...' (note 7-15) p 325

## Notes chapter 8

8-1    Weiner, Charles 1970 'Physics in the Great Depression' *PT* **23** 31–38 on p 34

8-2    Linus Pauling to Samuel Goudsmit May 1933 cited in *ibid* p 33

8-3    Rider, Robin E 1984 'Alarm and opportunity: emigration of mathematicians and physicists to Britain and the United States 1933–1945' *HSPS* **15** 107–176 on p 125

8-4    Ibid p 126. United States Atomic Energy Commission 1971 *In the Matter of J Robert Oppenheimer* (Cambridge, MA: MIT Press)

8-5    R T Birge to H L Johnston 26 October 1932 RTB Box 33

8-6    Heilbron and Seidel *Lawrence...* (note 4-38) p 28

8-7    Kevles, Daniel J 1995 *The Physicists: The History of a Scientific Community in Modern America* (Cambridge, MA, and London: Harvard University Press) fourth printing p 237

8-8    Tuve to Van de Graaff 21 June 1932 MAT Box 4

8-9    Cornell *Merle A Tuve and His Program...* (note 3-1) p 374

8-10   *New York Times* 24 February 1934 in Heilbron and Seidel *Lawrence...* (note 4-38) p 31

8-11   *New York Times* 11 March 1934 in Weiner 'Physics...' (note 8-1) p 35

8-12   Jewett, Frank B 1932 'Social effects of modern science' *Science* **76** 24; also in Heilbron and Seidel *Lawrence...* (note 4-38) p 32

8-13   Heilbron and Seidel *Lawrence...* (note 4-38) pp 32–33

8-14   Hartcup and Allibone *Cockcroft...* (note 2-47) p 46

## Notes chapter 9

9-1    Rutherford 'Nuclear constitution...' (note 2-24)

**9-2**     Rutherford and Chadwick 1924 'On the origin and nature of the long-range particles observed with sources of radium C' *PM* **48** 509–526 (CPR **3** 120–135 on p 122); Stuewer 'Rutherford's Satellite Model...' (note 2-30) pp 333–334

**9-3**     Brown *The Neutron*... (note 2-17) p 53; Badash 'Nuclear physics...' (note 2-13) p 886

**9-4**     The concept of isotopy, linking chemical identities of elements of different atomic weights, was introduced independently in 1912 by Frederick Soddy in England and by Kasimir Fajans in Germany, and held sway until Chadwick's discovery of the neutron in 1932. From that time on, the isotopes of an element were viewed, not as inseparable bodies, but as consisting of atoms which have the same nuclear charge (number of protons) but different masses (number of neutrons)

**9-5**     Birge R T and Menzel D H 1931 'The relative abundance of the oxygen isotopes and the basis of the atomic weight system' *PR* **37** 1669. An early hint that something was peculiar about the composition of water was the difficulty of coming up with a definite value for the density of pure water in 1913 by Arthur B Lamb and Richard E Leen at New York University

**9-6**     Gilbert Lewis's students receiving the Nobel Prize in Chemistry were the following: Harold C Urey in 1933; William F Giauque (1949); Glenn T Seaborg (1951); Willard F Libby (1960); Melvin Calvin (1961)

**9-7**     Dahl, Per F 1999 'Heavy water' ch 4 in *Heavy Water and the Wartime Race for Nuclear Energy* (Bristol and Philadelphia: Institute of Physics Publishing)

**9-8**     Urey H C, Brickwedde F G and Murphy G M 1932 Abstract no 34 *PR* **39** 854; Urey, Brickwedde and Murphy 1932 'A hydrogen isotope of mass 2' *PR* **39** 164–165; Urey, Brickwedde and Murphy 1932 'A hydrogen isotope of mass 2 and its concentration' *PR* **40** 1–15. The last paper appeared back-to-back with Lawrence and Livingston's paper announcing 1.2 MeV protons from the 11-inch cyclotron

**9-9**     Aston F W 1934 in 'Discussion on heavy hydrogen' *PRS* A**144** 1–28 on p 9

**9-10**   'Discussion...' (note 9-9)

**9-11**   Compton effect: the scattering of X-ray quanta, by impacting with free electrons, first observed by Arthur Holly Compton in 1923

**9-12**   Curie I and Joliot F 1932 'Émission de protons de grande vitesse par les substances hydrogénées sous l'influence des rayons gamma très pénétrants' *CR* **194** 273

**9-13**   Chadwick J 1964 'Some personal notes on the search for the neutron' *Proc 10th Int Congress on the History of Science (Ithaca 1962)* (Paris: Hermann) p 161; CHAD V/1 (original handwritten manuscript in ink)

**9-14**   Chadwick J 1932 'Possible existence of a neutron' *NA* **129** 312. A full-length paper, 'The existence of a neutron,' was dispatched to the Royal Society three months later and appeared in *PRS* **136** 692–708

**9-15**   Glasson J 1921 *PM* **42** 596–600

**9-16**   Bainbridge K T 1932 'The isotopic weight of $H^2$' *PR* **42** 1; Pais *Inward Bound*... (note 4-11) p 233

**9-17**   Dahl *Flash of the Cathode Rays*... (note 2-1)

**9-18**   The *positive rays* or canal rays, studied as well by J J Thomson and others, were *ions*, not positive electrons. The erroneous experiments referred to were those of the Polish physicist Julius Edgar Lilienfeld and Jean

Becquerel of France; by 1910 their evidence for positive electrons in glow discharge tubes had been convincingly discredited by fellow physicists, as had certain indications in magneto-optic experiments

**9-19**  Anderson C D 1932 'The apparent existence of easily deflectable positives' *Science* **76** 238–239

**9-20**  Anderson C D, Transcript, oral history interview with C Weiner, 30 June 1966 (OH16)

**9-21**  Cockcroft and Walton 1932 'Experiments with high velocity positive ions I—Further developments of the method of obtaining high velocity positive ions' *PRS* A**136** 619–630. For early thoughts of Cockcroft and Walton on high-voltage accelerating systems, and their final apparatus used in the Li experiment, see also CKFT 31/10, *High Voltage Laboratory – Development of Apparatus*, and CKFT 31/2, *350 KV Transformer*

**9-22**  Allibone 'Metropolitan-Vickers...' (note 2-47) p 159. For Birch pumps and their oils, see also CKFT 31/7, *Birch Pumps*

**9-23**  Cockcroft writing home from USSR August 1931 CKFT 11/1

**9-24**  Ibid

**9-25**  Rutherford to Bohr, telegram 9 September 1931; Rutherford to Bohr, telegram 10 September 1931; Rutherford to Bohr, 10 September 1931; Bohr to Rutherford, 14 September 1931, BSC 1930–1945 Reel 25 Rosseland–Teller. See also Programme for the Clerck Maxwell Celebration, 30 Sept–2 Oct 1931; CUL, Add 7653, PA 143

**9-26**  Hartcup and Allibone *Cockcroft...* (note 2-47) p 49

**9-27**  Walton to Cockcroft, 9 December 1937, CKFT 1/4. The story went around the Cavendish that Rutherford called for Cockcroft and Walton and said: 'If you don't put up a scintillation screen by tomorrow I'll sack both of you.' Brown *The Neutron and the Bomb...* (note 2-17) p 114. This is confirmed in Cockcroft's letter to Karl Darrow in February 1938, in which he added that Chadwick had already advised them to use a scintillation screen. Cockcroft to Karl Darrow, 1 Februry 1938, CKFT 20/28

**9-28**  In Hartcup and Allibone *Cockcroft...* (note 2-47) p 52, the date is given as 14 April, as it is in a letter, Walton to E M McMillan, 1 September 1977 in McMillan 'Early history...' (note 2-42) p 146; the same date appears in WLTN 1/10, as well as in Walton's Notebook No 4, WLTN 2/4. However, according to Cockcroft's notebook (also No 4), the first scintillations were observed on Thursday 13 April; CKFT 1/4

**9-29**  Walton to McMillan 14 May 1977 in McMillan 'Early history...' (note 2-42) p 145

**9-30**  Walton's recollections in Oliphant *Rutherford...* (note 2-26) p 86. The entry for 15 April in Walton's notebook No 4 shows the number of $\alpha$-particle scintillations per minute, including 'Counts by E.W.' and 'Counts by Prof,' with the following pencilled in the margin in Walton's hand: 'Data verified as (Rutherford), E.T.S.W. 20/10/76.' WLTN 2/4

**9-31**  Cockcroft and Walton 1932 'Disintegration of lithium by swift protons' *NA* **129** 649. For Cockcroft's and Walton's laboratory notebooks covering the historic experiment, see CKFT 1/4 (the Li experiment is written up on pp 27–33) and WLTN 2/4 (pages for April 14–18). In a letter to Wilson long after these events, Walton explains that his notebooks leading up to the first disintegration experiment (WLTN 2/2 and 2/3) are very sketchy, due to the effort to develop vacuum tubes able to withstand the requisite high voltages; failures more often than not were not recorded. Walton to A D Wilson, 29 Oct 1979; WLTN 1/10

**9-32**   Rutherford to Bohr 21 April 1932, BSC 1930–1945 Reel 25 Rosseland–Teller

**9-33**   Allibone 'Metropolitan-Vickers...' (note 2-47) p 167

**9-34**   Bohr to Rutherford 2 May 1932 BSC 1930–1945 Reel 25 Rosseland–Teller

**9-35**   Hartcup and Allibone *Cockcroft...* (note 2-47) p 53

**9-36**   Rutherford to Bohr 26 May 1932 BSC 1930–1945 Reel 25 Rosseland–Teller

**9-37**   Cockcroft and Walton 1932 'Experiments with high velocity positive ions. II – The disintegration of elements by high velocity protons' *PRS* A**137** 229–242

## Notes chapter 10

**10-1**   Childs, *Genius...* (note 3-2) p 181; Birge to Lawrence 7 May 1932 RTB Box 33

**10-2**   Childs *ibid*

**10-3**   Heilbron and Seidel *Lawrence...* (note 4-38) p 140

**10-4**   Childs *Genius...* (note 3-2) p 253

**10-5**   Lawrence to Cockcroft 20 August 1932 EOL Reel 5 Item 010415

**10-6**   Heilbron and Seidel *Lawrence...* (note 4-38) p 136

**10-7**   Lawrence to Cottrell 17 July 1931 (note 6-7)

**10-8**   Heilbron and Seidel *Lawrence...* (note 4-38) p 109; Lawrence to Cottrell 20 July 1931 (note 6-9)

**10-9**   Livingston 1959 'History of the cyclotron, Part 1' *PT* **12**(10) 18–23 on p 21

**10-10**   Livingston *ibid* pp 21–22

**10-11**   Lawrence to Tuve 13 January 1932 (note 6-14)

**10-12**   Lawrence to Curtis R Haupt, 11 March 1932. EOL Reel 13 Item 027403. In his reply Haupt takes Lawrence somewhat to task for working too hard and neglecting Molly. 'After all, you are an extremely lucky individual to have won the girl you did, and she is deserving of all the consideration and attention possible. Of course she is cognizant of the strain under which you are working on the research and is willing not to make any unnecessary demands on your time, but you will find after you are married you will not be able to continue research at the enormous pace you have maintained for the past few years.' Haupt to Lawrence, 15 March 1932, EOL Reel 13 Item 027404. Needless to say Lawrence did not take Haupt's advice

**10-13**   Childs *Genius...* (note 3-2) p 177; Livingston 1932 'High-speed hydrogen ions' *PR* **42** 441–442

**10-14**   White, Harvey E 'A brief history of the origins of the Lawrence Hall of Science 1959–1969' unpublished manuscript, archives LHS

**10-15**   Lawrence to Tuve 19 September 1932 EOL Reel 26 Item 056680; Lawrence to Cottrell 22 September 1932 Reel 7 Item 013990

**10-16**   Lawrence, Livingston and White 1932 'The disintegration of lithium by swiftly-moving protons' *PR* **42** 150–151

**10-17**   Cockcroft to Lawrence 3 October 1932 EOL Reel 5 Item 010418

**10-18**   Lawrence to Cockcroft 4 November 1932 EOL Reel 5 Item 010420

**10-19**   Cockcroft and Walton 'Experiments with high velocity positive ions' (note 9-37) p 241

**10-20**  Gamow to Cockcroft 7 September 1932. Hartcup and Allibone *Cockcroft . . .* (note 2-47) p 55

**10-21**  Cockcroft to Gamow 29 September 1932, CKFT 20/10A and in Hartcup and Allibone *ibid*

**10-22**  Cockcroft to Lawrence 17 September 1932 EOL Reel 5 Item 010415

**10-23**  Cockcroft to Lawrence 3 October 1932 EOL Reel 5 Item 010418

**10-24**  Tuve, report for December 1931, 6 January 1932, MAT Box 15

**10-25**  Tuve, Report for May 1932, 7 June 1932, MAT Box 15

**10-26**  Tuve, Hafstad and Dahl 1935 'High voltage technique for nuclear physics studies' *PR* **48** 315–337 on p 317

**10-27**  Tuve, Report for June 1932, 2 July 1932, MAT Box 15

**10-28**  Tuve, Special Report on the High Voltage Work, 27 May 1932, MAT Box 15

**10-29**  Tuve, Hafstad and Dahl 'High voltage technique . . .' (note 10-26) pp 318–321

**10-30**  Ibid p 330

**10-31**  Tuve, Report for June 1932, 2 July 1932, MAT Box 15

**10-32**  Breit to Tuve 14 July 1932 MAT Box 4

**10-33**  Tuve, Report for August 1932, 12 September 1932, MAT Box 15

**10-34**  Tuve to Breit 17 January 1933 MAT Box 8; Tuve to Compton 24 January 1933 MAT Box 4; J A Fleming 1933 'Studies in nuclear physics' *Science* **77** 298–300; Tuve 1933 'The atomic nucleus and high voltages' *Journal of the Franklin Institute* **216**(1) 1–38

**10-35**  Breit to Tuve 20 January 1933 MAT Box 4

**10-36**  Tuve, Hafstad and Dahl 1933 'Disintegration-experiments on elements of medium atomic number' *PR* **43** 942; Tuve, Report for March 1933, 7 April 1933, MAT Box 15; Report for April 1933, 5 May 1933, MAT Box 15

**10-37**  Cockcroft to Elizabeth Cockcroft in Hartcup and Allibone *Cockcroft . . .* (note 2-47) p 60

**10-38**  *Time* 'Complementarity in Chicago' **22** (3 July 1933) p 48. The title refers to Bohr's fumbling attempt to discuss his ideas on the duality of nature before his sophisticated audience. Childs *Genius . . .* (note 3-2) p 200

**10-39**  *Time ibid*

**10-40**  Cockcroft to Walton and Dee, 24 June 1933. CUL, Add 7653, C58. On his way to Chicago, Cockcroft stopped in Cambridge, Mass., calling on Van de Graaff at Round Hill. 'I went up into a sphere and was raised to 2.5 million volts' *ibid*

**10-41**  Lawrence to GN Lewis *circa* 7 July 1933 GLP CU-30 Box 3 Item 425

**10-42**  Cockcroft to Lawrence 30 March 1933 EOL Reel 5 Item 010424

**10-43**  Lauritsen to Tuve 1 March 1933 MAT Box 12

**10-44**  Tuve to Lauritsen 30 March 1933 MAT Box 12

**10-45**  Tuve Memorandum . . . MAT Box 4 and Box 5

**10-46**  Tuve to Breit 17 January 1933; Fleming to W Gilbert 17 January 1933 MAT Box 8

**10-47**  Tuve, Report for August 1933, 17 October 1933, MAT Box 15

**10-48**  Tuve, Hafstad and Dahl 'High voltage technique . . .' (note 10-26), pp. 321–322. The two-meter generator was reassembled for the opening of the exhibit 'Atom Smashers: 50 Years' in July of 1980 at the Smithsonian Museum of History and Technology in Washington, DC. There the generator was remounted in a rather clever manner in a circular stairwell, allowing the public to view the machine as they descended from the open terminal to the target chamber at the bottom end of the acceleration tube. At the time of writing, the machine is no longer on view.

## Notes chapter 11

**11-1**   Dahl *Heavy Water...* (note 9-7) p 312. 'I think the reason the mouse survived [thought Melvin Calvin] was that Gilbert ran out of $D_2O$ because the mouse wouldn't have lived if he had been able to increase the dosage over a long period of time.' Calvin, Melvin 1976 'Gilbert Newton Lewis,' address presented before 'The Robert A Welch Foundation Conferences on Chemical Research. XX. American Chemistry – Bicentennial' which was held in Houston, Texas, 8–10 November 1976. In Lewis's theory, dating from the early 1900s, the constituents of the atom were quiescent, with electrons situated at the corners of a tetrahedron, in contrast to Bohr's dynamic theory; hence Lewis's became known as the 'static atom.' Calvin 'Gilbert Newton Lewis,' above, p 124; Branch, Gerald E K 'Gilbert Newton Lewis 1875-1946' pp 18–21 on p 20 in Appendix to Melvin Calvin 1984 'Gilbert Newton Lewis: his influence on physical-organic chemistry at Berkeley' *Journal of Chemical Education* **61** 14–21. Branch was the first of Lewis's appointees at Berkeley. Lewis's 'mature' theory appeared in 1923: Lewis G N 1923 *Valence and the Structure of Atoms and Molecules* (New York: The Chemical Catalogue Company)

**11-2**   Davis, Nuel Pharr 1968 *Lawrence and Oppenheimer* (New York: Simon and Schuster) p 53; Dahl *Heavy Water...* (note 9-7) p 53

**11-3**   Lawrence to Tuve 3 May 1933 EOL Reel 26 Item 056694

**11-4**   Lawrence to Cockcroft 4 May 1933 EOL Reel 5 Item 010426

**11-5**   *Science Service* 20 May 1933; Gilbert N, Lewis M, Stanley Livingston and Ernest O Lawrence 1933 'The emission of alpha-particles from various targets bombarded by deutons of high speed' *PR* **44** 55–56 (letter of June 10)

**11-6**   Rutherford 1934 'The new hydrogen' *Science* **80** (13 July 1934) 21–25 delivered before the Royal Institution of Great Britain March 23 1934

**11-7**   Hartcup and Allibone *Cockcroft...* (note 2-47) p 58

**11-8**   Oliphant and Rutherford 1933 'Experiments on the transmutation of elements by protons' *PRS* A**141** 259–281 (received June 16)

**11-9**   Oliphant *Rutherford...* (note 2-26) p 106. Oliphant and Rutherford soon switched from scintillation counting to the electronic counting techniques of Wynn-Williams. Particles were registered as sideways deflections of a spot of light reflected from the mirror of a moving iron oscillograph on a moving strip of photographic paper. Impatient for results, Rutherford insisted that the photographic record be developed as fast as possible. Pulling it out of the fixing bath, he sat at a table, dripping fixing solution upon the notebook pages and his own clothes, while dribbling pipe ash all over the wet photographic paper. He damaged it further by attempting to mark the messy material with the stump of a pencil. Oliphant *ibid* p 107

**11-10**   Lewis to Rutherford 15 May 1933 GLP Box 3 Item 857. Ralph Fowler was a visiting lecturer in the Berkeley Physics department during the Spring term of 1933. There he gave a course on ferro-magnetism, with an enrollment of five for credit and *sixty* auditors. Birge RT *History of the Physics Department* mimeograph 5 vols (Berkeley 1966–?) in vol 4 XI (14)

**11-11** Rutherford to Lewis 30 May 1933 GLP Box 3 Item 858

**11-12** Oliphant *Rutherford . . .* (note 2-26) p 105

**11-13** Oliphant Kinsey and Rutherford 1933 'The transmutation of lithium by protons and by ions of the heavy isotope of hydrogen' *PRS* A**141** 722–733

**11-14** Ibid

**11-15** Rutherford to Lewis 20 June 1933 GLP Box 3 Item 859

**11-16** Dee P I and Walton E T S 1933 'A photographic investigation of the transmutation of lithium and boron by protons and of lithium by ions of the heavy isotope of hydrogen' *PRS* A**141** 733–742

**11-17** Rutherford to Lewis 10 August 1933 GLP Box 3 Item 862

**11-18** Lewis to Rutherford 5 October 1933 GLP Box 3 Item 863

**11-19** Harteck P 1934 'Bericht über Meine Arbeit in Cambridge und Erfahrungen augemeiner Art' PHP Acc MC 17 Box 2 Folder 2

**11-20** Lewis to Harteck 23 June 1933 GLP Box 2 Item 329

**11-21** Lawrence to Cockcroft 2 June 1933 EOL Reel 5 Item 010429

**11-22** Lawrence, Livingston and Lewis 1933 'The emission of protons from various targets bombarded by deutons of high speed' *PR* **44** p 56

**11-23** Chadwick 'The existence of a neutron' (note 9-14) p 702

**11-24** Chadwick J 1933 'The Neutron' (Bakerian Lecture) *PRS* A **142** 1–25; Brown *The Neutron . . .* (note 2-17) p 117

**11-25** Bainbridge, Kenneth T 1933 'Comparison of the masses of $H^2$ and helium' *PR* **44** 57

**11-26** Livingston, Henderson and Lawrence 1933 'Neutrons from deutons and the mass of the neutron' *PR* **44** 781–782; Heilbron and Seidel *Lawrence . . .* (note 4-38) pp 155–156

**11-27** Childs *Genius . . .* (note 3-2) p 200

**11-28** Rutherford to Lewis 27 July 1933 GLP CU-30 Box 3 Item 861

**11-29** Lewis to Rutherford 5 October 1933 GLP CU-30 Box 3 Item 863

**11-30** Lawrence to Langevin 4 October 1933 cited in Mark L Oliphant 1966 'The two Ernests—I' *PT* **19** 35-49 on p 45. All papers to be presented were circulated in advance to those invited and all were asked to submit the points they wished to raise concerning the various papers. Mehra, Jagdish 1975 *The Solvay Conferences on Physics: Aspects of the Development of Physics Since 1911* (Dordrecht, Holland: D Reidel Publishing Co)

**11-31** Cockcroft, John D 1933 'Disintegration of elements by accelerated protons,' manuscript copy EOL Reel 56 Solvay Congress; Reports Item 123597–123657. Thus under the subheading 'Acceleration by multiple impulses' (the cyclotron), Cockcroft discusses the loss of proton current by diffusion from the beam; Lawrence has written <u>no loss</u> in the margin (Reel 56 Item 123621). On the same page Cockcroft gives $10^{-8}$ amperes as the maximum molecular ion current produced by Lawrence to date; Lawrence has corrected the number to $0.2 \times 10^{-6}$ amperes

**11-32** Oliphant 'The two Ernests . . .' (note 11-30) p 45. Chadwick J 1933 Manuscript copy, with marginal corrections by Lawrence, EOL Reel 56 Solvay Congress; Reports Item 123528–123561

**11-33** Heilbron and Seidel *Lawrence . . .* (note 4-38) p 164. Chadwick, for that matter, was not spared from Heisenberg's intervention while still speaking, but his rude interruption was squelched by Dirac

**11-34**   Evans R D 1955 *The Atomic Nucleus* (New York: McGraw-Hill Book Co) p 432

**11-35**   Joliot-Curie, Frédéric and Irène 1934 'Un noveau type de radioactivité' *CR* **198** 254; 'Artificial production of a new type of radioelement' *NA* **133** 201

**11-36**   Childs *Genius...* (note 3-2) p 207

**11-37**   Rutherford to Lewis 30 October 1933 GLP CU-30 Box 3 Item 865

**11-38**   Childs *Genius...* (note 3-2) p 210

**11-39**   Tuve, Report for November 1933, 3 January 1934, MAT Box 15

**11-40**   Lawrence to Cockcroft 20 November 1933 EOL Reel 5 Item 010439

**11-41**   Lawrence to Tuve 21 December 1933 EOL Reel 26 Item 056696

**11-42**   Lawrence and Livingston 1934 'The emission of protons and neutrons from various targets bombarded by three million volt deutons' *PR* **45** 220; Lewis, Livingston, Henderson and Lawrence 1934 'The disintegration of deutons by high speed protons and the instability of the deuton' *PR* **45** 242–244

**11-43**   Cockcroft to Lawrence 21 December 1933 EOL Reel 5 Item 010441

**11-44**   Lawrence to Cockcroft 12 January 1934 EOL Reel 5 Item 010443 Lawrence to Tuve 12 January 1934 EOL Reel 26 Item 056701 voicing the same sentiments

**11-45**   Cockcroft to Lawrence 28 February 1934 EOL Reel 5 Item 010445

**11-46**   Lawrence to Cockcroft 14 March 1934 EOL Reel 5 Item 010447

**11-47**   Fowler to Lawrence 14 March 1934 EOL Reel 5 Item 010448

**11-48**   Lauritsen and Crane 1934 'Gamma-rays from carbon bombarded with deutons' *PR* **45** 345–346

**11-49**   Tuve report for December 1933, 3 January 1934, MAT Box 15

**11-50**   Tuve report for January 1934, 9 February 1934, MAT Box 15

**11-51**   Tuve Report for February 1934, 3 March 1934, MAT Box 15

**11-52**   Tuve Report for March 1934, 20 April 1934, MAT Box 15

**11-53**   Tuve to Lawrence 6 January 1934 EOL Reel 26 Item 056699

**11-54**   Tuve to Lawrence 28 February 1934 EOL Reel 26 Item 056702

**11-55**   Lawrence to Tuve 14 March 1934 EOL Reel 26 Item 056704

**11-56**   Lewis, Livingston, Henderson and Lawrence 1934 'On the hypothesis of the instability of the deuton' *PR* **45** 497 (submitted 15 March 1934). See also Lawrence 'Lecture on the Mass of the Neutron,' undated, EOL Reel 61 Item 135012 where no mention is made of the 'light neutron'

**11-57**   Tuve to Lawrence 17 April 1934 EOL Reel 26 Item 056706; Tuve and Hafstad 1934 'The emission of disintegration-particles from targets bombarded by protons and by deuterium ions at 1200 kilovolts' *PR* **45** 651–653

**11-58**   Lawrence to Tuve 20 April 1934 EOL Reel 26 Item 056714

**11-59**   Fowler to Lawrence 14 March 1934 (note 11-47)

**11-60**   Rutherford to Lawrence 13 March 1934 EOL Reel 23 Item 050650

**11-61**   Lawrence to Rutherford 10 May 1934 EOL Reel 23 Item 050652

**11-62**   Tuve to Cockcroft 18 April 1934 MAT Box 16

**11-63**   Cockcroft to Tuve 30 April 1934 MAT Box 16. Cockcroft also brings up a disagreement with Tuve, as we explore in chapter 12

**11-64**   Oliphant 'The two Ernests...' (note 11-30) p 49

**11-65**   Heilbron and Seidel *Lawrence...* (note 4-38) pp 239, 350

**11-66**   'A third variety of hydrogen? Experiments at Cambridge' *The Times* London 22 March 1934

**11-67**  Tuve, Hafstad and Dahl 1934 'A stable hydrogen isotope of mass three' *PR* **45** 840–841

**11-68**  Rutherford to the Director, Norsk Hydro, 19 March 1935 NHAN K03 50n 1.11: Tungt vann general Hefte 1. Dahl *Heavy Water...* (note 9-7) pp 133–134

**11-69**  Urey to Dr S Bronne [sic] 28 February and 22 March 1935 NIAM Box 4F-D02 960: Tungt vann Ark Nr 704C

**11-70**  Axel Aubert to Rutherford 23 March 1935 NHAN KO3 50 n1.1: Tungt vann general Hefte 1

**11-71**  Rutherford to the Director Norsk Hydro 3 April 1936 NHAO K0350 n2.1: Tungt vann Hefte 4. Rutherford 1937 'The search for the isotopes of hydrogen and helium of mass 3' *NA* **140** pp 303-305

**11-72**  Laurence W L 1950 'Cosmic rays are found to produce tritium vital for hydrogen bomb' *New York Times* Friday 15 September 1950. He$^3$, in contrast to H$^3$, is stable but almost as rare, with a natural abundance of about $10^{-9}$%

**11-73**  Goldhaber, Maurice 1979 'The nuclear photoelectric effect and remarks on higher multipole transitions: a personal history' pp 83–110 in *Nuclear Physics in Retrospect* (note 2-42)

**11-74**  Ibid p 86. Rudolf Peierls, who would make important contributions to the nuclear photoelectric, effect writes that Goldhaber was exceptionally bright but a little naïve when he came to Cambridge and 'had not yet acquired a very polished manner. He caused raised eyebrows by telling everybody, including Rutherford and Chadwick, what experiments they ought to be doing. It was particularly aggravating that he was usually right.' Peierls, Rudolf 1985 *Bird of Passage: Reflections of a Physicist* (Princeton: Princeton University Press) p 119

**11-75**  As Goldhaber noted in the discussion following his retrospective talk, 'the masses of deuterium and hydrogen were known approximately; the neutron mass was also known approximately. Depending on who was right on the neutron mass, it looked either easy or just touch and go.' *Ibid* p 106

**11-76**  Chadwick J and Goldhaber M 1934 'A "nuclear photo-effect": disintegration of the diplon by $\gamma$-rays' *NA* **134** 237–238; CHAD III 1/6

**11-77**  Rudolf Peierls, in his retrospective paper on the development of our ideas on nuclear forces, quotes Chadwick saying that 'it is, of course, possible that the neutron may be an elementary particle. This view has little to recommend itself at present, except the possibility of explaining the statistics of such nuclei as N$^{14}$.' Peierls Rudolf 'The development of our ideas on the nuclear forces' pp 183–211 on p 184 in *Nuclear Physics in Retrospect* R H Stuewer ed (note 2-42). Chadwick's remarks are found in Chadwick 'The existence of a neutron' (note 9-14)

**11-78**  Goldhaber 'The nuclear photoelectric effect...' (note 11-73) pp 88–89

**11-79**  Heilbron and Seidel *Lawrence...* (note 4-38) p 175

## Notes chapter 12

**12-1**  Reported as 'Atomic structure. Progress in 21 years' *Times* (3 October) p 6 col a. In his opening address Rutherford noted that Tuve was among 'those pioneers ... who have opened up these new methods of attack.' Other pioneers on his list were Coolidge, Allibone, Lauritsen and Brasch and Lange. Cornell *Merle A Tuve and His Program*... (note 3-1) pp 425–426

12-2   Segrè Emilio 1970 *Enrico Fermi, Physicist* (Chicago: The University of Chicago Press) p 78. The subsequent discovery of the effect on $(n, \gamma)$ reactions by slow neutrons 'was a great relief to Amaldi and [Segrè] because it explained a serious puzzle and showed that there was no need to correct the minutes of the London conference' *ibid* p 80

12-3   Andrade E N da C 1978 *Rutherford and the Nature of the Atom* (Gloucester, MA: Peter Smith) p 201

12-4   Hartcup and Allibone *Cockcroft*... (note 2-47) p 66

12-5   Statement by the Soviet Embassy *The Times* 24 April 1935

12-6   Massey, Harrie and Norman Feather 1976 'James Chadwick, 20 October 1891–24 July 1974' Royal Society *Biographical Memoirs* **22** pp 11–70 on p 22

12-7   Lawrence to Chadwick 27 November 1935 cited in Andrew P Brown 1996 'Liverpool and Berkeley: the Chadwick–Lawrence letters' *PT* **49** 34–40 on pp 35–36

12-8   Wilson *Rutherford*... (note 2-4) p 587; Brown *The Neutron*... (note 2-17) p 152

12-9   Hughes, Jeff 2000 '1932: The annus mirabilis of nuclear physics?' *Physics World* July pp 43–48. Oddly, few copies of the pamphlet survive today; not even Rutherford's files contain a copy. According to Hughes, Eddington appropriated the term 'Annus Mirabilis' for his 1934 pamphlet from an article by the Oxford radiochemist and former student of Rutherford, Alexander Russell. It was written in the wake of the celebrations of 1931 commemorating the centenary of Maxwell's birth and other British scientific milestones (page 112); Russell A 1931 'Annus Mirabilis: 1831' *The Nineteenth Century and After*

12-10  Oliphant to Rutherford 25 August 1935 CUL Add 7653 O5

12-11  Cockcroft to Lawrence 25 February 1936 EOL Reel 5 Item 010488; Lawrence to Rutherford 21 March 1936

12-12  Brown *The Neutron*... (note 2-17) p 153

12-13  Heilbron and Seidel *Lawrence*... (note 4-38) p 339. CKFT 20/52, *Design of Cyclotrons*, contains much correspondence on the topic between Cockcroft and Bernards Kinsey (in Liverpool) and between Cockcroft and Walton

12-14  Ibid pp 342, 344

12-15  Ibid p 350

12-16  Hartcup and Allibone *Cockcroft*... (note 2-47) p 71

12-17  Anna Kapitza to Cockcroft 30 March 1939 CKFT 20/15

12-18  Cited in Andrade *Rutherford*... (note 12-3) p 206

12-19  Cochran, William 1995 'Solid-state structure analysis' pp 421–519 on p 429 in *Twentieth Century Physics Volume I* edited by L M Brown, A Pais and Sir Brian Pippard (Bristol and Philadelphia: Institute of Physics Publishing). Bragg, we might add, had a strong interest in the history of physics, and founded the Cavendish Museum of experimental apparatus (note 2-21), originally installed in the Austin wing

12-20  Maurice Goldsmith, in his biography of Joliot, claims that Joliot thought the term 'artificial radioactivity' poorly chosen. To him, the new radioactivity was just as spontaneous and natural as that of the natural radioactive elements. He would have retained the adjective 'artificial' for the radioisotopes produced, not for their radioactivity. Maurice Goldsmith 1976 *Frédéric Joliot-Curie: A Biography* (London: Lawrence and Wishart) p 53

**12-21**  Joliot-Curie 'Un noveau type de radioctivité...' (note 11-35)

**12-22**  Lauritsen C C and Crane H R 1934 'Gamma-rays from carbon bombarded with deutons' *PR* **45** 345–346 submitted 15 February

**12-23**  Lauritsen C C, Crane H R and Harper W W 1934 'Artificial production of radioactive substances' *Science* **79** 234–235 submitted February 27; Crane H R and Lauritsen C C 1934 'Radioactivity from carbon and boron oxide bombarded with deutons and the conversion of positrons into radiation' *PR* **45** 430–432 submitted 1 March

**12-24**  Tuve, Report for February 1934, 3 March 1934, MAT Box 15

**12-25**  Tuve, Report for March 1934, 20 April 1934, MAT Box 15

**12-26**  Tuve, Report for April 1934, 8 May 1934, MAT Box 15

**12-27**  Cockcroft to Tuve 30 April 1934 (note 11-63)

**12-28**  Dahl to Tuve 22 May 1934 MAT Box 16

**12-29**  Dahl *ibid*

**12-30**  Tuve to Dahl 8 June 1934 MAT Box 13

**12-31**  Crane H R and Lauritsen C C 1934 'Further experiments with artificially produced radioactive substances' *PR* **45** 497–498 submitted March 14

**12-32**  Tuve, Report for May 1934, 9 June 1934, MAT Box 15; Hafstad and Tuve 1934 'Artificial radioactivity using carbon targets' *PR* **45** 902–903 submitted June 1

**12-33**  Childs *Genius*... (note 3-2) p 219. In Tuve's own copy of his note rebutting Lawrence for not mentioning contamination effects, he has scribbled in the margin: 'This would make us guilty of incautious and premature criticism of the results of others plus the failure to retract such criticism in an adequate and proper way' MAT Box 16

**12-34**  Loeb LB and TR Reed 1934 *Science* **80** 48–49

**12-35**  Tuve 1934 'Nuclear-physics symposium, a correction' *Science* **80** 161–162. It seems that Lawrence, who also had followed up on the Joliot-Curie's suggestion of inducing artificial activities through deuteron bombardment — Lawrence's strong card — made excessive claims in that regard as well, even going so far as hinting, if not outright claiming, that he had discovered artificial radioactivity before Joliot and Curie: Heilbron and Seidel *Lawrence*... (note 4-38) p 181

**12-36**  Tuve to Hafstad 2 July 1934 MAT Box 16

**12-37**  Hafstad to Tuve 5 July 1934 MAT Box 16

**12-38**  Tuve, Report for September 1934, 6 October 1934, MAT Box 15

**12-39**  Hafstad, Report on the International Conference on Physics, London England 1-6 October 1934 MAT Box 15

**12-40**  Tuve, Report for December 1934, 7 January 1935, MAT Box 15. The resonance is actually at 459 keV. Brown *Golden Age*... (note 3-17) p 359

**12-41**  Tuve, Hafstad and Dahl 1935 'High voltage technique for nuclear physics studies' *PR* **48** 315–337; Tuve, Dahl and Hafstad 1935 'The production and focusing of intense positive ion beams' *PR* **48** 241–256; Hafstad and Tuve 1935 'Carbon radioactivity and other resonance transformations by protons' *PR* **48** 306–315. A figure on p 322 of the first paper (Tuve, Hafstad and Dahl) reproduced as our Figure 10.7 on p 136, caused Tuve much grief, as its publication made it seem as if he had 'engaged in an act of self-promotion.' Its inclusion as a 'full page newspaper picture' was due to 'Fleming's work on the manuscript'; Fleming himself thought it was a good picture that deserved to be printed in the large format. Tuve to J T Tate (editor of the *Physical Review*) 20 August 1935 MAT Box 3

**12-42**  Hans Bethe to John Fleming 5 May 1935 cited in Brown *Golden Age...* (note 3-17). George Gamow had stipulated the creation of the Washington conferences as a condition of his appointment at George Washington University; he modeled them after Bohr's annual gatherings in Copenhagen

**12-43**  Norton G A, Ferry J A, Daniel R E and Klody G M undated 'A retrospective of the career of Ray Herb' unpublished manuscript National Electrostatics Corp, Middleton, Wisconsin

**12-44**  Brown, Louis undated *Outline for DTM Centennial History* unpublished manuscript

**12-45**  In addition to the question of pressure insulation for Van de Graaff generators, there remained the vexing mechanical problem with their charging belts, which had to be made from material that was a good insulator, yet was strong enough to work as a belt driven over pulleys at high speed. Silk and paper with various coatings frayed and tore with distressing regularity. Instead, Dahl resurrected an old version of the electrostatic generator with rotating dielectric disks. Dahl 1936 'Note on disk-type electrostatic generators' *Review of Scientific Instruments* **7** 254–256. However, in due course they found that rubberized airship fabrics worked well enough for belts, and the scheme was dropped

**12-46**  Tuve, Heydenburg and Hafstad 1936 'The scattering of protons by protons' *PR* **50** 806–825; Tuve and Hafstad 1936 'The scattering of neutrons by protons' *PR* **50** 490–491. It goes without saying that the experimental technique used in the n–p scattering experiment was quite different from that used in p–p scattering. It suffices to note that reactions yielding energetic neutrons provided incident neutron beams, albeit at very low intensity; now DTM's long experience with Wilson cloud chambers paid off. Brown, Louis 'The nuclear force' ch 12 in *Outline for DTM Centennial History* (note 12-44). Note that the earlier model of the nucleus consisting of protons and electrons, held together by electric and magnetic forces, had held sway from 1913 (Rutherford) till 1934 (Heisenberg, Ettore Majorana, Eugene Wigner, and others), or well past the discovery of the neutron (Chadwick) in 1932. Brown and Rechenberg *The Origin...* (note 4-19) pp 19, 40; Pais *Inward Bound...* (note 4-11) pp 231, 296. It was Heisenberg's nuclear neutron–proton model of 1932–33 that cast the first serious doubts on the electron's purported nuclear role. Laurie M Brown 2001 'The electron and the nucleus' ch 9 of *Histories of the Electron: The Birth of Microphysics* ed by Jed Buchwald and Andrew Warwick (Cambridge, MA: The MIT Press)

**12-47**  Breit G, Condon E U and Present R D 1936 'Theory of scattering of protons by protons' *PR* **50** 825–845. Note also Rudolf Peierls's discussion of the subject from a theoretical point of view, including the analysis of Breit, Condon and Present, and by Breit and Feenberg, below. Peierls 'The development of our ideas...' (note 11-77) p 190

**12-48**  Rutherford to Tuve 17 Nov 1936 CIW Archives 'DTM–Director 1935–1940' cited by Cornell *Merle A Tuve and His Program...* (note 3-1) p 483

**12-49**  Breit G and Feenberg E 1936 'The possibility of the same form of specific interaction for all nuclear particles' *PR* **49** 850–856

**12-50** Brown, Laurie M 1995 'Nuclear forces, mesons, and isospin symmetry' ch 5 pp 357–419 in *Twentieth Century Physics Volume I* (note 12-19). Just as the electron, neutron, and proton have a *spin* that can be represented by a half-integer multiple of $h/2\pi$(where $h$ is Planck's constant), *isospin* emerged in the 1930s as an analogue to spin; it is a so-called internal quantum number characterizing strongly interacting particles. Brown *ibid* p 395. Briefly, in 1932 Heisenberg pointed out that, if we ignore the electric charge, the proton and the neutron could be viewed as a single particle in two different states. In 1937 Eugene P Wigner proposed that the two states could be regarded as some type of spin; he named Heisenberg's states 'isotopic spin,' nowadays shortened to 'isospin'

**12-51** Tuve 'Odd Dahl...' (note 4-23) p 46

**12-52** Henderson, Livingston, and Lawrence 1934 'Artificial radioactivity produced by deuton bombardment' *PR* **45** 428–429; letter submitted 27 February

**12-53** Lawrence 'Outline of lecture on artificial radioactivity induced by bombardment of $_1H^1$, $_1H^2$' undated EOL Reel 61 Item 135016

**12-54** Segrè, Emilio 1971 'Fermi Enrico' *DSB* **4** 576–583; Segrè, Emilio 1970 *Enrico Fermi: Physicist* (note 12-2)

**12-55** Fermi's chief theoretical contribution, just prior to his appointment in Rome, had been the discovery in 1926 of a new type of statistics applicable to particles obeying Pauli's exclusion principle. (Paul Dirac came up independently with the same statistics soon afterward, and connected it to the new quantum mechanics.) Fermi's crowning theoretical achievement would be his theory of beta decay, based on the neutrino hypothesis of Wolfgang Pauli, announced shortly after the Solvay Conference in 1933

**12-56** Noddack, Ida 1934 'Über das Element 93' *Z Angewandte Chemie* **47** 653–655. English translation in H G Graetzer and D L Anderson 1971 *The Discovery of Nuclear Fission: A Documentary History* (New York: Van Nostrand Reinhold). Wrote Noddack, co-discoverer of Rhenium with her husband Walter, 'It is conceivable that when heavy nuclei are bombarded by neutrons, these nuclei break up into several *larger* fragments, which would of course be isotopes of known elements but not necessarily of the irradiated elements.' Her suggestion of nuclear fission was ignored in Rome as well as in Berlin!

**12-57** Fermi E, Amaldi E, d'Agostino O, Rasetti F and Segrè E 1934 'Artificial radioactivity produced by neutron bombardment' *PRS* A**146** 483–500

**12-58** Rutherford to Fermi 23 April 1934. Fermi *Collected Papers* Vol I (Chicago: University of Chicago Press) p 641

**12-59** Segrè, *Enrico Fermi: Physicist* (note 12-54) p 77

**12-60** Cooksey to Lawrence 6 May 1933 EOL Reel 6 Item 012027 pointed out in Heilbron and Seidel *Lawrence...* (note 4-38) p 184

**12-61** Heilbron and Seidel *Lawrence...* (note 4-38) p 185

**12-62** Childs *Genius...* (note 3-2) p 216

**12-63** Lawrence to Poillon 3 March 1934 cited in Heilbron and Seidel *Lawrence...* (note 4-38) p 186

**12-64** Heilbron and Seidel *Lawrence...* (note 4-38) p 188

**12-65** Heilbron and Seidel *Lawrence...* (note 4-38) p 191

**12-66** Lawrence to Cooksey 4 November 1934 cited in Heilbron and Seidel *Lawrence...* (note 4-38) p 189 note 103

**12-67** Heilbron and Seidel *Lawrence...* (note 4-38) p 189

**12-68**   Heilbron and Seidel *Lawrence* . . . (note 4-38) p 283

**12-69**   Childs *Genius* . . . (note 3-2) p 294

**12-70**   Heilbron and Seidel *Lawrence* . . . (note 4-38) Table 65: US Cyclotrons by Size 1940 p 310

**12-71**   Hirosige, Tetu 1963 'Social conditions for the researches of nuclear physics in pre-war Japan' in *Japanese Studies in the History of Science* **2** 87–88

**12-72**   Heilbron and Seidel *Lawrence* . . . (note 4-38) Table 71: Foreign Cyclotrons by Size 1940 p 321. Both cyclotrons survived the firebombing of Tokyo, in which nearly two-thirds of the institute's buildings were destroyed, but were dismantled and dumped in Tokyo Bay by the US occupation authorities in November 1945

**12-73**   Fleming to James Cork (Michigan's chief cyclotron builder) 22 November 1939 MAT Box 23

## Notes chapter 13

**13-1**   A neutron entering a hydrogenous material at room temperature with an energy of 5 MeV, providing it does not combine with some nucleus, will be reduced to the energy of a molecule at room temperature, or about 0.025 eV, after some 20 collisions. Neutrons which have come into thermal equilibrium with matter are known as *thermal neutrons*

**13-2**   Fermi, Amaldi, Pontecorvo, Rasetti and Segrè 1962 'Influence of hydrogenous substances on the radioactivity produced by neutrons' translated by E Segrè. Fermi *Collected Papers* Vol I (note 12-58) pp 761–762

**13-3**   Segrè, *Enrico Fermi: Physicist* (note 12-2) p 83

**13-4**   Curie I, Von Halban H and Preiswerk P 1935 'Sur la création artificielle d'éléments appartenant à une familie radioactive inconnue lors de l'irradiation du thorium par les neutrons' *JPR* **6** 361

**13-5**   Curie I and Savitch P 1937 'Sur les radioéléments formés dans l'uranium irradié par les neutrons' *JPR* **8** 385–387

**13-6**   Curie I and Savitch P 1938 'Sur la nature du radioélément de périod 3.5 heures formé dans l'uranium irradié par les neutrons' *CR* **206** 1643. Translated as 'Concerning the nature of the radioactive element with 3.5-hour half-life formed from uranium irradiation by neutrons' in Graetzer and Anderson *The Discovery of Nuclear Fission* . . . (note 12-56) pp 37–38

**13-7**   Weart SR 1979 *Scientists in Power* (Cambridge, MA: Harvard University Press) p 55

**13-8**   Badash Lawrence 1972 'Hahn, Otto' *DSB* **6** 14–17

**13-9**   Frisch OR 1974 'Meitner, Lise' *DSB* **9** 260–263

**13-10**   Graetzer and Anderson 'Radiochemistry experiments (1935-1939)' ch 2 in *The Discovery of Nuclear Fission* . . . (note 12-56) pp 21–48

**13-11**   Graetzer and Anderson *ibid* p 40

**13-12**   Hahn and Strassmann 1938 'Über die Entstehung von Radiumisotopen aus Uran beim Bestrahlen mit Schnellen und Verlangtsamten Neutronen' *NW* **26** 755–756; translated as 'Concerning the creation in radium isotopes from uranium by irradiation with fast and slow neutrons' in Graetzer and Anderson *The Discovery of Nuclear Fission* . . . (note 12-56) pp 41–44

**13-13**   Hahn to Meitner 19 December 1938; copy, Hahn–Meitner correspondence MPG A538 Map 3. Nor is the original, only a copy, to be found in the

Hahn–Meitner correspondence (before 1940) preserved at Churchill College, Cambridge; MTNR 5/21A

**13-14** Hahn and Strassmann 1939 ' Über den Nachweis und das Verhalten der bei der Bestrahlung des Urans mittels Neutronen Entstehenden Erdalkalimetalle' *NW* **27** 11–15 received 22 December 1938, published 6 January. 1939 Translated as 'Concerning the existence of alkaline earth metals resulting from neutron irradiation of uranium' in Graetzer and Anderson *The Discovery of Nuclear Fission . . .* (note 12-56) pp 44–47

**13-15** Meitner to Hahn 21 December 1938; original Meitner–Hahn correspondence MPG Item 04873

**13-16** Hendry, John 1970 'Frisch, Otto' *DSB* **17** Suppl II pp 320–322

**13-17** Frisch O R 1979 *What Little I Remember* (Cambridge: Cambridge University Press) pp 115–116

**13-18** Frisch O R 1967 'The discovery of fission: how it all began' *PT* **20** 43–48

**13-19** Brink, David M 'Nuclear dynamics' ch 15 p 1213 in *Twentieth Century Physics Volume II* edited by L M Brown, A Pais and Sir Brian Pippard (Bristol and Philadelphia: Institute of Physics Publishing). Evans 'Modes of nuclei' ch 11 p 365 in *The Atomic Nucleus* (note 11-34). Wheeler, John A 1967 'The discovery of fission: mechanism of fission' *PT* **20** 278–281

**13-20** At the end of their article, Hahn and Strassmann point out that barium of mass 138 plus 'masurium' (technetium 101, then still undiscovered) add up to the mass of uranium 239. 'Masurium' lies in the middle of the platinum group, and would have chemical characteristics that could be reconciled with those of the 'transuranics' observed earlier. In fact, they should have worked with atomic *numbers* (charges), not masses, as the relevant parameters as Meitner and Frisch were quick to appreciate, and they should add up to Z = 92 for uranium The correct explanation for fission is

$$_{92}U \longrightarrow {}_{56}Ba + {}_{36}Kr$$

where Kr (krypton) is an inert gas. Hahn O 1962 *A Scientific Autobiography* translated and edited by W Ley (New York: Scribners) p 155

**13-21** Meitner to Hahn 3 January 1939; original Meitner–Hahn correspondence MPG Item 04875

**13-22** Krafft, Fritz 1983 'Internal and external conditions for the discovery of nuclear fission by the Berlin team' in William R Shea, editor, *Otto Hahn and the Rise of Nuclear Physics* (Dordrecht: D Reidel Publishing Co)

**13-23** The apparatus used in the fission discovery has been on permanent display in the chemistry section of the Deutsches Museum, Munich, since 1953. For years the accompanying text failed to mention Meitner in the discovery—a blunder only corrected in 1990. In 1945 Otto Hahn received the Nobel Prize in Chemistry for the discovery of nuclear fission; his two partners were not included. Reexamining the controversial decision by the Nobel committees, Elisabeth Crawford, Ruth Sime and Mark Walker conclude that Strassmann was ignored 'because he was not a senior scientist.' Meitner's exclusion is more complicated, seen as the result of four basic flaws in the decision process.' 'The difficulty of evaluating an interdisciplinary discovery, a lack of expertise in theoretical physics, Sweden's scientific and political isolation during the war, and a general failure of the evaluation committees to appreciate the extent to which German prosecution of Jews skewed the published scientific record.' Elisabeth Crawford, Ruth Lewin Sime and Mark Walker 1977 'A Nobel tale of postwar injustice' *PT* **50** 26–32

13-24    Stuewer, Roger H 1985 'Bringing the news of fission to America' *PT* **38** 49–56

13-25    Churchill, Winston 1948 *The Gathering Storm* (Boston: Houghton Mifflin) p 292

13-26    Hahn to Meitner 1 October 1938 cited in Ruth Lewin Sime 1996 *Lise Meitner: A Life in Physics* (Berkeley: University of California Press) p 448

13-27    Tuve to James Franck 28 March 1938 MAT Box 17. Tuve, Report for September 1938, 7 October 1938, MAT Box 15. The generator peaked at somewhat under 2 MV at atmospheric pressure

13-28    Tuve, Report for October 1938, 14 November 1938, MAT Box 15

13-29    Tuve, Report for December 1938, 11 January 1939, MAT Box 15

13-30    Frisch to Meitner 3 January 1939, MTNR 5/23 (Meitner–Frisch 1936–52) cited in Lemmerich, Jost ed 1988 *Die Geschichte der Entdeckung der Kernspaltung: Ausstellungskatalog* (Berlin: Technische Universität Berlin, Universitätsbibliotek) p 177

13-31    Frisch *What Little I Remember* (note 13-17) p 116

13-32    Frisch *ibid* p 117. Placzek, a theorist, was not above sharing a laboratory workbench with Frisch, and had earlier collaborated with him on the absorption of slow neutrons in gold and other metals. For gold, he used several Nobel Prize medals left for safe keeping at the Institute. When Denmark was invaded in 1940, Hevesy dissolved the medals in acid and stored them on a chemical shelf. After the war they were converted back to metal, and the Swedish Academy struck them back into medals

13-33    Frisch O R 1939 'Physical evidence for the division of heavy nuclei under neutron bombardment' *NA* **143** 276, submitted 16 January, published 18 February 1939

13-34    Frisch 'The discovery...' (note 13-18) p 48

13-35    Meitner L and Frisch O R 1939 'Disintegration of uranium by neutrons: a new type of nuclear reaction' *NA* **143** 239–240, submitted 16 January, published 11 February 1939; Frisch 'Physical evidence...' (note 13-33)

13-36    Bohr first met Rosenfeld in April 1929, when Rosenfeld, as a young postdoctoral student, attended the first of the conferences Bohr arranged in Copenhagen. Periods of intense collaboration followed, and by 1939 Rosenfeld was a full professor at his *alma mater*, the University of Liège. Stuewer 'Bringing news...' (note 13-24) p 49

13-37    Rosenfeld 1979 *Selected papers of Léon Rosenfeld* ed by R S Cohen and J J Stachels (Boston: D Reidel) p 343

13-38    Bohr to Frisch 20 January 1939 BSC 1930–1945 Reel 25

13-39    Fermi, Enrico 1955 'Physics at Columbia University: the genesis of the nuclear energy project' *PT* **8** 282–286

13-40    Anderson H L *et al.* 1939 'The fission of uranium' *PR* **55** 511–512, submitted 16 February 1939

13-41    Rhodes, Richard 1986 *The Making of the Atomic Bomb* (New York: Simon & Schuster) pp 268–270. Edward Teller with Allen Brown 1962 *The Legacy of Hiroshima* (Garden City New York: Doubleday & Co) p 8

13-42    Stuewer 'Bringing the news...' (note 13-24) p 54. The building today is George Washington University's Hall of Government, and a plaque outside Room 105 commemorates the historic meeting

**13-43** Tuve M A 1939 'Droplet fission of uranium and thorium nuclei' *PR* **55** 202–203

**13-44** Fowler R D and Dodson R W 1939 'Intensely ionizing particles produced by neutron bombardment of uranium and thorium' *PR* **55** 417–418

**13-45** Brown 'Golden age...' (note 3-17) figure 51

**13-46** Roberts, Richard Brooke 1979 *Autobiography*, unpublished manuscript, DTM archives

**13-47** Roberts R B, Laboratory notes, DTM archives. Roberts R B, Meyer R C and Hafstad L R 1939 'Droplet fission of uranium and thorium nuclei' *PR* **55** 416–417. (Basically the same text as Tuve's 'Droplet Fission...' note 13-43)

**13-48** Tuve, Roberts, Meyer, and Hafstad 'Droplet fission...' (notes 13-43 and 13-47)

**13-49** Teller *The Legacy*... (note 13-41)

**13-50** Bohr to Margrethe Bohr 29 January 1939 cited in Moore Ruth 1966 *Niels Bohr: The Man His Science and the World They Changed* (New York: Knopf) p 236

**13-51** Tuve 'Droplet fission...' (note 13-43) p 203

**13-52** Davis, Watson, Science Service, 30 January 1939

**13-53** Bohr 1939 'Disintegration of heavy nuclei' *NA* **143** 330, submitted 20 January, revised 3 February 1939

**13-54** Bohr 1939 'Resonance in uranium and thorium disintegrations and the phenomenon of nuclear fission' *PR* **55** 418–419, submitted 7 February

**13-55** Wheeler, John A 1967 'Mechanism of fission' *PT* **20** 278–281 on p 281

**13-56** Pais, Abraham 1991 *Niels Bohr's Times, in Physics, Philosophy, and Polity* (Oxford: Clarendon Press) p 457

**13-57** Nier A O *et al* 1940 'Nuclear fission of separated uranium isotopes' *PR* **57** 546; 'Further experiments on fission of separated uranium isotopes' *ibid* p 748

**13-58** Tuve, report for February 1939, 9 March 1939, MAT Box 15. Roberts R B, Meyer R C and Wang P 1939 'Further observations on the splitting of uranium and thorium' *PR* **55** 510–511; Roberts, Hafstad, Meyer and Wang 1939 'The delayed neutron emission which accompanies fission of uranium and thorium' *ibid* p 664

**13-59** Alvarz 'Berkeley in the 1930s' (note 5-18); Alvarez, Luis W 1982 'The early days of accelerator mass spectrometry' *PT* **35** 25–32

**13-60** As may be recalled, both $H^3$ and $He^3$ had been observed as high-speed reaction products from d–d collisions by Oliphant, Harteck and Rutherford in 1934, though a subsequent heroic attempt to isolate the mass-3 isotope of hydrogen by electrolysis ended an abject failure. During 1939, questioning the near-universal but incorrect view that $He^3$ was heavier than the supposedly stable $H^3$, Alvarez, with Bob Cornog, a graduate student, undertook to fuse deuterium by d–d reactions in the 37-inch cyclotron and feed the product into the 60-inch cyclotron acting as a giant mass spectrograph. The upshot was the demonstration that $He^3$ is stable, not radioactive as everybody had, held while $H^3$ was radioactive, contrary to the claim of Tuve, Hafstad and Dahl in 1934

**13-61** Alvarez, Luis W 1987 *Alvarez: Adventures of a Physicist* (New York: Basic Books) p 72

**13-62** Abelsen, Philip H 1975 'A graduate student with Ernest O Lawrence' pp 22–34 on pp 27–29 in *All in Our Time*... (note 5-18)

**13-63**  In 1913, building on earlier X-ray scattering work by Charles G Barkla and other evidence, Henry G J Moseley presented a systematic study of the characteristic radiations emitted by various targets, and found that the energy of the X-rays increased in a regular manner from element to element up through the periodic table. He also showed that X-rays were emitted when electrons 'fell' into a vacancy in the inner electron shells of atoms. Lines of shorter wavelength (from a given element) were associated with the innermost shell, the 'K shell' closest to the atomic nucleus; lines of longer wavelength were associated with the next shell, the 'L shell'

**13-64**  Abelson, Philip 1939 'Cleavage of the uranium nucleus' *PR* **55** 418, submitted 3 February

**13-65**  Green G K and Alvarez Luis W 1939 'Heavily ionizing particles from uranium' *PR* **55** 417, submitted 31 January

**13-66**  Corson D R and Thornton R L 1939 'Disintegration of uranium' *PR* **55** 509, submitted 15 February

**13-67**  McMillan, Edwin 1939 'Radioactive recoils from uranium activated by neutrons' *PR* **55** 510, submitted 17 February

**13-68**  Alvarez *Adventures* . . . (note 13-61) p 75

**13-69**  Ibid pp 75–76

**13-70**  Oppenheimer to George Uhlenbeck, 5 February 1939; Alice Kimball Smith and Charles Weiner 1980 *Robert Oppenheimer: Letters and Recollections* (Harvard University Press) p 209

**13-71**  Roberts *Autobiography* . . . (note 13-46) p 41

**13-72**  Roberts *ibid* p 37

**13-73**  Brown, Louis 'The proximity fuse and the war effort' ch 15 in *DTM Centennial History* (note 12-44); Louis Brown 1999 'The proximity fuse — the smallest radar' ch 44 in Brown *A Radar History of World War II: Technical and Military Imperatives* (Bristol and Philadelphia: Institute of Physics Publishing)

**13-74**  Booth E T, Dunning J R and Slack F G 1939 'Energy distribution of uranium fission fragments' *PR* **55** 981, submitted May 15

**13-75**  Joliot 1966 'Chemical evidence of the transmutation of elements' *Nobel lectures Including Presentation Speeches and Laureate's Biographies, Chemistry 1922-1941* (Amsterdam: Elsevier) pp 369–373 on p 373

**13-76**  Graetzer and Anderson *The Discovery of Nuclear Fission* . . . (note 12-56) p 68

**13-77**  Halban H, von Joliot F and Kowarski L 1939 'Liberation of neutrons in the nuclear explosion of uranium' *NA* **143** 470–471

**13-78**  Weart, Spencer R 1976 'Scientists with a secret' *PT* **29** 23–30

**13-79**  Graetzer and Anderson *The Discovery of Nuclear Fission* . . . (note 12-56) p 73. The value for $\nu$, released by both Russian and American scientists at the time of the Geneva Conference, was about 2.47 neutrons per fission

**13-80**  Perhaps the best is Rhodes *The Making of the Atomic Bomb* (note 13-41)

**13-81**  Segrè E 1939 'An unsuccessful search for transuranium elements ' *PR* **55** 1104, submitted June 1

**13-82**  Harteck P and Groth W to Army Ordnance, 24 April 1939, PHP Acc 85-22 Box 2 (copy including English translation)

**13-83**  Nineteen subcritical pile assemblies were completed under Werner Heisenberg in the German wartime project. The first, B-I (for 'Berlin'), was completed in Berlin in December 1940, using uranium-oxide and a paraffin moderator. L-IV, number 11, was completed in Leipzig in June 1942, and used uranium metal and (Norwegian) heavy water; it had the distinction of being the first model pile in the world exhibiting a positive neutron production coefficient, beating Fermi's subcritical assembly by the narrowest of margins. The last, B-VIII, was begun in Berlin but evacuated and reassembled in a cave in Haigerloch near Hechingen. It utilized 15 tons of uranium metal, 15 tons of heavy water, and had a graphite neutron reflector. It was started up in March of 1945, but was still short of self-sustaining performance. Simple calculations indicated a mere 50% more heavy water, and about the same amount of uranium, would have sufficed for it to reach criticality. Dahl *Heavy Water...* (note 9-7) p 258

**13-84** Cockcroft to Frisch, 30 Aug 1938, MTNR 5/23; Cockcroft to Chadwick, 11 March 1939, CKFT 20/52. Somewhat later, the possibility of inviting Meitner to Cambridge for a year was reconsidered but rejected. With most of the Cambridge nuclear physicists earmarked for radar, Cambridge no longer seemed the right place for her. Hartcup and Allibone *Cockcroft...* (note 2-47) p 87

**13-85** Meitner to O W Richardson 15 May 1940 cited in Hartcup and Allibone *Cockcroft...* (note 2-47) p 121

**13-86** A two-part memorandum, the technical part is reproduced on pp 389–393 in Gowing M 1965 *Britain and Atomic Energy 1939-1945* (New York: St Martin's Press). The second part, less technical, is reproduced on pp 215–218 in Clark R 1965 *Tizard* (Cambridge, MA: MIT Press)

**13-87** Their critical mass estimate, about a pound, was too low, but a useful order-of-magnitude estimate, indicating the possibility of an atomic bomb in principle

**13-88** Dahl *Heavy Water...* (note 9-7)

**13-89** Except for their neutron measurements, we have neglected the early studies in Paris, by Joliot and his team as well as by Francis Perrin at the Sorbonne, of exponential (neutron diffusion) assemblies of uranium plus a moderator up to June of 1940, as well as related work in other British groups; we have done so because their laboratories and programs are not central to the underlying theme of our chronicle: the pioneering accelerator experiments in Cambridge, in Berkeley, and in Washington

**13-90** Hartcup and Allibone *Cockcroft...* (note 2-47) p 126

**13-91** F-1, the Russian graphite reactor, similar to Fermi's first pile, went critical on Christmas Day 1946 and was the first self-sustaining chain reaction in Europe

## Notes chapter 14

**14-1** Cornell 'Tuve...' (note 3-1)

**14-2** Brown 'Explosion seismology' ch 20 in *DTM History...* (note 12-44)

**14-3** Brown 'Radio astronomy' ch 22 in *DTM History...* (note 12-44)

**14-4** Davis, Nuel Pharr 1968 *Lawrence and Oppenheimer* (New York: Simon and Schuster) p 112

**14-5** Hull, McAllister H 1983 'Gregory Breit' *PT* **35** 102–104

**14-6** In August 1952, Dahl, Frank Goward (an English accelerator physicist) and Rolf Wideröe (the inventor of the linear accelerator) visited Brookhaven to view their 3-GeV 'Cosmotron,' the first proton synchrotron, and discuss the new European project. In preparing to receive their guests, Brookhaven scientists, led by M Stanley Livingston, reviewed the design of the Cosmotron, and in so doing came up with the principle of 'strong focusing,' a new idea for focusing charged particles in accelerators. In effect, a specific arrangement of the beam focusing magnets made possible much higher beam energy for a given machine circumference. Returning to Geneva, Dahl had to make a difficult decision: either ignore the new, untried idea and proceed with the machine as planned (simply a scaled-up Cosmotron), or adopt it for a completely different design at considerable risk

14-7    Roberts *Autobiography* (note 13-46)

14-8    Ibid p 31. Tuve felt that such areas as terrestrial magnetism and electricity were now routine and better studied by government agencies

14-9    Brown 'Postwar nuclear physics' ch 17 in *DTM History*... (note 12-44)

14-10   Roberts *Autobiography* (note 13-46) p 43

14-11   Britten, Roy J 1993 'Richard Brooke Roberts, December 7, 1910–April 4, 1980,' National Academy of Sciences *Biographical Memoirs* vol **62** pp 326–348 on p 335

14-12   The tandem Van de Graaff utilizes *negative* ions. They are accelerated from ground potential to a positive terminal where the electrons are stripped off; the resulting positive ions are then accelerated back to ground, thereby doubling their energy

14-13   Fowler 'Charles Christian Lauritsen' (note 7-1) p 223

14-14   Ibid p 227

14-15   Alvarez, Luis W 1970 'Ernest Orlando Lawrence, August 1, 1901–August 27, 1958' National Academy of Sciences *Biographical Memoirs* vol xli pp 250–294 on p 273

14-16   Shortly after the war the Alpha calutrons, powered with solid silver bus bars on loan from the US Treasury Department, were shut down, and the more efficient gaseous diffusion plant took over the peacetime production of $U^{235}$

14-17   The success of this scheme relies on the principle of phase stability, which ensures that particles will tend to bunch at a particular phase of the r.f. voltage

14-18   In 1936 Seth Neddermeyer and Carl Anderson, using the same magnetic cloud chamber that Anderson had used in the discovery of the positron, found tracks of a particle with the charge of the electron but a greater mass. Subsequently, it was believed that the new, highly unstable particle was the meson postulated by Yukawa as the quantum of the strong nuclear force. Only in 1947, with Powell's discovery, was it realized that his particle, the $\pi$-meson, is Yukawa's particle, while Neddermeyer and Anderson's, the lighter $\mu$-meson, is not. The pion, as it is also called, decays into a muon and it in turn decays into a positron and two neutrinos. Donald H Perkins 1989 'Cosmic-ray work with emulsions in the 1940s and 1950s' Section II Part 5 pp 89–108 in *Pions to Quarks: Particle Physics in the 1950s*, Laurie M Brown, Max Dresden, Lillian Hoddeson eds (Cambridge: Cambridge University Press)

14-19   Heilbron JL, Seidel, Robert W and Wheaton, Bruce R 1981 *Lawrence and His Laboratory: Nuclear Science at Berkeley* (Lawrence Berkeley Laboratory and Office for the History of Science and Technology University of California) p 53

**14-20** Childs *Genius...* (note 3-2) p 401

**14-21** The first proton synchrotron, and the first billion-volt (GeV) accelerator, was actually the Brookhaven Cosmotron (note 14-6), which was brought into operation at 2.3 GeV in 1952. In 1954 the Brookhaven machine was pushed to 3 GeV, the same year that Berkeley's Bevatron first ran, at 5.4 GeV. The CERN PS, that Odd Dahl took charge of, was the first 'strong focusing' proton synchrotron; it was soon followed by one at Brookhaven. Electron synchrotrons are technically much the easier to build, and the first one was a small tabletop machine completed in 1946 by Frank Goward, who later accompanied Dahl on his fact-finding mission to Brookhaven (note 14-6). (Note that we are using G for 'giga' in GeV, replacing BeV, since 'billion' represents $10^{12}$ in British usage)

**14-22** Courant, Ernest D 1997 'Milton Stanley Livingston, May 25, 1905–August 25, 1986' National Academy of Sciences *Biographical Memoirs* vol 72 pp 265–283 on p 270. Courant, Livingston and Hartland Snyder were the three scientists who reviewed the Cosmotron design in preparing for the European visitors, and in so doing came up with the principle of strong focusing

**14-23** Crease, Robert P 1999 *Making Physics: A Biography of Brookhaven National Laboratory 1946–1972* (Chicago: The University of Chicago Press)

**14-24** Haworth volunteered to take the smaller accelerator, gambling that they would finish it first, which they did, and also that Brookhaven would be first in line for the next, even bigger accelerator in due course. And indeed, Brookhaven was the second laboratory, after CERN, to acquire a strong-focusing proton synchrotron, to be known as the Alternating Gradient Synchrotron, or AGS for short

**14-25** Oliphant, Sir Mark and Lord Penney 1968 'John Douglas Cockcroft' *Biographical Memoirs of Fellows of the Royal Society* vol 14 pp 139–188

**14-26** *Harwell: The British Atomic Energy Research Establishment 1946–1951* (New York: Philosophical Library 1952)

**14-27** Penney, Lord 1967 'Sir John Cockcroft' *NA* **216** 11 November. Unfortunately, when Mark Oliphant examined the 'Black Book,' no pages earlier than for 1938 could be found. CKFT 21/12

**14-28** McMillan to Guy Hartcup cited in Hartcup and Allibone *Cockcroft...* (note 2-47) p 150

**14-29** Ibid p 164

**14-30** Cited in Hartcup and Allibone *Cockcroft...* (note 2-47) p 164

**14-31** Oliphant and Penney 'John Douglas Cockcroft' (note 14-25) pp 181–182

**14-32** Ibid p 185

**14-33** Hartcup and Allibone *Cockcroft...* (note 2-47) p 253

**14-34** Weber, Robert L 1980 *Pioneers of Science: Nobel Prize Winners in Physics* (Bristol and London: The Institute of Physics) p 141

**14-35** Cockburn, Stewart and Ellyard, David 1981 *Oliphant: The Life and Times of Sir Mark Oliphant* (Adelaide: Axiom Books)

**14-36** Weart and Szilard *Leo Szilard...* (note 4-50) p 146

**14-37** Notes Cockcroft in a letter to Karl Darrow in 1938: 'It is not true in any sense to say that Gamow suggested the [high voltage disintegration] work, though his presence in the laboratory and frequent discussions of the $\alpha$ particle disintegration considerably focussed our attention on the subject.' Cockcroft to Karl Darrow, 1 February 1938; CKFT 20/28

# SELECT BIBLIOGRAPHY

---

The following bibliography lists many, but not all, books, journal articles, and other sources cited in the notes. Thus, bibliographical entries in the *Dictionary of Scientific Biography* are not included. Abbreviations are those used in the notes.

Abelson P 1939 Cleavage of the uranium nucleus *PR* **55** 418
—— 1975 A graduate student with Ernest O Lawrence pp 22–34 in *All in Our Time: The Reminiscences of Twelve Nuclear Pioneers* ed J Wilson (Chicago: The Bulletin of the Atomic Scientists)
Allibone T E 1964 The industrial development of nuclear power Rutherford Memorial Lecture 1963 *PRS* **282** 447–463
—— 1984 Metropolitan-Vickers Electrical Company and the Cavendish Laboratory in *Cambridge Physics in the Thirties* ed J Hendry (Bristol: Adam Hilger)
Alvarez L W 1970 Ernest Orlando Lawrence, August 1, 1901–August 27, 1958 NAS *Biographical Memoirs* **41** 250–294
—— 1975 Berkeley in the 1930's in *All in Our Time: The Reminiscences of Twelve Nuclear Pioneers* ed Jane Wilson (Chicago: The Bulletin of the Atomic Scientists)
—— 1982 The early days of accelerator mass spectrometry *PT* **35** 25–32
—— 1987 *Alvarez: Adventures of a Physicist* (New York: Basic Books)
Anderson C D 1932 The apparent existence of easily deflectable positives *Science* **76** 238–239
Anderson H L, Booth E T, Dunning J R, Glasoe G N and Slack F G 1939 The fission of uranium *PR* **55** 511–512
Andrade E N da C 1964 *Rutherford and the Nature of the Atom* (Garden City, NY: Doubleday)
Aston F W 1934 Discussion on heavy hydrogen *PRS* **144** 1–28
Badash L 1983 Nuclear physics in Rutherford's laboratory before the discovery of the neutron *AJP* **51** 884–889
Bainbridge K T 1932 The isotopic weight of $H^2$ *PR* **42** 1
—— 1933 Comparison of the masses of $H^2$ and helium *PR* **44** 57
Birge R T and Menzel D H 1931 The relative abundance of the oxygen isotopes and the basis of the atomic weight system *PR* **37** 1669
Birge R T *History of the Physics Department* Mimeograph 5 vols (Berkeley 1966–?)

Blackett PMS 1922 On the analysis of α-ray photographs *PRS* **102** 294–317

—— 1925 The ejection of protons from nitrogen nuclei photographed by the Wilson method *PRS* **107** 349–360

Bohr N 1913 On the constitution of atoms and molecules part I *PM* **26** 1–25; On the constitution... Part II. Systems containing only a single nucleus *PM* **26** 476–502; On the constitution... Part III. Systems containing several nuclei *PM* **26** 857–875

—— 1939 Disintegration of heavy nuclei *NA* **143** 330

—— 1939 Resonance in uranium and thorium disintegrations and the phenomenon of nuclear fission *PR* **55** 418–419

Boorse H A and Motz L (eds) 1966 *The World of the Atom* (New York: Basics Books)

Booth E T, Dunning J R and Slack F G 1939 Energy distribution of uranium fission fragments *PR* **55** 981

Brasch A and Lange F 1930 Künstliche γ-Strahlung Ein Vakuum- Entladungsrohr für 2,4 Million Volt *Die Naturwissenschaften* **18** 765-766

Brish S G and Gillmor C S 1995 Geophysics ch 26 in *Twentieth Century Physics* vol III by L M Brown, A Pais and B Pippard (Bristol: Institute of Physics Publishing; New York: American Institute of Physics Press)

Britten R J 1993 Richard Brooke Roberts, December 7, 1910–April 4, 1980 NAS *Biographical Memoirs* **62** 326–348

Breit G 1929 On the possibility of nuclear disintegration by artificial sources *PR* **34** 817–818

Breit G, Condon E U and Present R D 1936 Theory of scattering of protons by protons *PR* **50** 825–845

Breit G, Tuve M A and Dahl O 1927 Magnetism and atomic physics *Carnegie Institution of Washington Yearbook* **27** 208–209

—— 1929 A laboratory method of producing high potentials *PR* **35** 51–65

Brown A 1996 Liverpool and Berkeley: The Chadwick–Lawrence letters *PT* **49** 34–40

—— 1997 *The Neutron and the Bomb: A Biography of Sir James Chadwick* (Oxford: Oxford University Press)

Brown L *Report From a Golden Age of Physics* unpublished draft manuscript

—— 1999 The proximity fuse—the smallest radar in Brown *A Radar History of World War II: Technical and Military Imperatives* (Bristol: Institute of Physics Publishing)

Brown L M and Rechenberg H 1996 *The Origin of the Concept of Nuclear Forces* (Bristol and Philadelphia: Institute of Physics Publishing)

Burrill E A 1976 Van de Graaff, the man and his accelerators *PT* **20** 49–52

Chadwick J 1932 Possible existence of a neutron *NA* **129** 312

—— 1932 The existence of a neutron *PRS* **136** 692–708

—— 1933 The neutron (Bakerian Lecture) *PRS* **142** 1–25

—— 1964 Some personal notes on the search for the neutron *Proc 10th Int Congress on the History of Science (Ithaca 1962)* (Paris: Hermann)

Chadwick J and Goldhaber M 1934 A 'nuclear photo-effect': Disintegration of the diplon by γ-rays *NA* **134** 237–238

Childs H 1968 *An American Genius: The Life of Ernest Orlando Lawrence* (New York: E P Dutton & Co)

Clark R 1965 *Tizard* (Cambridge, MA: The MIT Press)

Cockburn S and Ellyard D 1981 *Oliphant: The Life and Times of Sir Mark Oliphant* (Adelaide: Axiom Books)

Cockcroft J D and Walton E T S 1930 Experiments with high velocity positive ions *PRS* **129** 477–489

—— 1932 Experiments with high velocity positive ions (I). Further developments in the method of obtaining high velocity positive ions *PRS* **136** 619–630

—— 1932 Experiments with high velocity positive ions (II). The disintegration of elements by high velocity protons *PRS* **137** 229–242

—— 1932 Disintegration of lithium by swift protons *NA* **129** 649

Compton A H 1956 *Atomic Quest: A Personal Narrative* (New York: Oxford University Press)

Compton K T 1933 High voltage II *Science* **78** 48–52

Cornell Thomas D 1986 *Merle A Tuve and His Program of Nuclear Studies at the Department of Terrestrial Magnetism: The Early Career of a Modern American Physicist* (PhD Dissertation, The Johns Hopkins University)

Corson D R and Thornton R L 1939 Disintegration of uranium *PR* **55** 509

Courant E D 1997 Milton Stanley Livingston, May 25, 1905–August 25, 1986 NAS *Biographical Memoirs* **72** 265–283

Crane R and Lauritsen C C 1934 Radioactivity from carbon and boron oxide bombarded with deutons and the conversion of positrons into radiation *PR* **45** 430–432

—— 1934 Further experiments with artificially produced radioactive substances *PR* **45** 497–498

Crawford E, Sime R L and Walker M 1977 A Noble tale of postwar injustice *PT* **50** 26–32

Crease R P 1999 *Making Physics: A Biography of Brookhaven National Laboratory 1946–1972* (Chicago: The University of Chicago Press)

Curie I *See* Joliot-Curie I

Dahl O 1936 Note on disk-type electrostatic generators *RSI* **7** 254–256

Dahl O, Hafstad L R and Tuve M A 1933 On the technique and design of Wilson cloud-chambers *RSI* **4** 373–378

Dahl P F 1997 *Flash of the Cathode Rays: A History of J J Thomson's Electron* (Bristol: Institute of Physics Publishing)

—— 1999 *Heavy Water and the Wartime Race for Nuclear Energy* (Bristol: Institute of Physics Publishing)

Davis N P 1968 *Lawrence and Oppenheimer* (New York: Simon and Schuster)

Dee P I and Walton E T S 1993 A photographic investigation of the transmutation of lithium and boron by protons and of lithium by ions of the heavy isotope of hydrogen *PRS* **141** 733–742

Du Bridge L A and Epstein P S 1959 Robert Andrews Millikan, March 22, 1868–December 19, 1953 NAS *Biographical Memoirs* **33** 241–282

Eve A S 1939 *Rutherford: Being the Life and Letters of the Rt Hon Lord Rutherford OM* (New York: Cambridge University Press)

Falconer I J 1980 *Apparatus from the Cavendish Museum* (Cambridge: pamphlet University of Cambridge)

Feather N 1940 *Lord Rutherford* (Glasgow: Blackie & Sons)

Fermi E 1955 Physics at Columbia University: The genesis of the nuclear energy project *PT* **8** 282–286

Fermi E, Amaldi E, d'Agostino O, Rasetti F and Segrè E 1934 Artificial radioactivity produced by neutron bombardment *PRS* **146** 483–500

Fermi E, Amaldi E, Pontecorvo B M, Rasetti F and Segrè E 1962 Influence of hydrogenous substances on the radioactivity produced by neutrons translated by E Segrè. Fermi *Collected Papers* vol I (Chicago: University of Chicago Press) 761–762

Fleming J A 1933 Studies in nuclear physics *Science* **77** 298–300

Fowler W A 1975 Charles Christian Lauritsen, April 4, 1892–April 13, 1968 NAS *Biographical Memoirs* **46** 221–239

Fowler W A and Ajzenberg-Selove F 1985 Thomas Lauritsen, November 16, 1915–October 16, 1973 NAS *Biographical Memoirs* **55** 385–394

Fowler R D and Dodson R W 1939 Intensely ionizing particles produced by neutron bombardment of uranium and thorium *PR* **55** 417–418

Frisch O R 1939 Physical evidence for the division of heavy nuclei under neutron bombardment *NA* **143** 276

—— 1967 The discovery of fission: How it all began *PT* **20** 43–48

—— 1979 *What Little I Remember* (Cambridge: Cambridge University Press)

Gamow G 1928 Quantum theory of the atomic nucleus *ZP* **51** 204–212

—— 1970 *My World Line: An Informal Autobiography* (New York: The Viking Press)

Gehrenbeck R K 1978 Electron diffraction: Fifty years ago *PT* **31** 34–41

Goldhaber M 1979 The nuclear photoelectric effect and remarks on higher multipole transitions: A personal history pp 83–110 in *Nuclear Physics in Retrospect* ed R H Stuewer (Minneapolis: University of Minnesota Press)

Goldsmith M 1976 *Frédéric Joliot-Curie: A Biography* (London: Lawrence and Wishart)

Gowing M 1965 *Britain and Atomic Energy 1939–1945* (New York: St Martin's Press)

Graetzer H G and Anderson D L 1971 *The Discovery of Nuclear Fission: A Documentary History* (New York: Van Nostrand Reinhold)

Green G K and Alvarez L W 1939 Heavily ionizing particles from uranium *PR* **55** 417

Gurney R W and Condon E U 1928 Wave mechanics and radioactive disintegration *NA* **122** 439

Hafstad L R and Tuve M A 1934 Artificial radioactivity using carbon targets *PR* **45** 902–903

—— 1935 Carbon radioactivity and other resonance transformations by protons *PR* **48** 306–315

Hahn O 1962 *A Scientific Autobiography* translated and edited by W Ley (New York: Scribners)

Hahn O and Strassmann F 1938 Über die Entstehung von Radiumisotopen aus Uran beim Bestrahlen mit Schnellen und Verlangtsamten Neutronen *Die Naturwissenschaften* **26** 755–756

—— 1939 Über den Nachweis und das Verhalten der bei der Bestrahlung des Urans mittels Neutronen Entstehenden Erdalkalimetalle *Die Naturwissenschaften* **27** 11-15

Halban H, von Joliet F and Kowarski L 1939 Liberation of neutrons in the nuclear explosion of uranium *NA* **143** 470–471

Hartcup G and Allibone T E 1984 *Cockcroft and the Atom* (Bristol: Adam Hilger)

Heilbron J L and Seidel R W 1989 *Lawrence and His Laboratory: A History of the Lawrence Berkeley Laboratory* vol 1 (Berkeley: University of California Press)

Heilbron J L, Seidel R W and Wheaton B R 1981 *Lawrence and His Laboratory: Nuclear Science at Berkeley* (Lawrence Berkeley Laboratory and Office for the History of Science and Technology University of California)

Henderson M, Livingston M S and Lawrence E O 1934 Artificial radioactivity produced by deuton bombardment *PR* **45** 428–429

Hirosige T 1963 Social conditions for the researches of nuclear physics in pre-war Japan *Japanese Studies in the History of Science* **2** 87–88

Holbrow C H 1981 The giant cancer tube and the Kellogg Radiation Laboratory *PT* **34** 42–49

Hull M H 1983 Gregory Breit *PT* **35** 102–104

Ising G 1924 Prinzip einer Methode zur Herstellung von Kanalstrahlen Hoher Voltzahl *Arkiv för Mathematik, Astronomi och Fysik* **18** 1–4. Translated by F W Brasse as The principle of a method for the production of canal rays of high voltage in *The Development of High-Energy Accelerators* ed M S Livingston (New York: Dover)

Joliot-Curie F 1966 Chemical evidence of the transmutation of elements *Nobel Lectures Including Presentation Speeches and Laureate's Biographies: Chemistry 1922–1941* (Amsterdam: Elsevier) 369–373

Joliot-Curie F and I 1934 Un noveau type de radioactivité *CR* **198** 254

—— 1934 Artificial production of a new type of radioelements *NA* **133** 201

Joliot-Curie I and F 1932 Émission de protons de grande vitesse par les substances hydrogénées sous l'influence des rayons gamma très pénétrants *CR* **194** 273

Joliot-Curie I, Halban H von and Preiswerk P 1935 Sur la création artificielle d'éléments appartenant à une familie radioactive inconnue lors de l'irradiation du thorium par les neutrons *JPR* **6** 361

Joliot-Curie I and Savitch P 1937 Sur les radioéléments formés dans l'uranium irradié par les neutrons *JPR* **8** 385–387

—— 1938 Sur la nature du radioélément de périod 3.5 heures formé dans l'uranium irradié par les neutrons *CR* **206** 1643

Jungnickel C and McCormmach R 1986 *Intellectual Masters of Nature: Theoretical Physics from Ohm to Einstein* (Chicago: The University of Chicago Press)

Kevles D J 1995 *The Physicists: The History of a Scientific Community in Modern America* (Cambridge, MA: Harvard University Press)

Kirsch G and Pettersson H 1923 Long-range particles from radium-active deposit *NA* **112** 394–395

—— 1924 The artificial disintegration of atoms *NA* **113** 603

Kraft F 1983 Internal and external conditions for the discovery of nuclear fission by the Berlin team in *Otto Hahn and the Rise of Nuclear Physics* ed W R Shea (Dordrecht: D Reidel)

Laurence W L 1931 Giant X-ray tube wins science award *The New York Times* 4 January 1931

—— 1950 Cosmic rays are found to produce tritium vital for hydrogen bomb *The New York Times* 15 September 1950

Lauritsen C C and Bennett R D 1928 A new high potential X-ray tube *PR* **32** 850–857

Lauritsen C C and Cassen B 1930 High potential X-ray tube *PR* **36** 988–992

Lauritsen C C and Crane H R 1934 Gamma-rays from carbon bombarded with deutons *PR* **45** 345–346

Lauritsen C C, Crane H R and Harper W W 1934 Artificial production of radio-active substances *Science* **79** 234–235

Lauritsen T, Lauritsen CC and Fowler WA 1941 Application of a pressure electrostatic generator to the transmutation of light elements by protons *PR* **59** 241–252

Lawrence E O 1952 The evolution of the cyclotron *Les Prix Nobel en 1951*, Imprimerie Royale (Stockholm: Norstedt & Sönen) reprinted in *The Development of High-Energy Accelerators* ed M S Livingston (New York: Dover)

Lawrence E O and Edlefsen N E 1930 On the production of high speed protons *Science* **72** 376–377

Lawrence E O and Livingston M S 1931 A method for producing high speed hydrogen ions without the use of high voltage *PR* **37** 1707

—— 1931 The production of high speed protons without the use of high voltages *PR* **38** 834

—— 1932 The production of high speed light ions without the use of high voltages *PR* **40** 19–35

—— 1934 The emission of protons and neutrons from various targets bombarded by three million volt deutons *PR* **45** 220

Lawrence E O, Livingston M S and Lewis G N 1933 The emission of protons from various targets bombarded by deutons of high speed *PR* **44** 56

Lawrence E O, Livingston M S and White M G 1932 The disintegration of lithium by swiftly moving protons *PR* **42** 150–151

Lawrence E O and Sloan D H 1931 The production of high speed cathode rays without the use of high voltages NAS *Proceedings* **17** 64–70

Lemmerich J ed 1988 *Die Geschichte der Entdeckung der Kernspaltung: Ausstellungs-katalog* (Berlin: Technische Universität Berlin, Universitätsbibliotek)

Lewis G N, Livingston M S, Henderson M C and Lawrence E O 1934 The disintegration of deutons by high speed protons and the instability of the deuton *PR* **45** 242–244

—— 1934 On the hypothesis of the instability of the deuton *PR* **45** 497

Lewis G N, Livingston M S and Lawrence E O 1933 The emission of alpha-particles from various targets bombarded by deutons of high speed *PR* **44** 55–56

Livingston M S 1931 *The Production of High Velocity Hydrogen Atoms Without the Use of High Voltages* PhD thesis, University of California, Berkeley, 14 April 1931

—— 1932 High-speed hydrogen ions *PR* **42** 441–442

—— 1959 History of the cyclotron, Part 1 *PT* **12** 18–23

—— 1980 Early history of particle accelerators *Advances in Electronics and Electron Physics* **50** ed L Marton and C Marton (New York: Academic Press)

Livingston M S, Henderson M C and Lawrence E O 1933 Neutrons from deutons and the mass of the neutron *PR* **44** 781–782

McMillan E M 1939 Radioactive recoils from uranium activated by neutrons *PR* **55** 510

—— 1979 Early history of particle accelerators *Nuclear Physics in Retrospect* ed R H Stuewer (Minneapolis: University of Minnesota Press)

Marsden E 1962 Rutherford at Manchester *Rutherford at Manchester* ed J B Birks (London: Heywood & Company)

Marsden E and Lantsberry W C 1915 The passage of $\alpha$-particles through hydrogen, II *PM* **30** 240–243

Massey H and Feather N 1976 James Chadwick, 20 October 1891–24 July 1974 Royal Society *Biographical Memoirs* **22** 11–70

Mehra J 1975 *The Solvay Conferences on Physics: Aspects of the Development of Physics Since 1911* (Dordrecht: D Reidel)

Meitner L and Frisch O R 1939 Disintegration of uranium by neutrons: A new type of nuclear reaction *NA* **143** 239–240

Mladjenovic M 1998 *The Defining Years in Nuclear Physics 1932-1960s* (Bristol and Philadelphia: Institute of Physics Publishing)

Moore R 1966 *Niels Bohr: The Man, His Science, and the World They Changed* (New York: Knopf)

Nier A O, Booth E T, Dunning J R and von Grose A 1940 Nuclear fission of separated uranium isotopes *PR* **57** 546

—— 1940 Further experiments on fission of separated uranium isotopes *PR* **57** 748

Noddack I 1934 Über das Element 93 *Z Angewandte Chemie* **47** 653–655

Norton G A, Ferry J A, Daniel R E and Klody G M undated *A Retrospective of the Career of Ray Herb* unpublished manuscript, National Electrostatics Corporation, Middleton, WI

Oliphant M L E 1966 The two Ernests—I *PT* **19** 35–49

—— 1972 *Rutherford: Recollections of the Cambridge Days* (Amsterdam: Elsevier)

Oliphant M L E, Kinsey B B and Rutherford E 1933 The transmutation of lithium by protons and by ions of the heavy isotope of hydrogen *PRS* **141** 722–733

Oliphant M L E and Penney W 1968 John Douglas Cockcroft *Biographical Memoirs of Fellows of the Royal Society* **14** 139–188

Oliphant M L E and Rutherford E 1933 Experiments on the transmutation of elements by protons *PRS* **141** 259–281

Pais A 1986 *Inward Bound: Of Matter and Forces in the Physical World* (Oxford: Clarendon Press)

—— 1991 *Niels Bohr's Times in Physics Philosophy and Polity* (Oxford: Clarendon Press)

Peierls R 1985 *Bird of Passage: Reflections of a Physicist* (Princeton: Princeton University Press)

Pettersson H 1924 On the structure of the atomic nucleus and the mechanism of its disintegration Physical Society *Proceedings* **36** 194–202

Rhodes R 1986 *The Making of the Atomic Bomb* (New York: Simon and Schuster)

Roberts R B 1979 *Autobiography* unpublished manuscript, DTM Archives, Carnegie Institution of Washington

Roberts R B, Hafstad L R, Meyer R C and Wang P 1939 The delayed neutron emission which accompanies fission of uranium and thorium *PR* **55** 664

Roberts R B, Meyer R C and Wang P 1939 Further observations on the splitting of uranium and thorium *PR* **55** 510–511

Rose P H 1968 In memoriam: Robert Jemison Van de Graaff *Nuclear Instruments and Methods* **60** 1–3

Rutherford E 1919 Collision of $\alpha$ particles with light atoms. IV. An anomalous effect in nitrogen *PM* **37** 581–587

—— 1920 Nuclear constitution of atoms *PRS* **97** 374–400

—— 1924 The natural and artificial disintegration of the elements Franklin Institute *Journal* **198** 725–744

—— 1927 Scientific aspects of intense magnetic fields and high voltages *NA* **120** 809–811

—— 1929 Discussion of the structure of atomic nuclei Royal Society of London *Proceedings* **123** 373–382

—— 1934 The new hydrogen *Science* **80** 21–25

—— 1937 The search for the isotopes of hydrogen and helium of mass 3 *NA* **140** 303–305

Rutherford E and Chadwick J 1921 The artificial disintegration of light elements *PM* **42** 809–825

—— 1924 On the origin and nature of the long-range particles observed with sources of radium C *PM* **48** 509–526

Rutherford E, Chadwick J and Ellis C D 1930 *Radiations from Radioactive Substances* (Cambridge: Cambridge University Press)

Segrè E 1939 An unsuccessful search for transuranium elements *PR* **55** 1104

—— 1970 *Enrico Fermi, Physicist* (Chicago: The University of Chicago Press)

Sime R L 1996 *Lise Meitner: A Life in Physics* (Berkeley: University of California Press)

Sloan D H and Lawrence E O 1931 Production of heavy high speed ions without the use of high voltages *PR* **38** 2021–2032

Smith A K and Weiner C 1980 *Robert Oppenheimer: Letters and Recollections* (Cambridge, MA: Harvard University Press)

Stuewer R H 1985 Bringing the news of fission to America *PT* **38** 49–56

—— 1986 Rutherford's satellite model of the nucleus *HSPS* **16** 321–352

Tuve M A 1933 The atomic nucleus and high voltages *Journal of the Franklin Institute* **216** 1–38

—— 1934 Nuclear-physics symposium, a correction *Science* **80** 161–162

—— 1939 Droplet fission of uranium and thorium nuclei *PR* **55** 202–204

—— 1968 Odd Dahl at the Carnegie Institution 1926–1936 *Festskrift til Odd Dahl I Anledning av Hans Fylte 70 År, 3 November 1968* (Bergen: John Griegs Boktrykkeri)

Tuve M A, Breit G and Hafstad L R 1929 Application of high potentials to vacuum tubes *PR* **35** 66–71

Tuve M A, Dahl O and Hafstad L R 1935 The production and focusing of intense positive ion beams *PR* **48** 241–256

Tuve M A and Hafstad L R 1934 The emission of disintegration-particles from targets bombarded by protons and by deuterium ions at 1200 kilovolts *PR* **45** 651–653

—— 1936 The scattering of neutrons by protons *PR* **50** 490–491

Tuve M A, Hafstad L R and Dahl O 1931 Experiments with high-voltage tubes *PR* **37** 469

—— 1932 High-speed protons *PR* **39** 384–385

—— 1933 Disintegration-experiments on elements of medium atomic number *PR* **43** 942

—— 1934 A stable hydrogen isotope of mass three *PR* **45** 840–841

—— 1935 High voltage technique for nuclear physics studies *PR* **48** 315–337

Tuve M A, Heydenburg N P and Hafstad L R 1936 The scattering of protons by protons *PR* **50** 806–825

Urey H C, Brickwedde F G and Murphy G M 1932 A hydrogen isotope of mass 2 *PR* **39** 164–165

—— 1932 A hydrogen isotope of mass 2 and its concentration *PR* **40** 1–15

Van Atta L C, Northrup D L, Van Atta C M and Van de Graaff R J 1936 The design operation and performance of the Round Hill electrostatic generator *PR* **49** 761–776

Van de Graaff R J 1931 A 1,500,000 volt electrostatic generator *PR* **38** 1919–1920

Van de Graaff R J, Compton K T and Van Atta L C 1933 The electrostatic production of high voltage for nuclear investigations *PR* **43** 149–157

Waloschek P ed 1994 *The Infancy of Particle Accelerators: Life and Work of Rolf Wideröe* (Braunschweig: Vieweg & Sohn)

Walton E T S 1984 Personal recollections of the discovery of fast particles *Cambridge Physics in the Thirties* ed J Hendry (Bristol: Adam Hilger)

Weart S R 1976 Scientists with a secret *PT* **29** 23–30

—— 1979 *Scientists in Power* (Cambridge, MA: Harvard University Press)

Weart S R and Weiss Szilard G eds 1978 *Leo Szilard: His Version of the Facts: Selected Recollections and Correspondence* (Cambridge MA: The MIT Press)

Weber R L 1980 *Pioneers of Science: Nobel Prize Winners in Physics* (Bristol: The Institute of Physics)

Weiner C 1970 Physics in the Great Depression *PT* **23** 31–38

—— 1972 1932 — Moving into the new physics *PT* **25** 332–339

Wheeler J A 1967 The discovery of fission: Mechanism of fission *PT* **20** 278–281

—— 1979 Some men and moments in the history of nuclear physics: the interplay of colleagues and motivations in *Nuclear Physics in Retrospect* ed R H Stuewer (Minneapolis: University of Minnesota Press)

Wideröe R 1928 Über ein neues Prinzip zur Herstellung hoher Spannungen *Archiv für Elektrotechnik* **21** 387–406. Translated by G E Fischer, F W Basse, H Kumpfert and H Hartmann 1966 as A new principle for the generation of high voltages in *The Development of High-Energy Accelerators* ed M S Livingston (New York: Dover)

—— 1984 Some memories and dreams from the childhood of particle accelerators *Europhysics News* **15** 9–11

Wilson C T R 1965 On the cloud method of making visible ions and the tracks of ionizing particles *Nobel Lectures Including Presentation Speeches and Laureate's Biographies: Physics 1922–1941* (Amsterdam: Elsevier)

Wilson D 1983 *Rutherford: Simple Genius* (Cambridge, MA: The MIT Press)

# NAME INDEX

*Note*: Years of birth and death (if known) are recorded after each name. Endnotes are indicated by 'n' and a note number.

Abelson, Philip Hauge [1913– ]
  and Alvarez, 205, 206
  biophysics at DTM, 218
  delayed neutrons from uranium, 207
  and DTM cyclotron, 207
  fission verification experiment, 205–206
  isolating neptunium and plutonium, 210
  joins DTM, 207
  joins Radiation Laboratory, 205
  learns of fission, 205
  radiochemistry experiments, 205–206, 210
  and Roberts, 207
Adams, John Bertram [1920–1984], 218, 225
Aebersold, Paul, 183
Allibone, Thomas Edward [1903–1948]
  background, 19
  on Burch, 112
  and Cockcroft, 20
  and Ellis, 19
  and Fleming, 20
  and Gamow, 43
  and Goodlet, 20, 58
  at Metropolitan-Vickers, 19, 24, 140
  and Rutherford, 20, 140
  Tesla coil experiments, 20, 21
  thermionic rectifiers, 58, 108
  and Walton, 19
Allier, Jacques [1928– ]: Oslo mission, 213
Alvarez, Luis Walter [1911–1989]
  and Abelson, 205
  background, 204
  characteristic x-rays from radioisotopes, 205

  finds original cyclotrons, n.5–18
  fission verification experiment, 206
  hydrogen-3 and helium-3 experiments, 204, n.13–60
  and Lawrence, 204
  learns of fission, 204–205
  at MIT Radiation Laboratory, 220
  and Oppenheimer, 206
  and 60-inch cyclotron, 183, 184
Alvarez, Walter Clement (father of Luis Alvarez), 204
Amaldi, Edorado [1908–1989], 180, 184, 224
  neutron bombardment studies, 160, 180, 186
  professor in Rome, 187
Ames, Joseph Sweetman [1864–1943], 28, 30
Amundsen, Roald Engebreth Gravning [1872–1928], 44
Anderson, Carl David [1905–1991], 169
  background, 106
  on Dirac's electron theory, 107
  meson discovery, n.14–18
  positron discovery, 2, 107, 147
Anderson, Herbert Lawrence [1914–1988]
  and Bohr, 198
  fission demonstration at Columbia, 198
Appleton, Sir Edward Victor [1892–1965]
  ionosphere, 30
  radio receiving apparatus, n.3–14
Arnold, William Archibald [1904– ], 196
Aston, Francis William [1877–1945], 102, 130, 157, 210, 227, n.2–8

α-bombardment, radiation from, 103–104, 180
artificial radioactivity, 147–148, 160, 169, 178, 187, n.12–20
background, 103
and Bohr, 148
on censorship, 209
chair at Collège de France, 187
and Hahn, 188
and Halban and Kowarski, 208–209, 213
laboratory equipment, 103
and Lawrence, 148
and Madame Curie, 103
and Meitner, 147–148
and neutron discovery, 107
neutron mass, 147
Nobel Prize, 162, 187
and Norwegian heavy-water escapade in France, 213
nuclear chain reaction studies, 213, n.13–89
and Paris cyclotron, 148, 166
photograph of electron–positron pair, 108
and polonium source, 103
and positron discovery, 108
recoil protons from neutral 'beryllium radiation,' 2, 104
and Rutherford, 162
secondary neutrons from fission, 208, 209
at Solvay Conference of 1933, 147–148
Joliot-Curie, Irène [1897–1956], 194
α-ray studies, 103–104
artificial radioactivity, 147, 160, 178, 187–188
background, 103
and Bohr, 148
Hahn on, 188, 190
health, 187
and Joliot, 103–104, 187
and Marie Curie, 103, 187
Meitner on, 188
and neutron discovery, 107
neutron mass, 147
Nobel Prize, 162, 187
photograph of electron–positron pair, 108
and polonium source, 103
and positron discovery, 108
professorship at the Sorbonne, 187
and Radium Institute, 103, 187
recoil protons from neutral 'beryllium radiation,' 2, 103–104
and Savitch, 188
at Solvay Conference of 1933, 147–148

and Strassmann, 190
thorium investigation, 188
uranium investigation, 188
Jordan, Pascual [1902–1980], 39

Kaempffert, Waldemar, 3
Kapitza, Anna, 161, 167, 226
Kapitza, Peter (Pyotr Leonidovich) [1894–1984], 42, 49, 227
background, 22
and Cavendish Laboratory, 22–23
character, 23
detained in USSR, 161–162
equipment transferred to USSR, 161–162
generation of intense magnetic fields, 23
and Mond Laboratory, 23, 161–162
and Moscow Institute, 161, 162
Rutherford's opinion on, 22, 161
Rutherford supports, 161–162
visit to Churchill College, 226
work with Cockcroft, 23, 36, 114
writes to Rutherford, 161
Karman, Theodore von [1881–1963], 87
Kay, William [1879–1961], 6, 12, n.2–14
Kellogg, Will Keith: and Kellogg Radiation Laboratory, 89
Kelvin, Lord (William Thomson) [1824–1907], 10, 11, 168
Kendrew, Sir John Cowdery [1917– ], 168
Kennelly, Arthur Edwin [1861–1939]
ionospheric layer, 29
radio wave propagation, 29
Kerst, Donald William [1911–1993], 175, 177
Kikuchi, Ken, 185
Kinsey, Bernard B., 142, 163
Kirsch, Gerhard Theodor [1890–1956]: disintegration experiments, 14–16, 71, 192
Kowarski, Lew [1907–1979], 224
background, 209
chain-reaction studies in Cambridge, 213
chain-reaction studies in Paris, 209
and Halban, 213
and ZEEP, 214
Kruger, Peter Gerald [1902–1978], 207
Krutkov, Yuri Alexsandrovich [1890–1952], 36
Kurie, Franz, 119, 124

Ladenburg, Rudolf Walter [1882–1952], 200
Lamb, Horace [1849–1934], 167
Lamb, Willis Eugene, Jr. [1913– ], 198

# SUBJECT INDEX

304